AF614969

Mechanics of fracture

VOLUME 5

Stress analysis of notch problems

Mechanics of fracture

edited by GEORGE C. SIH

VOLUME 1

Methods of analysis and solutions of crack problems

VOLUME 2

Three-dimensional crack problems

VOLUME 3

Plates and shells with cracks

VOLUME 4

Elastodynamic crack problems

VOLUME 5

Stress analysis of notch problems

Mechanics of fracture 5

Stress analysis of notch problems

Stress solutions to a variety of notch geometries used in engineering design

Edited by

G. C. SIH

Professor of Mechanics and Director of the Institute of Fracture and Solid Mechanics

Lehigh University Bethlehem, Pennsylvania

NOORDHOFF INTERNATIONAL PUBLISHING
ALPHEN AAN DEN RIJN

ISBN: 90 286 0166 X

PRINTED IN THE NETHERLANDS

Contents

Editor's preface

The magnification of stresses at geometric stress concentrations is of great importance in engineering design. In particular, problems of fracture are all the result of local concentrations of stresses arising from abrupt changes of shape or geometrical discontinuities. Typical examples are notches, grooves, fillets, flaws and cracks. Their shape and degree of sharpness can greatly influence the load bearing capacity of structural members. From the designer's point of view, a fracture analysis will involve two major concerns: (1) a stress analysis of the geometrical disturbance, and (2) a postulate predicting the event of fracture itself. Comprehensive treatments of the elastic stress distributions around a wide variety of geometric cavities are available in the open literature and will not be repeated in this volume. The number of proposed failure criteria is equally exhaustive. However, there is still no coherent treatment of failure for all geometric cavities regardless of their shapes and sizes. This fifth volume of the series on Mechanics of Fracture attempts to join the behavior of sharp discontinuities such as cracks to that of notches or cavities and to provide a few typical analytical methods of stress solutions. The finite element method has been purposely left out, for it is now becoming common procedure in cases too complex for analysis.

One of the major applications for fracture mechanics analyses is an assessment of the influence of defects on the strength of structural components. The classical approach involving energy release rates or stress intensity factors is restricted to sharp cracks and relies on the assumption that stable cracks are first initiated from notch sites before incipient fracture. This, however, may not always be the case. Most important of all, it is essential to correlate the fundamental concepts of sharp crack fracture criteria to blunted cracks or notches so that no special consideration is required when the sharpness of the notch is altered. The means by which this is accomplished involves the use of the strain energy density function as part of the failure criterion as discussed in the Introductory Chapter. The basis of the arguments rests on the hypothesis that fracture occurs when a small element of material in the bulk of the solid near the notch front has absorbed a critical amount of energy, and then releases it to allow crack extension. The critical value of the strain energy density functions is also characteristic of the material and serves as a measure of the fracture toughness. This strain energy density theory is also extended to notch boundaries where in addition the energy in a surface layer is calculated and the location of failure initiation is determined.

The concept of a core region near the notch front, and its consequences, are also examined in detail in the Introductory Chapter. This small region absorbs the unknown variations in material behavior, such as localized anisotropy or inhomogeneity, at the crack or notch tip while a valid elastic solution external to this region presumably provides an accurate measure of the failure behavior of material. The size of the core region leads to a length constant as an additional material characteristic. Thus, two material parameters are required to determine the strength of a notched solid: the critical strain energy density and the characteristic length. The examples treated are an internal elliptical notch and two external hyperbolic notches in a large isotropic, homogeneous medium, subjected to loads applied at an angle to the line of notch symmetry. The results are shown to approach those of the crack solutions for narrow ellipses or hyperbolas, and to display satisfactory agreement with published experimental data under both tensile and compressive loading conditions for the case of an elliptical notch. Results also indicate that in globally unstable configurations, the original loading and notch geometry are sufficient to predict the subsequent crack trajectory with considerable accuracy.

Special precautions should be taken when approximate methods of solution are used because in smoothing out stress peaks, they may underestimate stress concentrations. The number of boundary value problems which have been exactly solved is not large, and is mostly confined to two-dimensional problems of elasticity. The degree of difficulty increases very quickly when a notch interacts with neighboring notches or free surfaces. In this volume, several analytical methods of solution are presented and some two- and three-dimensional examples of geometrical discontinuities are provided as a means of showing the kinds of stress concentration that can arise.

Chapter 1 presents a method for solving two- and three-dimensional notch problems by making use of a point force solution in the region to be considered without any discontinuities such as notches or cracks. The method of solution involves the superposition of stress fields derived from a system of continuously embedded point forces referred to as body forces. The density of the body force at the location of the notch or crack site is adjusted so as to satisfy the specified boundary conditions. In the simpler problems, the density of the body force may be determined in closed form while numerical procedures must be adopted for the more complicated notch geometries. The local stresses for a wide variety of notch geometries are tabulated and presented graphically.

The application of MMC (modified mapping-collocation) in the stress analysis of two-dimensional notch problems is considered in Chapter 2. Conformal mapping is applied to subregions involving only the notch tip and its vicinity. This is in contrast to the classical approach of global mapping of the entire physical geometry on a rigidly prescribed parameter region. When the MMC method is combined with partitioning, separate expansions can be made for two or more subregions thus improving the convergence of the solution. Examples for several notch geometries are provided to show that only a few terms of these expansions are necessary for the cataloging of sufficient information for the calculation of the local notch and subsurface stresses. Displacement or mixed boundary conditions can also be adopted in the MMC plane. For those regions remote to the notch that are awkward to represent by series expansions, e.g., narrow strips, the regions can be partitioned, or alternatively represented by finite elements.

The primary emphasis of Chapter 3 is to offer a mathematical treatment of the stresses in a semi-infinite medium with one or more edge notches. The method of solution consists of converting the Airy's stress function into two analytic functions of the complex variable $z = x + iy$. Mapping is then used to transform the notch boundary in the physical plane into a portion of the circumference of a unit circle in the mapped plane. The resulting set of equations is then truncated to satisfy the boundary conditions of no tractions on the notch surface. A suitable method of minimizing the inaccuracy owing to truncations is then devised. Stress solutions are provided for a variety of edge notch shapes, namely semi-circular, almost circular, semi-elliptical, U-shaped and V-shaped. When there are more than one notch on the edge of a semi-infinite medium, the solution can be constructed in the same way by superposition from the fundamental solution of a single notch. If the notches are not of the same size and/or shape, the boundary conditions of each notch must then be adjusted individually and the method of solution becomes exceedingly complicated.

Chapter 4 is concerned with the triaxial character of the stress field in the neighborhood of a three-dimensional notch embedded in an infinite elastic solid. The formulation makes use of various derivatives of the harmonic function often used in potential theory associated with the ellipsoid. The displacements and stresses near the notch front are derived in terms of a convenient set of spherical coordinates measured from the edge of a focal plane of an ellipsoid. The singular stresses for a sharp flat crack are recovered in the limit. Such information is pertinent to the formulation of fracture mechanics theories for three-dimensional cracks with blunt edges or notches.

It is fitting here to express gratitude to my two secretaries Mrs. Barbara DeLazaro and Mrs. Constance Weaver who were most helpful in typing and in assisting the details associated with the completion of this volume.

Lehigh University G. C. SIH
Bethlehem, Pennsylvania

September, 1977

Contributing authors

O. L. Bowie
Army Materials and Mechanics Research Center, Watertown, Massachusetts

C. E. Freese
Army Materials and Mechanics Research Center, Watertown, Massachusetts

M. K. Kassir
Department of Civil Engineering, The City College of the City University of New York, New York, New York

Chih-Bing Ling
Department of Mathematics, Virginia Polytechnic Institute and State University, Blacksburg, Virginia

H. Nisitani
Faculty Engineering, Kyushu University, Fukuoka, Japan

G. C. Sih
Institute of Fracture and Solid Mechanics, Lehigh University, Bethlehem, Pennsylvania

G. C. Sih

Introductory chapter:

Strain energy density and surface layer energy for blunt cracks or notches

I Background information

Although much effort has been spent in the past to develop failure theories, there is still considerable uncertainty involved in determining the strength of a material. A central problem facing the designer today is optimizing the use of the multivarious materials now known to him and available for application under various loading and environmental conditions. This calls for an a priori knowledge of the kind of failure expected before design limits can be established. To acquire this knowledge, it is essential to know, at a minimum, the nature of loading, the prevailing environments, the material behavior, and whether there are any damaging cracks, notches or other mechanical imperfections present in the solid.

In recent years, a bewildering number of theories have been proposed on the grounds that they 'fit the data' rather than that they rely on self-consistency, and hence it has not been difficult to find other sets of data for which the theories failed to apply. This has caused a great deal of confusion in the field. On the other hand, very little has been done to coordinate these theories so that the failure of solids can be explained on a more unified basis. Today, separate theories are still used for the failure of notches and cracks. A case in point is that of the failure condition of solid with notches which is usually based on the local maximum normal stress criterion. Griffith [1] employed such a stress criterion to predict the failure of a solid with an internal elliptical notch instead of the now well-known concept of energy release rate as he

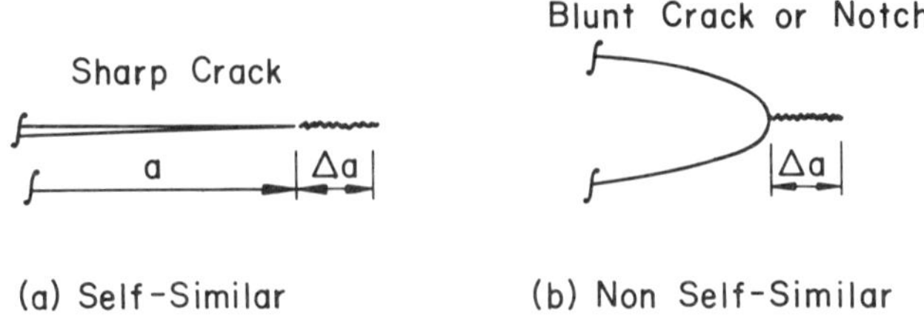

Figure 1. Crack and notch failure.

originally proposed [2] for a perfectly sharp crack, Figure 1a. This is because of the conceptual difficulties* involved in extending blunt cracks or notches in a self-similar manner. Figure 1b shows that the material ahead of the notch front fails by cracking and hence the notch geometry before and after material separation is not preserved. Therefore, Δa can be considered as crack extension only for the case of a sharp crack (Figure 1a) where the crack tip radius of curvature has a negligible influence.

A coherent theory of fracture that can consistently account for the fracture behavior of both sharp cracks and blunt notches has been advanced only recently [3, 4]. The basis of the theory is the hypothesis that fracture occurs when a small element of material near the notch or crack tip has absorbed a critical amount of energy, and then releases it to cause material separation. This material element is always kept at a finite distance, say r_0, from the crack tip, Figure 2a, or notch front, Figure 2b. The concept of a local core region in Figure 2 stems from a concern that in the immediate vicinity of the crack or notch tip, inhomogeneity of the material due to grain boundaries, microcracks, and dislocations preclude an accurate analytical solution, whereas an analysis presuming a valid continuum mechanics solution external to the core region can provide a sufficiently accurate measure of the failure behavior of the material. The continuum description requires the size of the core region to be limited to the smallest macroscopic element. For the notch geometry, a surface layer failure criterion [4] is dealt with as a preliminary requirement to locate the position (ϕ) where fracture may

* The application of the energy release rate concept to the notch geometry in Figure 1b would require the general solution of a small crack Δa extending from the notch boundary. Such an approach is inappropriate, for it delves into a complicated mathematical analysis whose accuracy will always be doubtful in the limit as $\Delta a \rightarrow 0$.

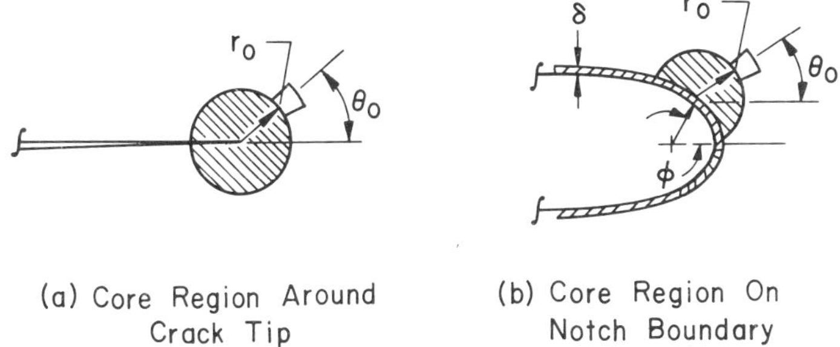

Figure 2. Core region near crack and notch front.

be expected to initiate from the notch surface. Irregularities at the microscopic scale are bounded within this layer of thickness δ as indicated in Figure 2b. Once the angle ϕ in Figure 2b is established, the knowledge of the energy stored in an element outside the core region is used as a means for establishing the location of failure at some point in the bulk of the solid. The angle θ_0 determines the initial direction of the surface of material separation. The calculation involves the determination of the stationary values of the strain energy density function, dW/dV.

The application of the strain energy density theory, involving an element outside the core region, makes possible the joining of the fracture analysis of cracks and notches. From the analysis point of view, the fracture of a sharp crack can be regarded simply as the limiting case of the elliptical notch, and needs, for the most part, no special consideration. Figures 3 and 4 provide some information as to how the initial direction of crack propagation would be affected by the crack tip radius of curvature, ρ. For an ellipse $\rho = a(b/a)^2$ where a and b are respectively the semi-major and semi-minor axes of the ellipse. Suppose that an elliptical notch in an infinite medium is stretched by a uniform stress p in a direction that makes an angle β with reference to the major axis of the ellipse. Figure 3 gives the fracture angle θ_0 (θ_0 is negative because it is in the clockwise direction) for tensile loading as a function of the notch (or load) angle β for a sharp crack ($b/a = 0$) and a blunt crack ($b/a = 0.07$). The radius of the core region is selected to be $r_0 = 0.05a$. Note that predictions based on the strain energy density theory show very little difference in θ_0 for the two cases when β is kept below 40°. Deviations are seen for β between 50° and 80°. A different set

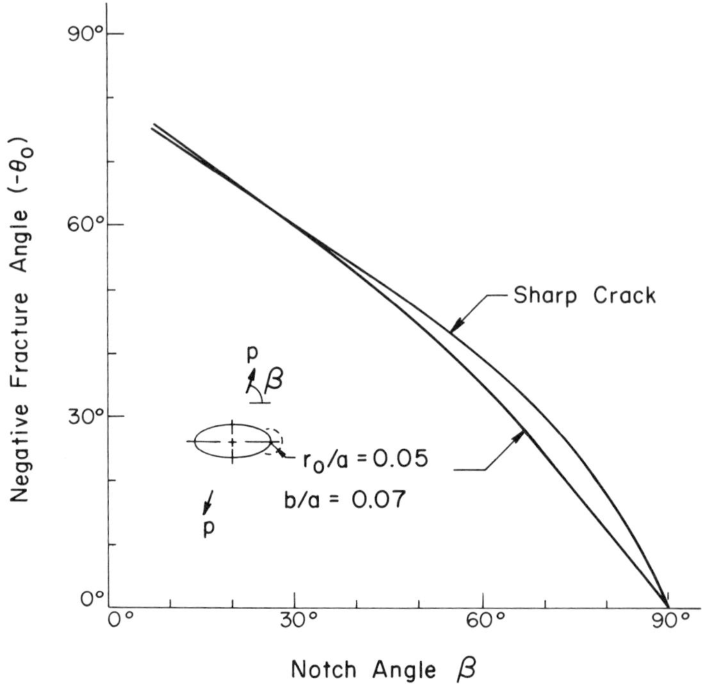

Figure 3. Fracture angle versus notch angle for elliptical notch in tension.

of results is obtained when the applied stress is compressive as shown in Figure 4. First of all, the direction of crack initiation tends toward the applied compressive stress and the difference between θ_0 for the sharp crack and notch is approximately the same for $20° \leqslant \beta \leqslant 80°$. Available experimental data [5, 6] on plexiglass plates with cracks show striking agreement with the results in Figure 3. This will be discussed in more detail subsequently.

The use of the critical value of strain energy density function, $(dW/dV)_c$, as a material constant has also been demonstrated by others [7, 8]. In particular, Gillemot [8] and his co-workers have performed many experiments to measure $(dW/dV)_c$ for numerous engineering materials. Sih has pointed out in [4, 9] that $(dW/dV)_c$ is equivalent to S_c/r where S_c is the critical strain energy density factor while r can be used as a radius vector to identify the location of failure.

The fundamental assumption at hand is that although failure initiates at a microscopic or atomic level, the continuum description is sufficient to establish design limits for the instability of structural components.

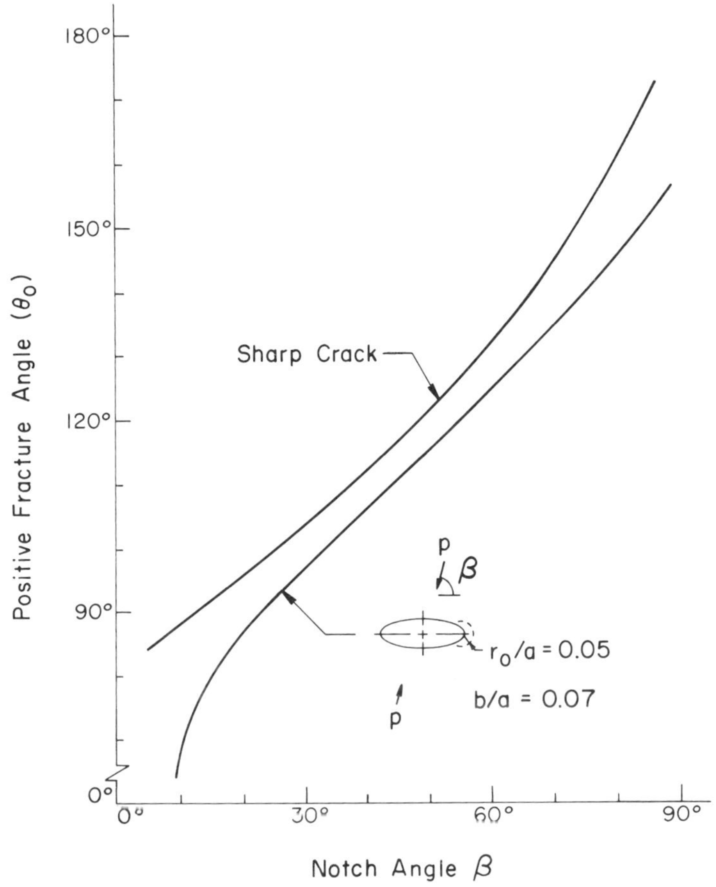

Figure 4. Fracture angle versus notch angle for elliptical notch in compression.

This has been the aim of the development of structural mechanics in which attention is focused not on modeling or understanding of the physical structure of the material, but on prediction. The desired properties of the material will be characterized by certain measurable parameters determined from test specimens in the laboratory.

II Surface layer energy

It is known from common experience that a physical boundary is not a sharp mathematically defined curve, but a rough surface composed of microscopic ridges, voids, cracks, etc. In machine parts, these boun-

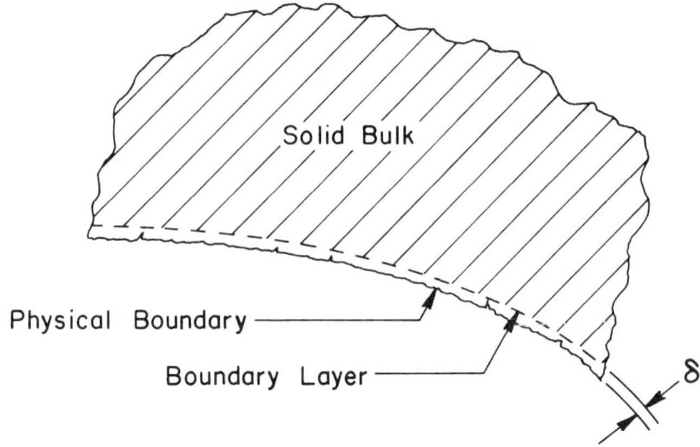

Figure 5. Free surface boundary layer.

daries may be formed by rolling, drawing or machining resulting in layers of materials that may be less porous or harder than those in the interior.

On the continuum level, modeling of the surface irregularities mentioned above is largely inhibited by lack of physical understanding and by the very nature of the continuum approximations. However, as noted from experience, it is these very surface features that can ultimately determine the failure strength of a solid and hence it is necessary to incorporate the surface effects into a continuum in an averaged way. To this end, a formal boundary layer of thickness δ is proposed as indicated in Figure 5 to separate surface and interior regions. To a first approximation, this layer is assumed to be a continuum such that its gross properties may be treated as an isotropic and homogeneous medium. Sih [4] has suggested a surface layer energy* failure criterion which assumes that failure will initiate at some location or region along the boundary layer of a solid when the local energy caused by loading will exceed some experimentally determined material constant.

Mechanics of the surface layer. The energy associated with the surface layer can be derived from the mechanics of the thin layer shown in Figure 6. A small element, $\rho \Delta s$, of the surface layer for a free

* This quantity is conceptually different from the surface tension for a liquid, nor is it related to the specific surface energy used in the Griffith criterion [2].

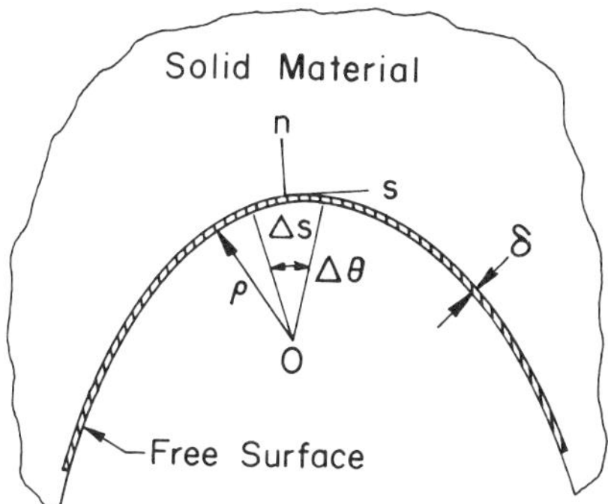

Figure 6. Boundary layer near free surface.

boundary is isolated and shown in Figure 7. If δ is of the order of magnitude of a continuum element, then σ_r and $\tau_{r\theta}$ are zero at both $r = \rho$ and $r = \rho + \delta$. Thus the element is under uniaxial tension, $\sigma_\theta = \sigma_t$, in the tangential direction. The corresponding tangential strain is

$$\varepsilon_t = \frac{1 - \nu^2}{E}\sigma_t \tag{1}$$

for plane strain where E is the Young's modulus and ν the Poisson's ratio for an elastic isotropic material. The strain energy density is therefore $\sigma_t\varepsilon_t/2$ and the strain energy per unit surface layer area becomes

$$\gamma_e = \frac{1}{2}\sigma_t\varepsilon_t\delta = \frac{1 - \nu^2}{2E}\delta\sigma_t^2 \tag{2}$$

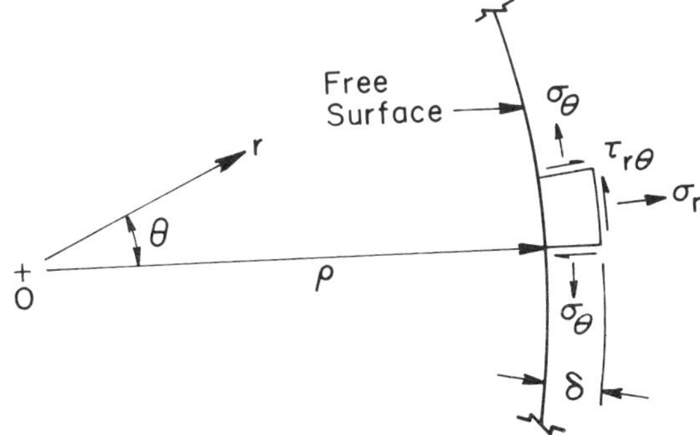

Figure 7. Continuum element on a free curved surface.

Since neither γ_e nor δ has explicitly been given, it is convenient to form the quantity $2E\gamma_e/\delta(1-v^2)$ and treat it as a material parameter. A notched specimen may be loaded to failure and from equation (2), $2E\gamma_e/\delta(1-\nu^2)$ computed; then for any notch geometry of this material, the failure load may be obtained. The location on the notch boundary at which failure occurs is assumed to coincide with the maximum value of γ_e.

For the simple case of an elliptical notch subjected to applied stress p in a direction normal to the major axis of the ellipse, the local tangential stress along the line of expected failure is

$$\sigma_t = p\left(1+\frac{2a}{b}\right)^2 \tag{3}$$

Using the relation $\rho = a(b/a)^2$ and substituting σ_t in equation (3) into (2) gives

$$\gamma_e = \frac{\delta(1-\nu^2)p^2}{2E}\left(1+2\sqrt{\frac{a}{\rho}}\right)^2 \tag{4}$$

If a/ρ is large in comparison with unity as in the case of a crack, equation (4) further reduces to

$$\gamma_e = \frac{2(1-\nu^2)p^2a}{E}\left(\frac{\delta}{\rho}\right), \quad \rho \ll a \tag{5}$$

Note that γ_e is not the Griffith surface energy term and is derived from an entirely different physical basis. It emphasizes a characteristic length δ with reference to the radius curvature of the crack and is somewhat similar to what Neuber [10] has used in his work on notches. However, his argument is completely different from the concept of the surface layer energy introduced here.

Although the form of equation (2) may suggest that the γ_e-criterion is equivalent to the commonly recognized failure criterion of the maximum surface tangential stress, there is a major conceptual difference that must be emphasized. The maximum stress criterion is basically applied in a layer of zero thickness, at surface points, whereas the surface layer energy has been applied over an element of non-zero thickness, and resulted in an additional parameter δ, in equation (2). As indicated earlier, the presence of δ provides a means of absorbing the local surface inhomogeneities into the energy theory; that is, different values of δ may be assigned to quantify different surface conditions, although the bulk material properties remain unaffected. The use of the surface tangential stress alone is unable to include such local phenomena. Brittle materials are especially sensitive to surface characteristics, and unless the continuum model can in some way include parameters reflecting such characteristics, the prediction of failure is likely to be quite inaccurate.

An experimental knowledge of the surface characteristic can indeed improve the use of the maximum surface layer energy by assigning material parameters to the thin boundary layer different from those of the bulk material. This step removes some of the arbitrariness associated with assigning $2E\gamma_e/\delta(1-\nu^2)$ as a new parameter, when it appears at first glance merely to bear the image of the maximum tangential stress criterion.

Elliptical notch under tension. A numerical example illustrating the behavior of the surface layer energy will now be given for the geometry of an elliptical notch in a large plate loaded at an angle as shown in Figure 8a. Around the notch surface, which is assumed to be rough, is depicted a boundary layer of uniform thickness, δ. At each point along the layer, represented by the equation of the ellipse in parametric form

$$x = a \cos \eta, \qquad y = b \sin \eta \tag{6}$$

where η is the eccentric angle for the ellipse in Figure 8b, the magnitude of the surface layer energy, γ_e, may be calculated from equation (2) once the local tangential stress is known. Without going into the details, σ_t for the problem in Figure 8a can be obtained from Muskhelishvili [11] and hence

$$\gamma_e = \frac{\delta(1-\nu^2)p^2}{2E} \times \left[\frac{(a+b)^2(\sin\eta\cos\beta - \cos\eta\sin\beta)^2 - a^2\sin^2\beta - b^2\cos^2\beta}{a^2\sin^2\eta + b^2\cos^2\eta}\right]^2 \tag{7}$$

For $\beta = 90°$, the largest value of γ_e is found at the ends of the major axis $\eta = 0$ and equation (7) reduces to the form in equation (4).

In its present form, equation (7) obscures the difference between tensile and compressive loading. Along the boundary of an elliptical

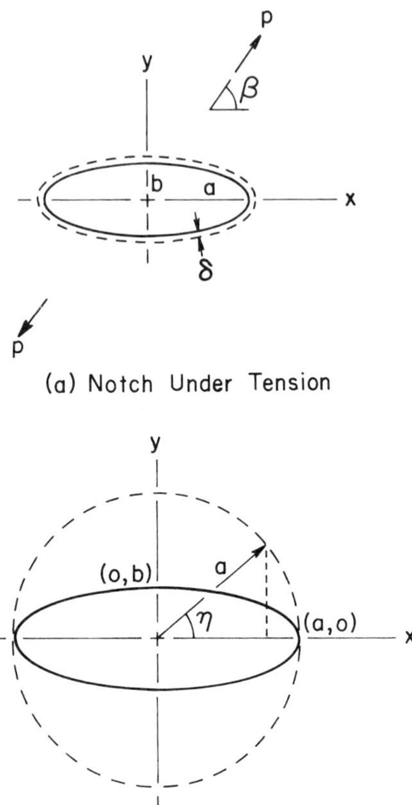

Figure 8. Elliptical notch.

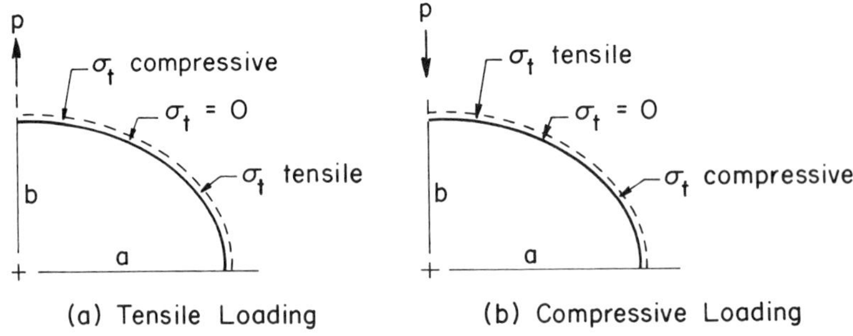

Figure 9. Tensile and compressive normal loading on an elliptical notch.

cavity, σ_t may be tensile or compressive, as indicated in Figure 9. However, γ_e remains positive at all points along the boundary, and hence, requires that the nature of the loading be dictated in order to select the region in which the surface layer is in tension. As the angle of loading varies, the regions of tensile and compressive surface stresses also shift positions.

If equation (7) is normalized to the form $2E\gamma_e/\delta(1-\nu^2)p^2$, where p is

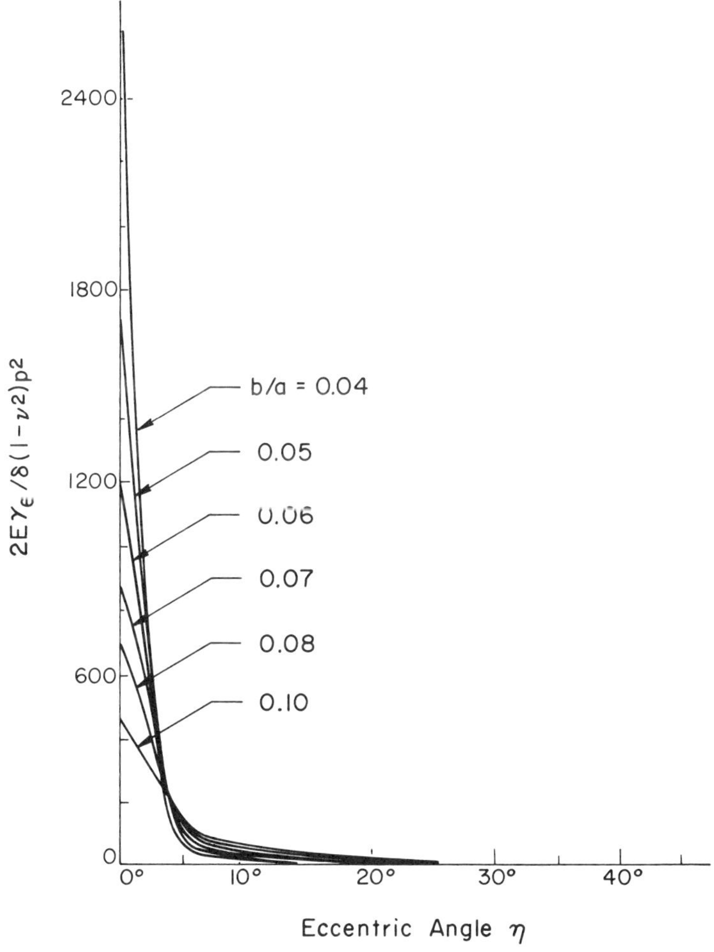

Figure 10. Normalized surface layer energy with eccentric angle for $\beta = 90°$.

the applied uniform stress, then this quantity may be represented as a function of the loading angle β, semi-minor to semi-major axes ratio b/a, and the eccentric angle η.

Fixing $\beta = 90°$, and applying a uniform tensile load results in the surface layer energy varying as a function of the eccentric angle η as seen in Figure 10, for several values of b/a. It is observed that for small values of b/a, the surface layer energy displays a sharp peak near $\eta = 0$ (an angle of symmetry), and this peak rapidly disappears as b/a increases. This is in part due to the nature of the eccentric angle. The same values of the surface layer energy may be plotted against the local normal angle ϕ as defined in Figure 11. The relation between η and ϕ is given by

$$\tan \phi = \frac{a}{b} \tan \eta \tag{8}$$

Referring to the results in Figure 12 in which $2E\gamma_e/\delta(1 - \nu^2)p^2$ is plotted against ϕ, it is seen that the pronounced peaks of Figure 10 have given way to more gentle slopes in Figure 12. In compression, at $\beta = 90°$, the stresses around an ellipse are always such that the stress at $\eta = 90°$, equal to the negative of the applied stress, is unchanged for all ratios b/a.

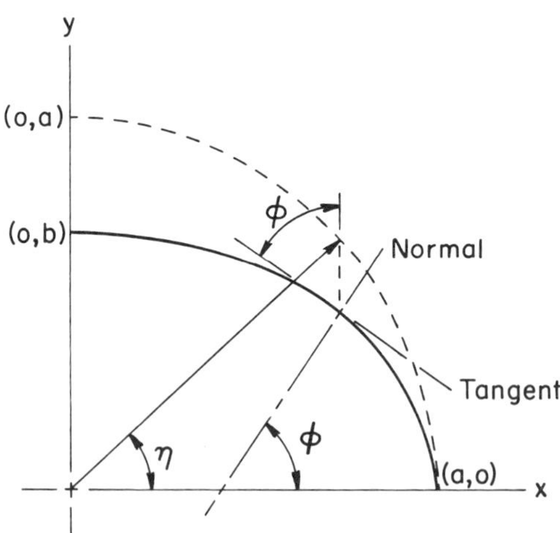

Figure 11. Relationship of normal and eccentric angles.

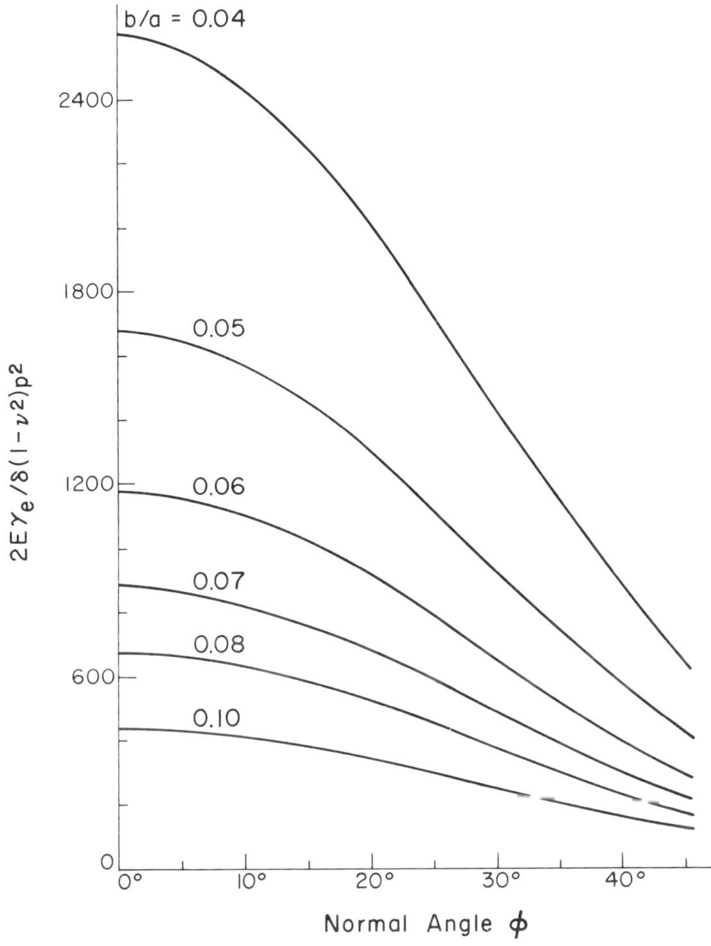

Figure 12. Normalized surface layer energy versus normal angle for $\beta = 90°$.

It has been assumed that failure will occur at the location of maximum surface layer energy (in tension), and both the magnitude and location along the surface vary with angle of loading. In tension, the maximum values of $2E\gamma_e/\delta(1-\nu^2)p^2$ vary with β as shown in Figure 13, for several values of b/a. In contrast, under compressive loading, for the same ratios b/a, the maximum values vary as in Figure 14, displaying a peak at some angle β near 45°. The importance of these results may be more

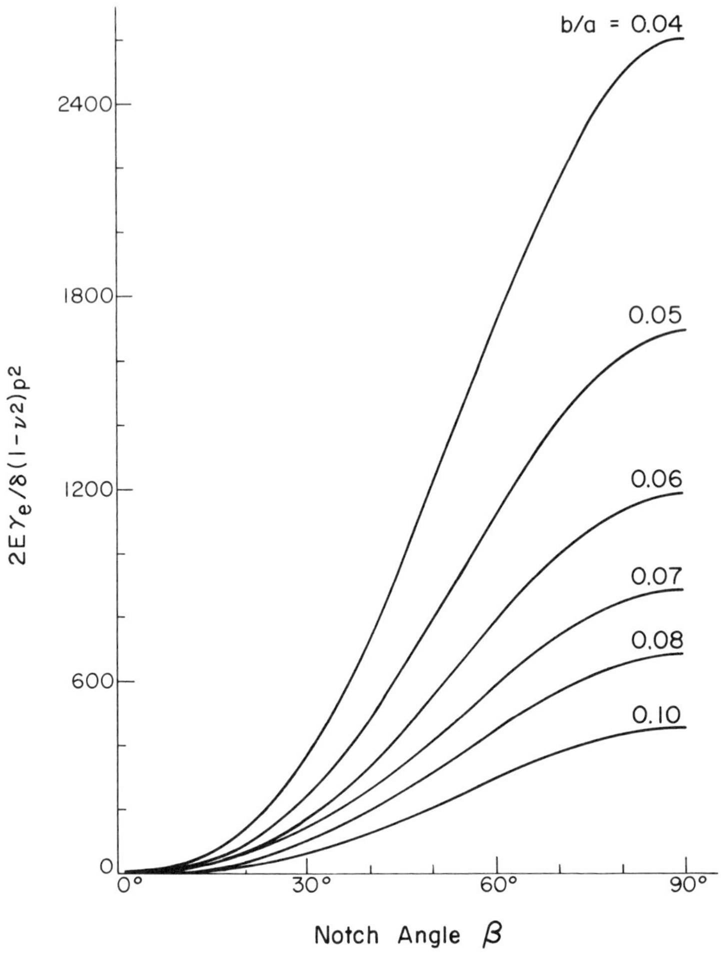

Figure 13. Variation of maximum surface layer energy with notch angle for the tension case.

clearly observed if the curves in Figures 13 and 14 are replotted as $p\sqrt{\delta(1-\nu^2)/2E\gamma_e}$, so that the permissible loads appear on the ordinate. Here the quantity $\delta(1-\nu^2)/2E\gamma_e$ is assumed to be characteristic of the material. In Figure 15, under tension the minimum load to failure is at $\beta = 90°$, while under compression (Figure 16), the minimum load to failure occurs in the region $\beta \simeq 45°$. Since the notch size enters into the problem, there is no need to assume surface contact under compression

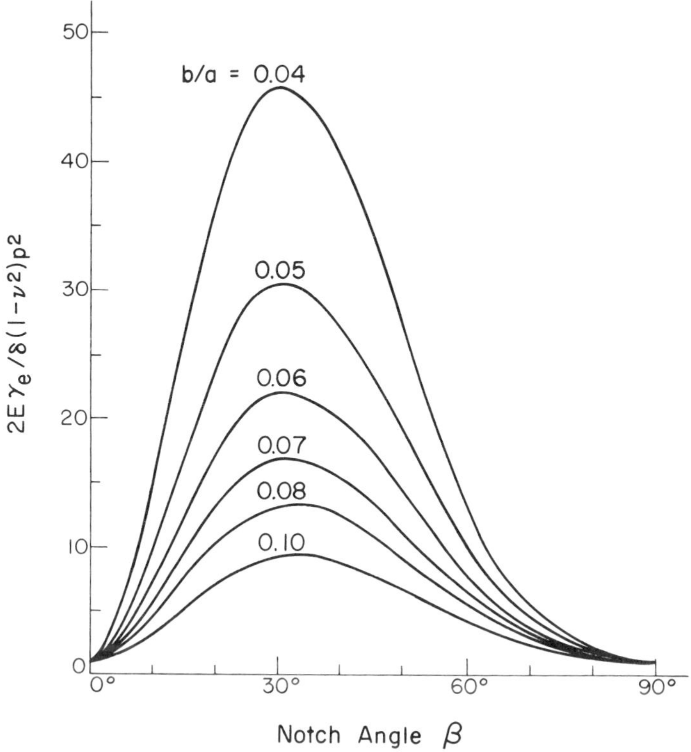

Figure 14. Variation of maximum surface layer energy with notch angle for the compression case.

as in the case of the crack configuration. This influence, however, has been exaggerated [12] in the past and does not lead to any significant correction on the continuum solution* for a crack.

III Strain energy density theory

Having discussed the surface layer energy criterion that applies only to the surface of a notch, this section moves into the strain energy density criterion local to points interior to the surface. A knowledge of the

* Experimental data on cracks under compression will show considerable influence due to surface contact as the crack surfaces are not mathematically smooth and parallel but notched with a zig-zag pattern.

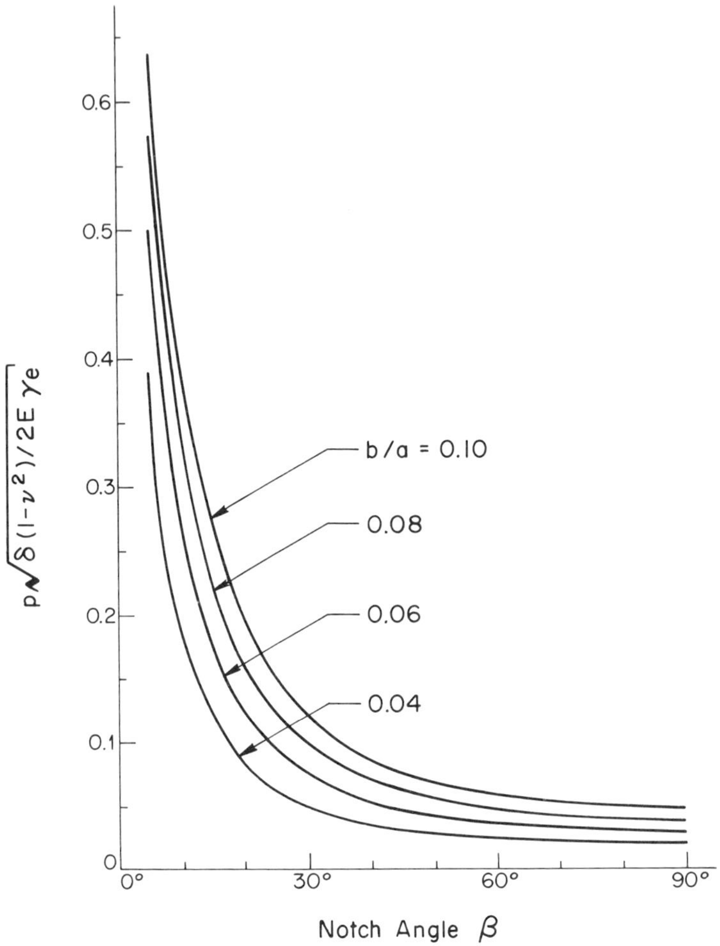

Figure 15. Maximum loading stress as a function of notch angle; tension case.

location of failure at points on the notch surface and in the bulk of the solid sheds light on the trajectory of crack propagation or failure path.

The inability of existing failure theories to adequately explain yield phenomena on the basis of maximum principle strain, maximum principle stress, or maximum shear stress hypotheses prompted Haigh [13] to expound a theory based upon the elastic energy absorbing capacity of a material. This strain energy function was defended rationally on the basis of thermodynamic considerations, and for materials with rather well

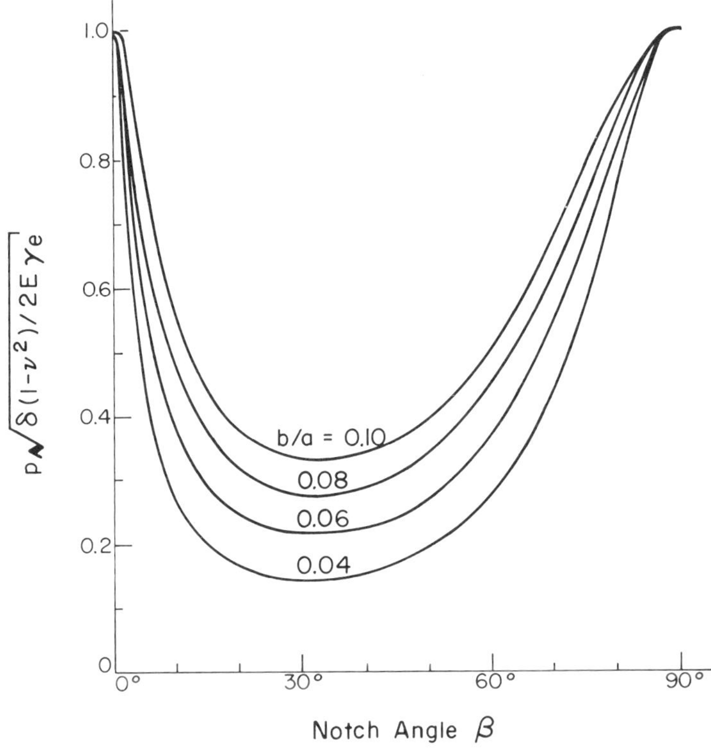

Figure 16. Maximum loading stress as a function of notch angle; compression case.

defined yield points, was substantiated by already existing data. The principal objective was to determine the 'limiting strain energy' that could be absorbed per unit volume of material uniformly strained to its elastic limit, and to use this quantity as a material constant. It was also noted that the principles set forth probably applied only to materials exhibiting ductile behavior, relegating brittle fracture phenomenon to be governed by other principles.

Recently, Sih [3, 4] has proposed a theory of fracture based on the field strength of the local strain energy density in an element of material ahead of the crack or notch. This theory can simultaneously account for yielding and fracture. The fundamental parameter in the new theory, the 'strain energy density' factor S, is direction sensitive in that it predicts the direction of crack initiation and/or propagation and yielding. This is accomplished by calculating the stationary value of S or dW/dV, where

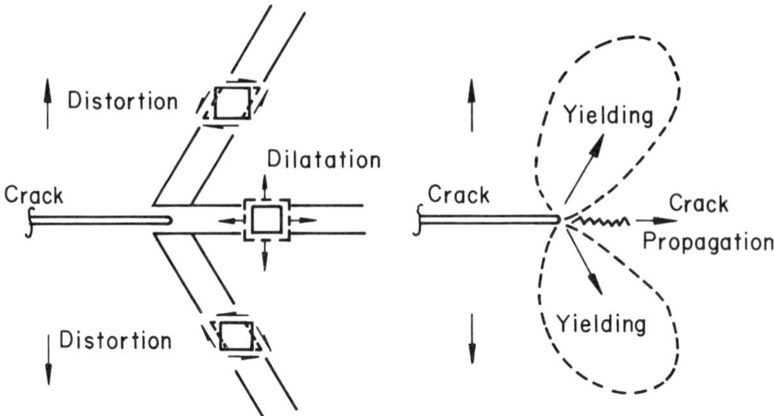

Figure 17. Dilatation and distortion of material elements ahead of a crack.

$\mathrm{d}W/\mathrm{d}V = S/r$ with r being the radial distance measured from the crack front and $\mathrm{d}W/\mathrm{d}V$ is the strain energy density function. The direction of $S_{\max}$ determines maximum distortion while $S_{\min}$ relates to dilatation. Figure 17 shows that distortion of material elements is associated with yielding which is to the sides of a crack under symmetric loading. Dilatation or volume change of material elements tends to be associated with the creation of free surface or fracture and occurs along the line of expected crack extension. As a fracture criterion, the critical value of $S_{\min}$, S_c, is used as a material constant to evaluate the failure load of a material, whether at a crack tip, notch front, re-entrant corners, or in an unflawed structure.

Strain energy density. In its original form, the theory makes use of the strain energy density function

$$\frac{\mathrm{d}W}{\mathrm{d}V} = \frac{\mathrm{d}W}{\mathrm{d}V}(\varepsilon_{ij}) = \tfrac{1}{2}(c_0 + c_{ij}\varepsilon_{ij} + c_{ijkl}\varepsilon_{ij}\varepsilon_{kl} + \cdots) \tag{9}$$

where ε_{ij} are the rectangular components of the strain and the coefficient c_0 can usually be disregarded if the body is not pre-stressed. For a linear elastic material, the stress components can be obtained from

$$\sigma_{ij} = \frac{\partial}{\partial \varepsilon_{ij}}\left(\frac{\mathrm{d}W}{\mathrm{d}V}\right) = \tfrac{1}{2}\sigma_{ij}\varepsilon_{ij}, \quad i, j = 1, 2, 3 \tag{10}$$

The components ε_{ij} in equation (10) may be eliminated by using the relation for an isotropic, homogeneous elastic medium:

$$\varepsilon_{ij} = \frac{1+\nu}{E}\sigma_{ij} - \frac{\nu}{E}\sigma_{kk}\delta_{ij} \tag{11}$$

This gives

$$\frac{dW}{dV} = \frac{1}{2E}(\sigma_{kk})^2 - \frac{\nu}{E}(\sigma_{11}\sigma_{22} + \sigma_{22}\sigma_{33} + \sigma_{11}\sigma_{33}) + \frac{1}{2\mu}(\sigma_{12}^2 + \sigma_{23}^2 + \sigma_{13}^2) \tag{12}$$

in which the Young's modulus E is related to the shear modulus μ as $E = 2(1+\nu)\mu$.

In order to determine the initiation of crack propagation or fracture, it suffices to take the singular terms of the stresses near a line crack tip, and substitute them into dW/dV in equation (12). This results in a quadratic form for the strain energy density function,

$$\frac{dW}{dV} = \frac{1}{r}(a_{11}k_1^2 + 2a_{12}k_1k_2 + a_{22}k_2^2 + a_{33}k_3^2) + \cdots \tag{13}$$

Here the quadratic

$$S = a_{11}k_1^2 + 2a_{12}k_1k_2 + a_{22}k_2^2 + a_{33}k_3^2 \tag{14}$$

represents the amplitude of the energy density field, where the coefficients $a_{ij}(i, j = 1, 2, 3)$ vary with a polar angle, θ, measured from the crack tip, and for the case of plane strain they are given by:

$$a_{11} = \frac{1}{16\mu}[(3\text{–}4\nu - \cos\theta)(1+\cos\theta)] \tag{15a}$$

$$a_{12} = \frac{1}{16\mu}2\sin\theta[\cos\theta - (1\text{–}2\nu)] \tag{15b}$$

$$a_{22} = \frac{1}{16\mu}[4(1-\nu)(1-\cos\theta) + (1+\cos\theta)(3\cos\theta - 1)] \tag{15c}$$

$$a_{33} = \frac{1}{4\mu} \tag{15d}$$

The stress intensity factors k_1, k_2, k_3 are dependent only upon loading and geometric conditions.

In general, for a notch, the form of the strain energy density will not be as concise as that of equation (13), but the evaluation of the energy field will be the same as that for the crack, and the resulting failure loads and angles also determined on the basis of the strain energy density assuming a critical constant value in a given material. For a relatively sharp elliptical notch, the stresses near the notch tip may be expanded in a series in terms of a set of local polar coordinates r and θ as follows:

$$\sigma_{11} = \frac{1}{\sqrt{2r}}\left[k_1 \cos\frac{\theta}{2}\left(1 - \sin\frac{\theta}{2}\sin\frac{3\theta}{2}\right) - k_1\left(\frac{\rho}{2r}\right)\cos\frac{3\theta}{2} - k_2 \sin\frac{\theta}{2}\left(2 + \cos\frac{\theta}{2}\cos\frac{3\theta}{2}\right) + k_2\left(\frac{\rho}{2r}\right)\sin\frac{3\theta}{2}\right] \tag{16a}$$

$$\sigma_{22} = \frac{1}{\sqrt{2r}}\left[k_1 \cos\frac{\theta}{2}\left(1 + \sin\frac{\theta}{2}\sin\frac{3\theta}{2}\right) + k_1\left(\frac{\rho}{2r}\right)\cos\frac{3\theta}{2} + k_2 \sin\frac{\theta}{2}\cos\frac{\theta}{2}\cos\frac{3\theta}{2} - k_2\left(\frac{\rho}{2r}\right)\sin\frac{3\theta}{2}\right] \tag{16b}$$

$$\sigma_{12} = \frac{1}{\sqrt{2r}}\left[k_1 \sin\frac{\theta}{2}\cos\frac{\theta}{2}\cos\frac{3\theta}{2} - k_1\left(\frac{\rho}{2r}\right)\sin\frac{3\theta}{2} + k_2 \cos\frac{\theta}{2}\left(1 - \sin\frac{\theta}{2}\sin\frac{3\theta}{2}\right) - k_2\left(\frac{\rho}{2r}\right)\cos\frac{3\theta}{2}\right] \tag{16c}$$

$$\sigma_{33} = \nu(\sigma_{11} + \sigma_{22}),\ \sigma_{13} = \sigma_{23} = 0 \tag{16d}$$

which correspond to a state of plane strain. The radial distance r in equations (16) is measured from the focal point of the notch and always remain nonzero and θ is measured from a line extended from the major axis of the notch. The ratio ρ/r is such that it is not negligible in comparison with unity. Although the stress field in equations (16) differs from that for a sharp crack where $\rho = 0$, the stress intensity factors k_1, k_2 and k_3 are unchanged. For plane strain problems, equation (12) simplifies to the form

$$\frac{dW}{dV} = \frac{1}{2\mu}\left[\frac{1-\nu}{2}\left(\sigma_{11}^2 + \sigma_{22}^2\right) - \nu\sigma_{11}\sigma_{22} + \sigma_{12}^2\right] \tag{17}$$

where $\mathrm{d}V = \mathrm{d}A \cdot 1$ for two-dimensional problems. Substituting the stresses in equations (16) into (17) yields a strain energy density factor

$$S = r\frac{\mathrm{d}W}{\mathrm{d}V} = S_1 + \frac{S_2}{r} + \frac{S_3}{r^2} \tag{18}$$

in which

$$S_1 = a_{11}k_1^2 + 2a_{12}k_1k_2 + a_{22}k_2^2 \tag{19a}$$

$$S_2 = b_{11}k_1^2 + 2b_{12}k_1k_2 + b_{22}k_2^2 \tag{19b}$$

$$S_3 = c_{11}k_1^2 + 2c_{12}k_1k_2 + c_{33}k_2^2 \tag{19c}$$

The coefficients a_{ij} are the same as those in equations (15) for the crack problem while b_{ij} and c_{ij} are given by

$$b_{11} = 0, \qquad b_{12} = -\frac{\rho}{8\mu}\sin\theta, \qquad b_{22} = -\frac{\rho}{4\mu}\cos\theta \tag{20a}$$

$$c_{11} = \frac{\rho^2}{16\mu}, \qquad c_{12} = 0, \qquad c_{22} = \frac{\rho^2}{16\mu} \tag{20b}$$

For more details, refer to the work of Sih [14] for the example of an elliptical notch subjected to in-plane shear loading.

Core region. A fundamental difficulty in fracture mechanics arises in the interpretation of high stresses at sharp corners in the analytical models which are physically unattainable. In part, this difficulty arises because analytically, one cannot expect the solution to be accurate much closer to the tip of a crack or notch than the minimum continuum element size. It must also be recognized that in a very local region at the crack tip, the physical behavior is unknown, and cannot be incorporated into a mathematical model. In this region, the material being highly strained, may become inhomogeneous, and in general, is not conducive to modeling. In polycrystalline materials, for example, the orientation of a grain may be significant, but unknown from a continuum viewpoint. These same limitations hold for sharp notches as well, and likely for blunted notches. But it must be emphasized that both the analytical and physical aberrations mentioned are confined to a very small region near the notch tip, and for this reason are not expected to significantly perturb the analytic solution external to this region.

The initial approach taken here is that of postulating the previously stated aberrant material behavior to be limited to a region local to the notch tip, Figure 2b, characterized by a length dimension, r. This core region will remain temporarily unspecified in size. The main purpose is to avoid the requirements of an explicit continuum solution everywhere around the notch and hence evaluation of the fracture toughness will be done on the basis of material behavior external to the core region.

Unstable crack extension is assumed to occur when some small element just outside the core region (Figure 2b) has absorbed as much elastic energy as possible and releases it to allow material separation*. The failure of the core region remains unexplained in view of the lack of a mathematical description of its material properties. It is also necessary to state that the 'boundary' separating the elastic material from the core region is a fictitious one. Continuity must prevail at the interface. Rather, the core dimension is simply a measure stating that on a continuum level, closer approach to the notch tip renders the analytical model inaccurate. In reality, the size of this core region will reflect a basic property of the material at the microscopic scale level.

As discussed earlier, the location on the free surface of a notch of the maximum surface layer energy may be postulated to be the region of initial failure of the notch. Subsequent behavior, such as the immediate post-failure crack direction, may be in part described by the local features of the strain energy density function. A radius vector may be attached to the point of maximum surface layer energy. The vector is of length r, and its position relative to a fixed axis is given by the angle θ, measured positive counterclockwise as shown in Figure 2b. The origin of the vector is a function of both loading and notch geometry. Further, for each notch and loading geometry, there is usually a separate origin for the tension and compression cases. In the degenerate case, when the notch becomes a line crack, the concept of surface layer energy is inapplicable, and the origin moves to the crack tip, as in the S-theory [3].

Prediction of failure path. Prior to becoming involved in the details of an example, a few additional observations and conjectures on failure

* Brittle behavior is assumed in that the material and notch geometry do not deform appreciably prior to instability triggered by crack propagation.

at notches and cracks will be advanced for consideration. Having already noted that while in principle the strain energy density theory may be extended with suitable constitutive and kinematic relationships to account accurately for behavior near the notch tip, the primary limitation placed on the present discussion of the theory, including that of the core region, is that the material or structural member under consideration behaves in a brittle fashion. Within such a framework, it does not seem unreasonable to assert that the initial notch and loading geometries (and material properties) determine not only the load required to precipitate fracture and the initial fracture angle, but also the subsequent path the fracture will follow. That is, upon being loaded, the notch must seek a path of release, through knowledge of the local properties of the strained material, seen from its *present configuration.* And in view of the work of Kipp and Sih [15], it appears possible that in tension the fracture path follows the trajectory of points of minimum strain energy density remarkably closely. That is, once fracture is initiated, it is assumed that its path can be determined from the condition of the uncracked state. These are, however, displacement controlled situations for which crack growth is globally stable. This is because the input energy remains constant and the crack will initially grow, then stop, and remain stationary until a further increase in load or input energy occurs. Under such conditions, long range path predictions would not be possible. In the situations where local and global instability occur almost simultaneously, such as notches loaded uniformly at distances far from the notch, crack growth is quite sudden and unstable. Then the material has no time to readjust to a new configuration, and as suggested above, quite possibly this path can be predetermined with sufficient accuracy. A primary argument in favor of such behavior comes from an understanding of the global energy field. For brittle behavior of blunt notches, or line cracks at low angles of loading, the loads required for initial fracture are relatively high (when compared to that of the line crack in symmetric loading) so that there already exists a high density energy field in the regions at both near and far distance from the point of incipient failure. Such a field will do much to contribute to rapid fracture and failure, when few available physical resources exist to inhibit the fracture from propagating.

In order to determine a path for the fracture, the family of curves generated for a wide range of radius vectors provides the locations of the sequence of points that indicated stationary values of the strain energy density function. It is suggested that this trajectory should be in close agreement with the actual fracture path* taken from a notch that behaves in a brittle fashion. The subsequent discussion for the specific example considered will bear out at least locally the accuracy of the path prediction capabilities.

IV The embedded elliptical notch

For the purpose of illustration, a specific notch geometry has been chosen to demonstrate the methods of analysis and the particular uses for which the resulting information about the strain energy density near the notch tip can be employed. The choice of a uniformly loaded solid containing an embedded elliptical notch is based primarily on the availability of an analytic solution for this geometry, and also on the availability of recent experimental work involving this geometry under both tensile and compressive loading conditions. Additionally, the ease with which the solution is reduced to that of the finite crack geometry permits comparison with this limiting case of blunted notches.

The details of the stress field around the elliptical cavity under uniform loading are provided by Muskhelishvili [11] for an isotropic and homogeneous material under plane strain. The elliptical notch is defined by the ratio of minor to major axis half-lengths, b/a, with all other length

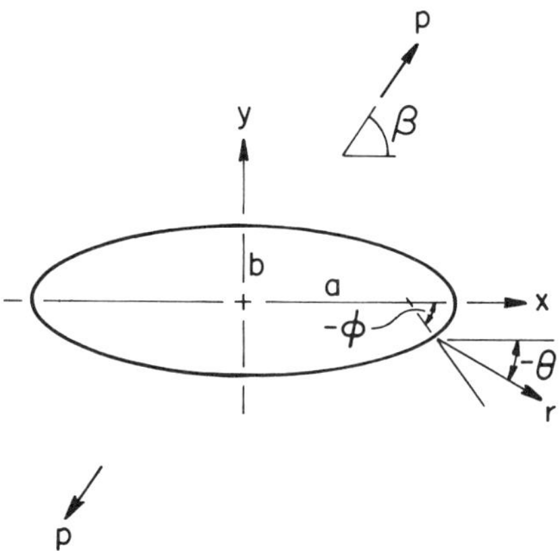

Figure 18. Elliptical notch under angle loading.

* Hartranft and Sih [16] have predicted the fracture surface emanating from an embedded three-dimensional flat elliptical crack in a combined stress field. This surface is assumed to coincide with that of minimum strain energy density.

parameters normalized with respect to the major axis half-length, a. The uniform loading (at infinity) forms an angle of β with the major axis of the cavity, and the radial distance r is referenced from the point on the notch surface that forms the normal angle ϕ with the major axis and is measured positive counterclockwise. The position of r with respect to the horizontal line is determined by the angle θ (positive counterclockwise). Figure 18 indicates the geometry described. Only the right hand notch tip will be considered, as antisymmetry of the geometry implicitly allows description of the other notch tip as well.

A brief exposition on the strain energy density function in linear elasticity theory has already been given in Section III. As mentioned earlier, the stresses for the elliptical notch problem in Figure 18 can be obtained from Muskhelishvili [11]:

$$\sigma_{11} + \sigma_{22} = 2[\phi'(z) + \overline{\phi'(z)}] \tag{21a}$$

$$\sigma_{22} - \sigma_{11} + 2i\sigma_{12} = 2[\bar{z}\phi''(z) + \psi'(z)] \tag{21b}$$

in which σ_{11}, σ_{22} and σ_{12} are the rectangular stress components and $\phi(z)$ and $\psi(z)$ are complex functions of the variable $z = x + iy$. The subscripts 1 and 2 correspond to the directions x and y, respectively. These functions are given by

$$4\phi'(z) = \frac{p}{c^2}\left\{(a+b)^2 \cos 2\beta + [c^2 - (a^2 + b^2)e^{2i\beta}]\frac{z}{\sqrt{z^2 - c^2}}\right\} \tag{22a}$$

$$4\psi'(z) = p(a^2 + b^2 - c^2 \cos 2\beta)\frac{z}{(z^2 - c^2)^{3/2}} - \frac{(a+b)^2}{c^4}\left\{2(a^2 + b^2)\cos 2\beta + 4iab \sin 2\beta + [2ab \cos 2\beta + i(a^2 + b^2)\sin 2\beta]\frac{z(2z^2 - 3c^2)}{(z^2 - c^2)^{3/2}}\right\} \tag{22b}$$

where $c^2 = a^2 - b^2$. Putting equations (22) into (21) and the subsequent results into equation (17) yields the strain energy field surrounding an elliptical notch. All stationary values of the strain energy density are obtained numerically, as the contours followed do not lend themselves to analytic evaluation.

Since the behavior of the strain energy field is load dependent, the cases of compression and tension are treated separately. There are some fundamental differences in the two cases, and it is intended that separation will clarify these departures without redundancy of explanation.

The strain energy density function is also dependent upon the value of Poisson's ratio, and it will be assigned a value of 0.250, unless otherwise stated. It is clear from the work of Sih [3, 4] that there are significant deviations attributable to Poisson's ratio change. These deviations are restricted to magnitudes only, however, and trends remain unchanged. Hence, subsequent discussion will center not on this effect, but on basic observable behavior, restricting changes in the Poisson's ratio to matching experimental specimen material where necessary.

The choice of specific values for the ratio b/a has been governed to large extent by the experimental data available. In other cases, the choice is simply for illustrative purposes, and any further comparison with new data requires new calculations to be performed. It is intended that the choices reflect fairly completely the range of possible behavior for this geometry of notch and loading.

Origin of failure on notch. The application of the local strain energy density field requires a knowledge of the position of maximum surface layer energy, γ_e, which determines the point on the notch boundary where failure may occur. For the elliptical notch, this position may be obtained analytically by setting the first derivative of γ_e in equation (7) with respect to eccentric angle η equal to zero, i.e., $\partial\gamma_e/\partial\eta = 0$. The resulting equation, quadratic in $\tan\eta$, may be solved to give

$$\tan\eta = \frac{b[a\sin^2\beta - b\cos^2\beta \pm \sqrt{a^2\sin^2\beta + b^2\cos^2\beta}]}{a(a+b)\sin\beta\cos\beta} \tag{23}$$

The positive root provides the location for failure initiation under compressive loading, and the negative root for failure initiation under tensile loading (Figure 19).

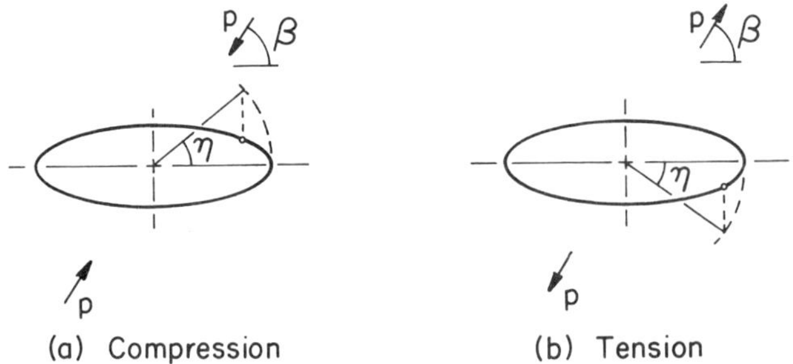

Figure 19. Location of failure points on notch in tension and compression.

It will be convenient to have the normal angle to the surface at both these points; the transformation is provided in equation (8). These equations are meaningful only for $b/a \neq 0$. The special case of $b/a = 0$ corresponds to that of a line crack, and in such a case, the radius vector is attached to the crack tip. This has been treated to considerable extent by Sih [3], where the asymptotic expansion is considered. Rather than repeat the results of that work, it will be instructive here to discuss possible behavior further from the crack tip.

Elliptical notch in tension. If an elliptical cavity in an isotropic, homogeneous, linearly elastic solid is uniformly loaded as shown in Figure 18, at an angle β to the major axis, the maximum surface layer energy occurs in the surface region in the fourth quadrant, where $-90° \leqslant \phi \leqslant 0°$, (considering right hand tip only). Loading behavior as a function of crack angle, notch size, and position in the medium is the subject of the initial discussion. Then considerations will focus on the initial and subsequent crack trajectories. This order is necessary in light of the earlier criteria established for the expected nature of failure and its relationships to the core region size.

For purposes of initial analysis, the applied uniform stress, p, will be assumed to be small, with no core region present, to establish the behavior of the strain energy field. Although normalized slightly differently, the strain energy behavior at the surface coincides with that shown in Figure 15, since at the surface, the strain energy is a function only of the one non-zero stress component, the tangential stress, the same one involved in the surface layer energy. As the radius (normalized to the major axis half-length) is made to increase, the stationary values of the strain energy begin to decrease, indicating that analytically, (exclusive of surface layer energy considerations) the strain energy density assumes its absolute maximum on the surface of the notch, (Figures 20–22). In each of the figures, a family of curves appears, each curve representing the possible loading behavior of a particular notch at a different distance from the notch surface. It may be observed that in all cases, the minimum load required for fracture occurs in the normal loading case ($\beta = 90°$). Intuitively, such a result appears to be reasonable, although some other fracture criteria have been proposed in which the minimum load appears in the vicinity of $\beta = 70°$. Palaniswamy [17] attempted an energy release rate procedure for complex loading, in which a small extension to the main crack was varied in direction until the maximum strain energy release rate for small crack extension was

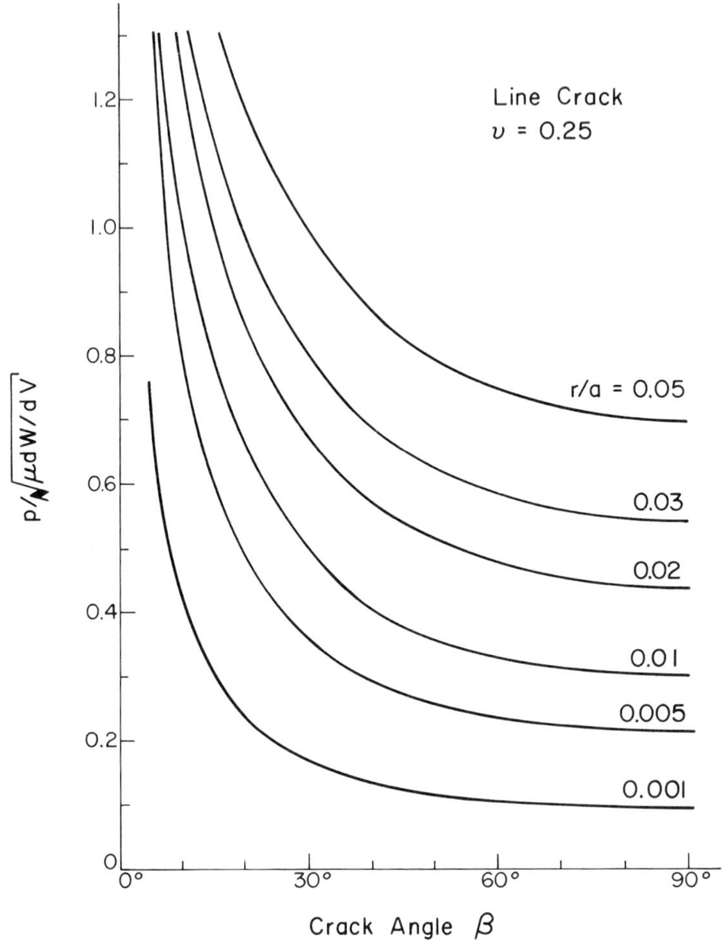

Figure 20. Variation of load with crack angle: line crack in tension.

determined; in this case, a minimum was located near 72°. It must be mentioned that in order to solve the stated problem, approximations in the numerical procedures altered the strain energy calculation by unknown amounts, assumed small, casting some doubt on the validity of the results, and the extent to which any implications can be made concerning fracture load behavior. A second case in which the minimum load is predicted to appear near $\beta = 70°$ occurs in the criterion of maximum tangential stress near the crack tip. Unmentioned in the

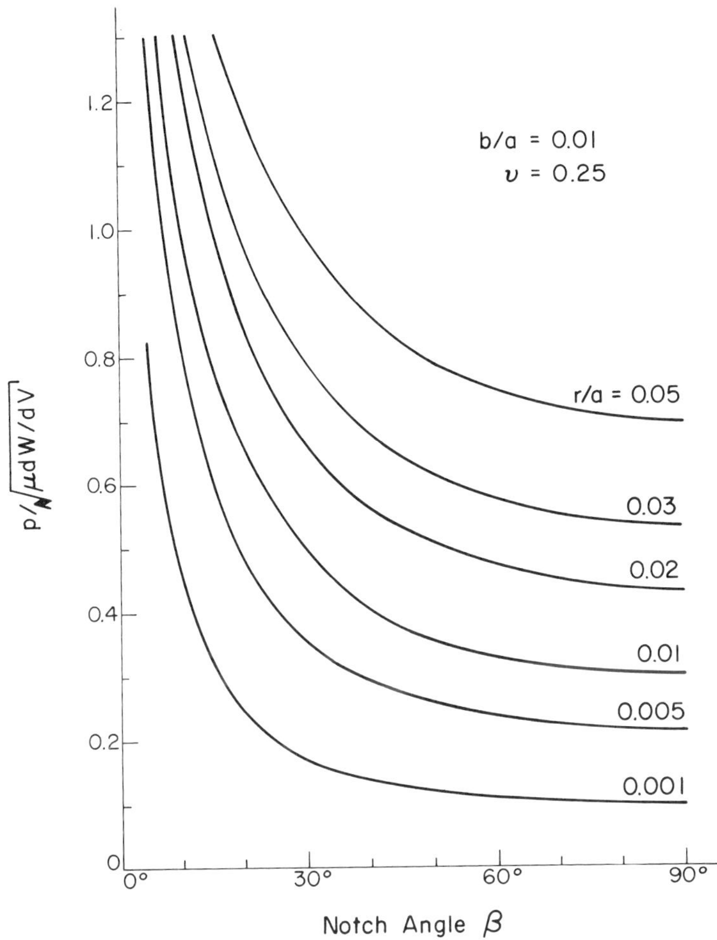

Figure 21. Variation of load with notch angle: $b/a = 0.01$ and tensile loading.

original paper on the inclined loading of a crack by Erdogan and Sih [18], Williams and Ewing [6] pointed out this effect in the loading response. This latter work included an attempt to avoid the local material behavior by applying the criterion a short distance from the crack tip using an additional term in the asymptotic solution. Unfortunately, truncation errors are the cause of the discrepancy, as is manifested when compared to the exact solution (Sih and Kipp [5] and Kassir [19]). The data presented in [6] appear to display a trend towards

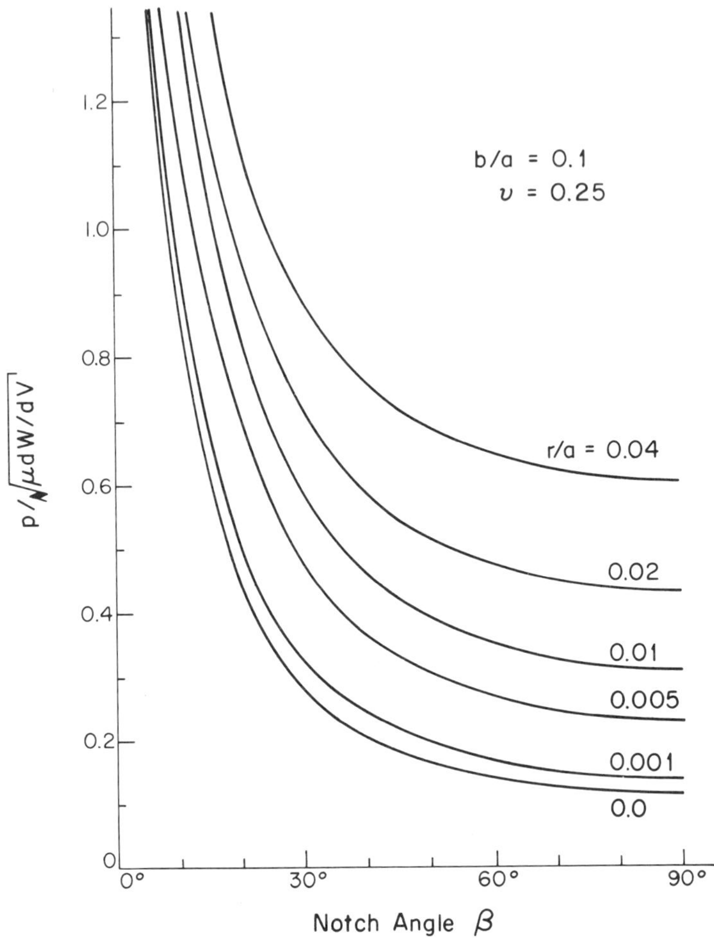

Figure 22. Variation of load with notch angle: $b/a = 0.1$ and tensile loading.

minimum load near $\beta = 70°$; however, much of the data had to be questioned because of inability to distinguish the initial direction of crack growth. In addition to the interaction of Mode I and II stress intensity factors, Mode III may also be present if the crack propagates at an angle in the thickness direction, and so the data [6] is not truly representative of the failure load as a function of the angle of loading. As Sih [9] has shown, the minimum failure would indeed be near $\beta = 70°$ when Mode III is present in addition to in-plane extension.

Figures 20–22 also reflect the trend of the increased loading permitted as the notch tip becomes more blunted. If an elliptical notch is experimentally loaded to failure at several crack angles, then from the corresponding family of curves for the same size notch, the curve that matches the data may be extracted, and that choice will provide the characteristic dimension of the core region, assuming that the elastic strain energy density constant is known. A knowledge of these two parameters has been earlier postulated to be sufficient to describe the failure of this material in the presence of any geometric configuration, whether line crack or notch. In the case at hand, given the core dimension, a family of curves may now be established that indicates the change in loading patterns as the geometry changes. For example, should $r_0/a = 0.01$, then for various values of b/a, the variation in loading appears as in Figure 23. The value of the angled crack problem should now be more clearly seen: the failure criterion is based not on a single load to failure, but a function, and while more difficult to obtain, in principle, the prediction reliability should be considerably greater than presently is the case.

Although interest from a safety standpoint centers on loading capacity, and more specifically, on the worst case if possible, the actual post-failure behavior of the fracture provides another insight into the validity of the theory under study. As briefly mentioned earlier, Griffith had suggested that a crack would extend in a direction normal to the maximum tangential stress, and Erdogan and Sih [18], in applying this criterion obtained striking agreement with their experimental data. In a discussion of this paper, McClintock [20] suggested the use of the normal angle from the ellipse surface as the directional property, but the result was not in the slightest agreement with the observed behavior. The published data of Williams and Ewing [6] for initial crack angle (with respect to the plane of the crack) corraborates that of [18]. Figure 24 illustrates the predictions of the strain energy density function for the core region dimensions suggested by the experimental data. At $\beta = 90°$, the crack is expected to propagate in its own plane, but as β becomes small, the direction becomes less well defined. At $\beta = 0$, the material reacts as if (in theory) no crack at all were present, and while the material would be expected to break at a normal to the load, the crack solution cannot predict this. Before proceeding further, it is necessary to be more precise in stating how the crack will extend from the tip of the crack or notch. When the stationary values of the strain energy density were determined for various radius vectors, in addition to the load, an

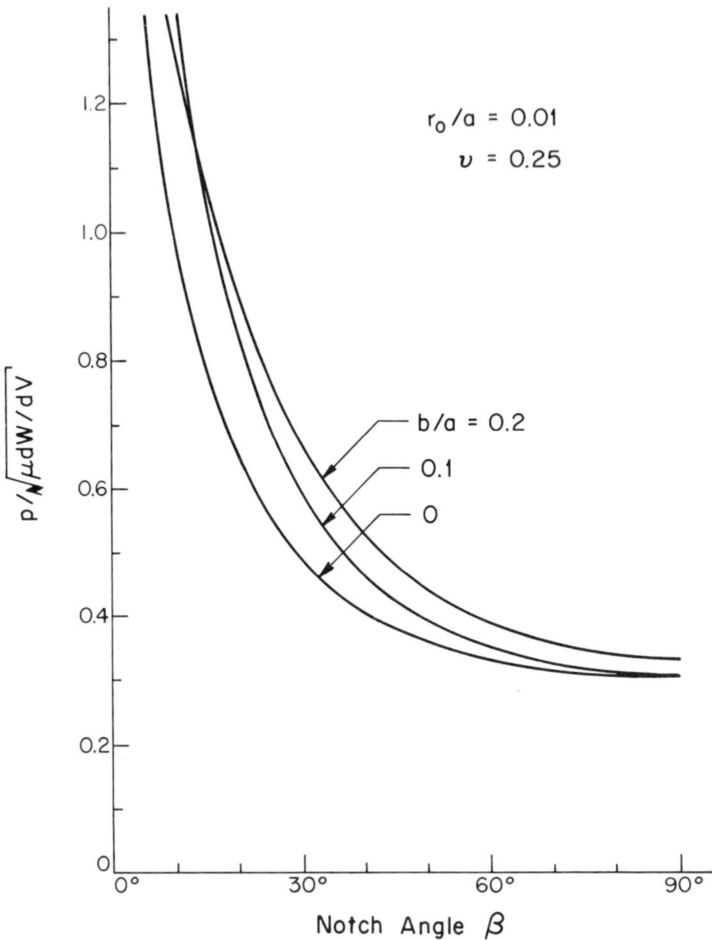

Figure 23. Variation of load with notch angle β for fixed core dimension: $r_0/a = 0.01$ and tensile loading.

angle θ was found corresponding to the load. So for each radius r/a, a curve is generated in the angle θ; Figures 25–27 reflect this angle for the same ratios of b/a and r/a as used for the load variations of Figures 20–22. Once the parameter r_0/a has been established from loading considerations, the angle of fracture, θ_0, is determined from the assumption that there is separation of the material from the surface to

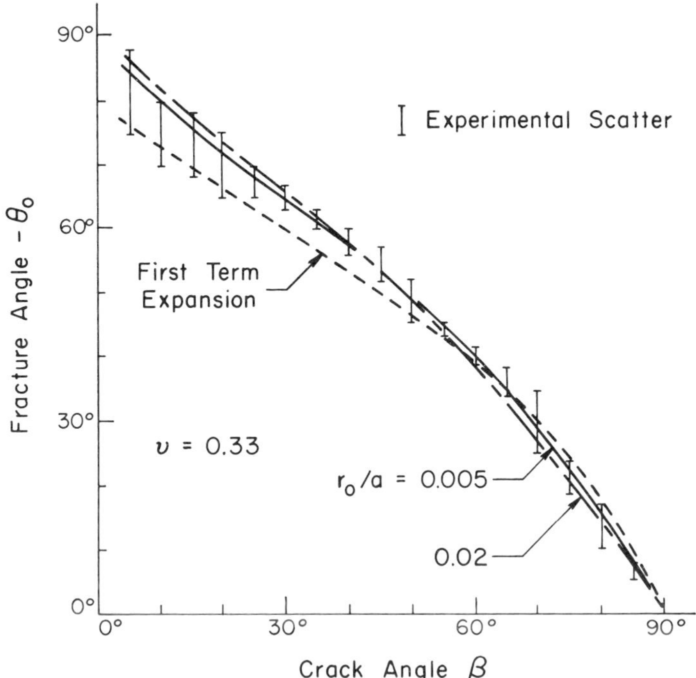

Figure 24 Variation of fracture angle with crack angle: strain energy density solution for line crack in tension.

the edge of the core region in the direction θ_0, with subsequent fracture proceeding from there. If r_0/a is taken to be 0.01, corresponding to the load example, then for varying ratios of b/a, the resulting alterations in initial fracture angle would appear as in Figure 28.

One must retain cognizant of the non-stationary position of fracture initiation along the notch surface. As the loading angle changes, so does the position of initial failure. One result is that now, when $\beta = 0$, there is a well defined point of failure (because the geometry is now non-trivial), and it is to be expected that the fracture will extend from the minor edge of the cavity along the minor axis direction. The path is, of course, perpendicular to the load, and the behavior, apart from load magnitude, corresponds exactly to that from the major axis notch tip at $\beta = 90°$.

Cotterell [21] has observed that for notches, rather than immediately

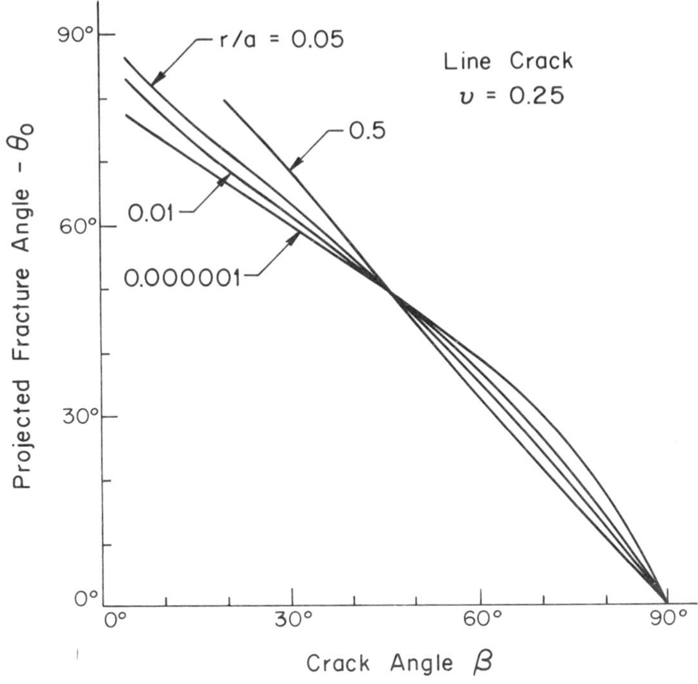

Figure 25. Variation of fracture angle with crack angle: line crack in tension.

turning into a path of fracture coinciding with that of the line crack, the fracture trajectory extends forward from the notch tip, roughly in the same plane, for a distance of approximately one notch tip radius, before turning into the path direction matching that for the line crack. The ideal line crack has, of course, no tip radius, and would immediately assume its characteristic direction of failure. If it is assumed that the crack path can be predetermined for unstable crack propagation, then the path of minimum strain energy density should trace out the trajectory along which fracture is expected to occur. In effect, the trajectory will follow a path that restores symmetry to the geometry of fracture. For a line crack, Figure 29 illustrates the projected paths for several angles of loading. Photographs of actual crack trajectories [15] are available for angles of loading $\beta = 30°$ and $60°$ which agree very well with the predicted path shown in Figure 30, out to a distance of at least one-half crack length.

If the behavior of a rather blunt notch, say $b/a = 0.1$, is examined, an

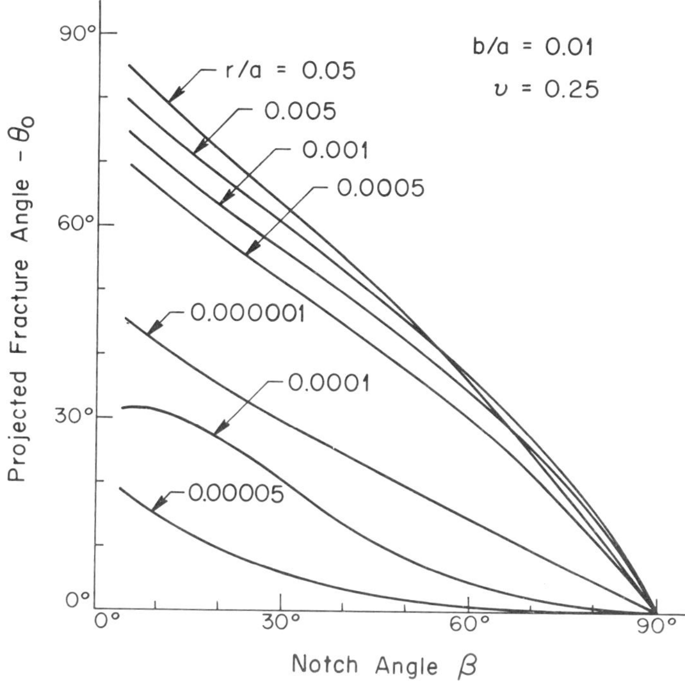

Figure 26. Variation of fracture angle with notch angle: $b/a = 0.01$ and tensile loading.

overall view (Figure 30) suggests that there have been at least some local changes in behavior from that of the line crack. Upon close observation, the trajectories near the surface reveal a behavior that connects some previously advanced, but analytically unsupported, ideas. When the surface of the notch separates, the fracture begins its course tangentially to the normal angle at the surface; i.e., initial fracture is perpendicular to the surface, coinciding with McClintock's [20] hypothesis. But this is only a local tangent, for the path then bends sharply towards the horizontal (with respect to the major axis of the notch), and extends in this way for a short distance. Again, a sharp turn occurs, now into a direction whose tangent matches that for the initial fracture angle of the line crack. The short extension distance is approximately equal to the notch tip radius, in accordance with the observations of Cotterell [21]. Subsequent to turning onto the path of fracture coinciding with that of the line crack, the trajectory continues in much the same manner as that of the line crack. Experimentally, surface conditions likely would ob-

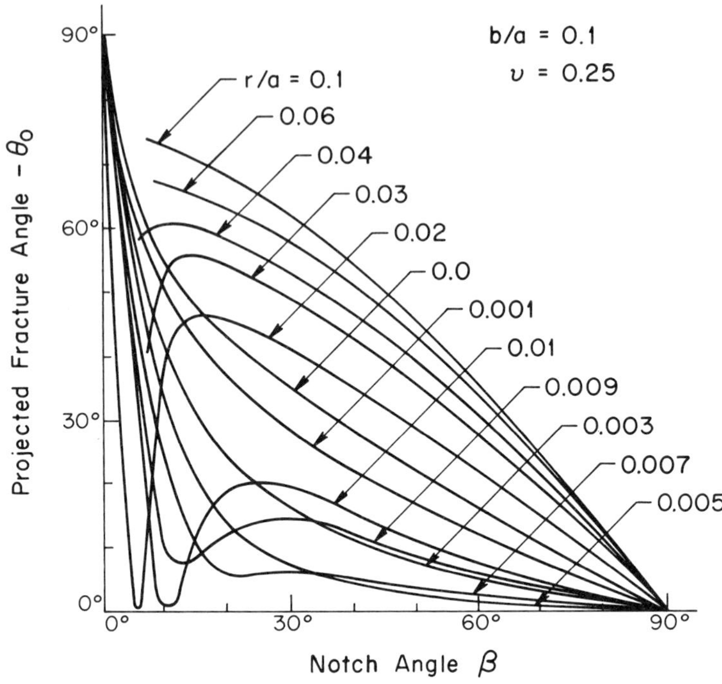

Figure 27. Variation of fracture angle with notch angle: $b/a = 0.1$ and tensile loading.

scure much of this local behavior, yet clearly, for small core regions, the separation would occur linking two regions somewhere before the trajectory turns into that of the line crack, and the local horizontal extension is still expected to occur.

These latter observations have been made assuming prevailing linear elastic response both locally and globally. Should there be extensive localized inhomogeneity, or unaccountable grain size effects, it is expected that few of the above phenomena will be observable, and a more refined model must be created to determine the local behavior. Also, if the fracture is slow, then the material has time to react to the new geometry, and the details listed above are no longer valid.

Other fracture trajectory data predicted on the basis of the strain energy density theory for various notch geometries and loading angles can be found in Appendix A.

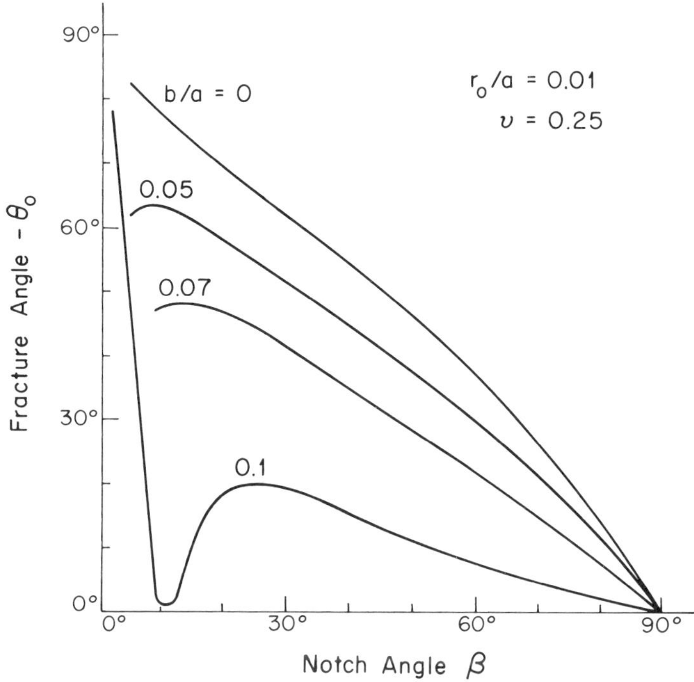

Figure 28. Variation of fracture angle with notch angle for fixed core dimension: r_0/a = 0.01 and tensile loading.

Elliptical notch in compression. Although much of the behavior to be described in this section corresponds to that of the tensile case, there are a few additional difficulties that must be dealt with in compression. Rather than reiterating the arguments already presented in detail, emphasis will be placed upon the features that tend to complicate the compressive analysis. Caution must be exercised in requiring that no interpenetration of notch faces occurs. The line crack, for instance, must be reformulated to ensure that sufficient normal stresses occur on the faces to prevent penetration, and this has been attempted by McClintock and Walsh [12]. In addition, it is clear that functional stresses will be developed on the faces if any slip is to occur to allow fracture to progress. In the development here, the notch geometry will be restricted to such dimensions that contact between opposing faces cannot occur.

As before, assuming a small applied uniform stress, the surface

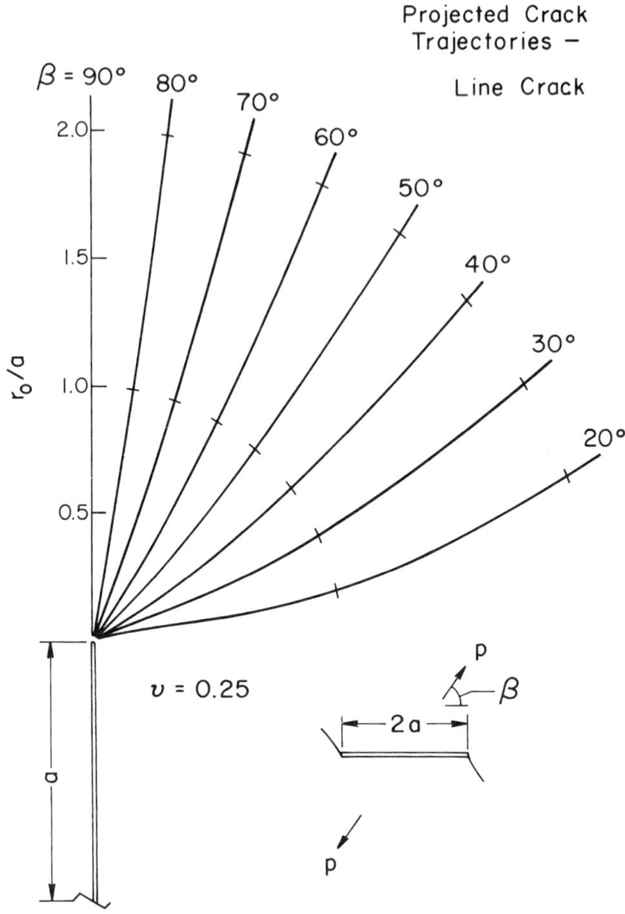

Figure 29. Fracture trajectories predicted from the strain energy density theory for a line crack under tension.

position of maximum surface layer energy may be located, and an origin established to evaluate the load behavior as a function of radial distance from the surface. Figures 31 and 32 describe the load behavior as a function of notch angle, for various constant radii, r/a. The curves for $r/a = 0$ correspond to those of the maximum surface layer energy in Figure 16, apart from a different normalization of the load.

Once a core dimension has been established, say $r_0/a = 0.01$, the variation in load with geometry may be described, as in Figure 33. For a

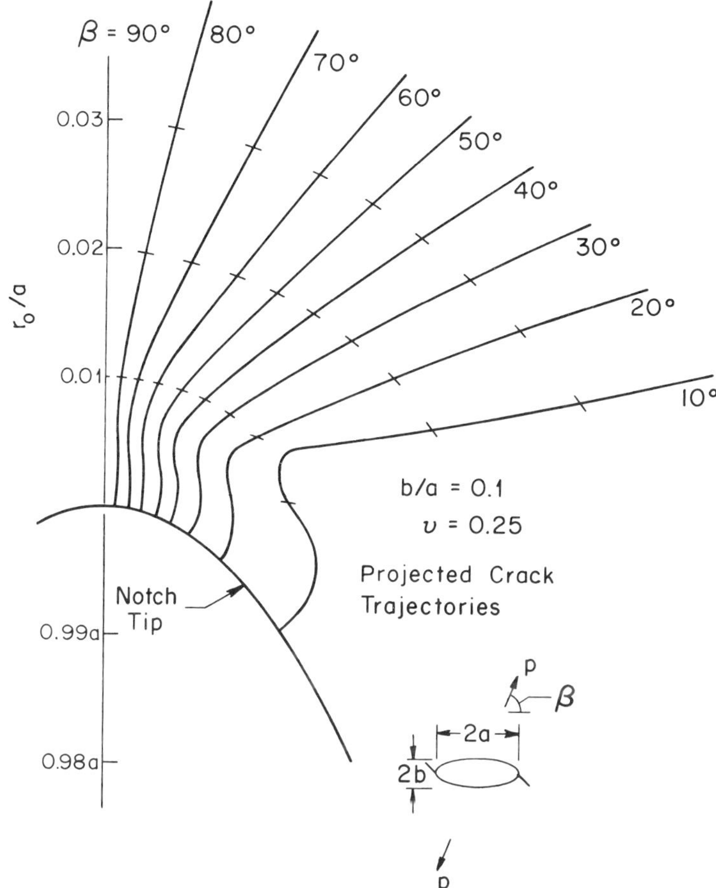

Figure 30. Fracture trajectories for a notch with $b/a = 0.1$ under tension.

given material, and the present analysis, it is expected that a new core dimension, r_0/a, must be determined, since in general, the behavior of a material in compression differs considerably from that in tension. That is to say, another sequence of tests would be required to obtain the core dimension.

A characteristic of compressive loading is that the minimum load to fracture occurs at some angle between $\beta = 0°$ and $90°$ depending on the size of the core region. Cotterell [22] has published the experimental data that appear in Figure 34 for an elliptical cavity, $b/a = 0.1$, in glass.

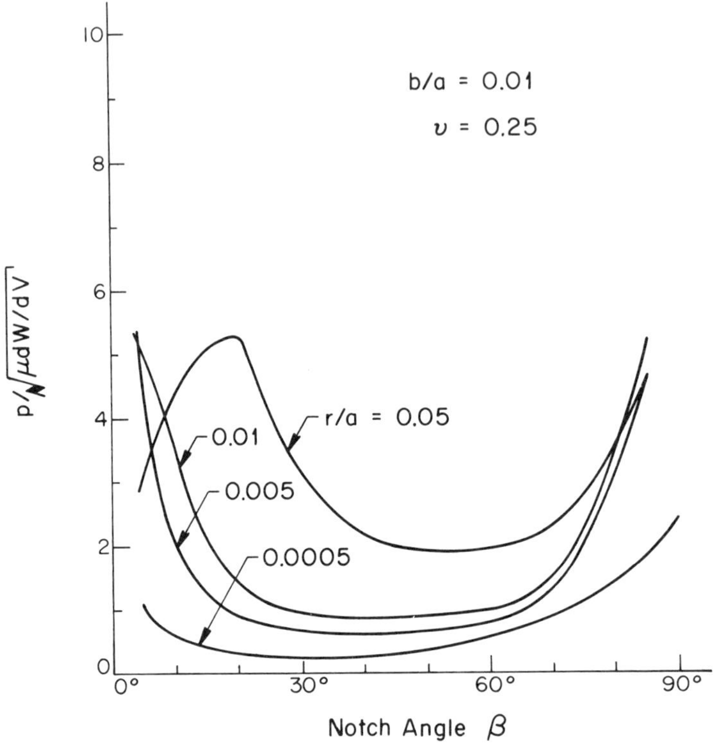

Figure 31. Variation of load with notch angle for $b/a = 0.01$ and compressive loading.

The curves in this figure are those of Figure 32, normalized to the load at $\beta = 90°$. The curve $r/a = 0$ is that of the maximum surface layer energy criterion. As the position of possible failure is allowed to move deeper into the material, there appears quite satisfactory agreement between $r/a = 0.005$ and 0.008. Hence, the energy criterion may be expected to provide quite reasonable failure predictions. On the basis of the data, one could choose a critical core dimension of, say, $r_0/a = 0.005$, and then use this dimension as set forth in previous sections for load predictions for other geometries in this material.

Under uniform loads, it has been observed that propagating cracks tend to restore symmetry by running normal to tensile loads, and parallel to compressive loads. Then, it is expected that for the elliptical cavity under compression, for loads parallel to the major axis ($\beta = 0$), the fracture will initiate at the notch tip on the major axis, and extend along

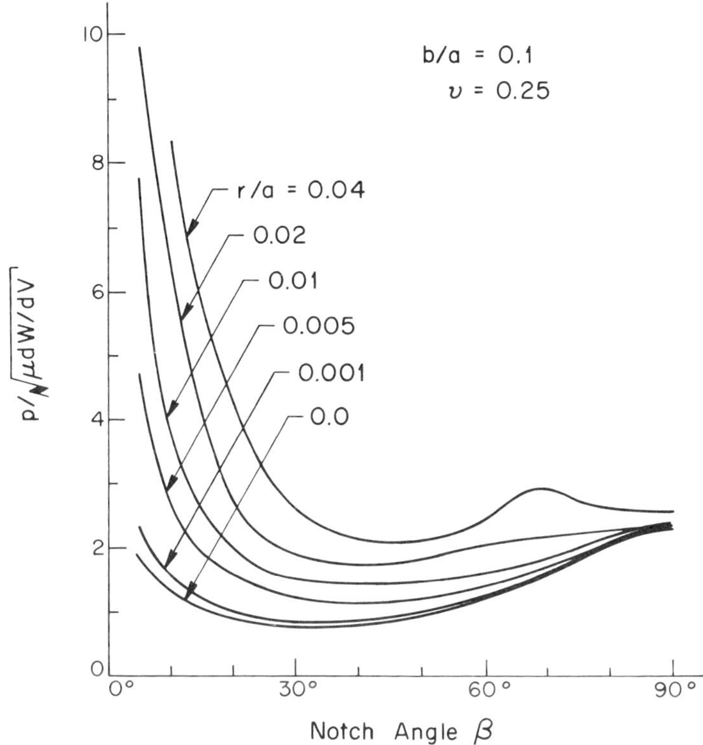

Figure 32. Variation of load with notch angle for $b/a = 0.1$ and compressive loading.

that same axis towards the applied load, while for loads parallel to the minor axis, fracture will initiate at the face intersection with the minor axis, and again extend towards the applied load; i.e., for $\beta = 0°$, $\theta_0 = 0°$ and for $\beta = 90°$, $\theta_0 = 90°$. Between these two limiting cases, the behavior is determined as in the tensile loading case: the stationary values of θ are plotted as functions of crack angle β, for various magnitudes of r/a (Figures 35 and 36). It must be emphasized that not until a specific core dimension has been obtained can any of these curves represent the initial fracture angle. If, for example, $r_0/a = 0.01$, then the initial fracture angle, as it varies with geometry, is given by Figure 37. The data shown in Figure 38 is from Cotterell [22] for elliptical notches in glass plates loaded in compression. The curves are those from Figure 36, since the data is for cavities of ratio $b/a = 0.1$.

The lower curves presented in Figure 38 are for small values of r/a,

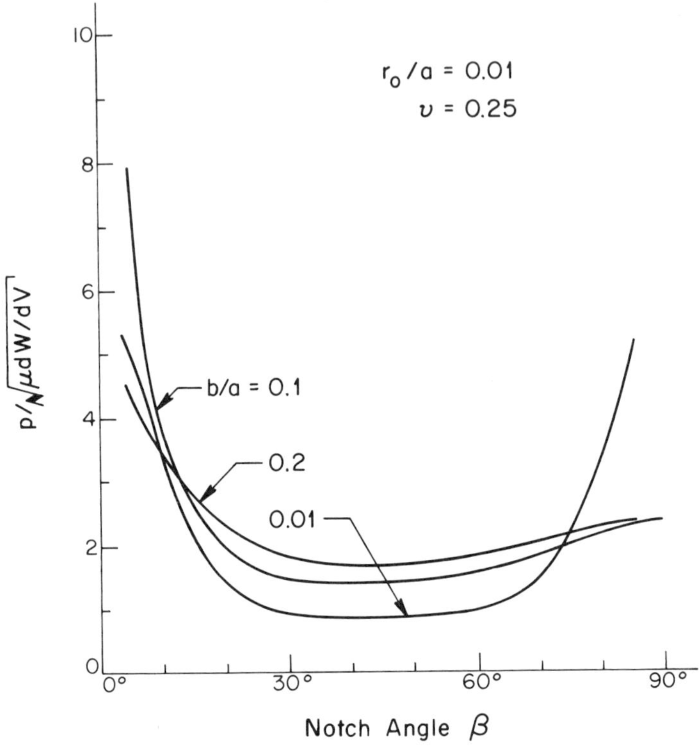

Figure 33. Variation of load with notch angle β for fixed core dimension: $r_0/a = 0.01$ and compressive loading.

since the core region of necessity must be small compared to the notch length. The upper curve represents the tangent angle taken from the plots in Figures 39 and 40, of the crack trajectory after it makes the sharp turn into its final path.

In general, the loads required to cause failure in compression are substantially higher than those in tension, and there results throughout the member severely strained material, so that there is prevailing a high energy density; once a crack is initiated, it is expected to propagate catastrophically in this particular geometry. Then the locus of stationary values of the strain energy density may be plotted to attempt to predict the subsequent path taken by the fracture (Figures 39 and 40). As in the tension case, the path is initially tangent to the local normal before

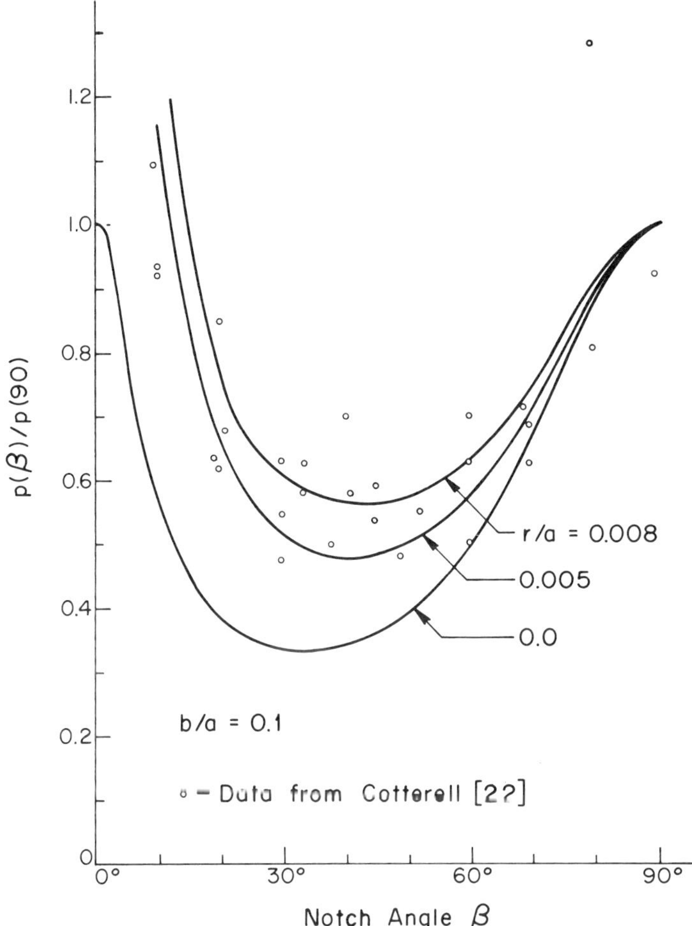

Figure 34. Variation of load with notch angle for $b/a = 0.1$ and compressive loading data [22].

bending away; then there is a quick return into the general trend leading the path towards the uniform load. This describes the behavior for most of the angles of loading; apparently, as β increases, so does the length of the initial trajectory before its final turn into the load. This is also evident by following changes in r/a for constant β in Figures 35 and 36.

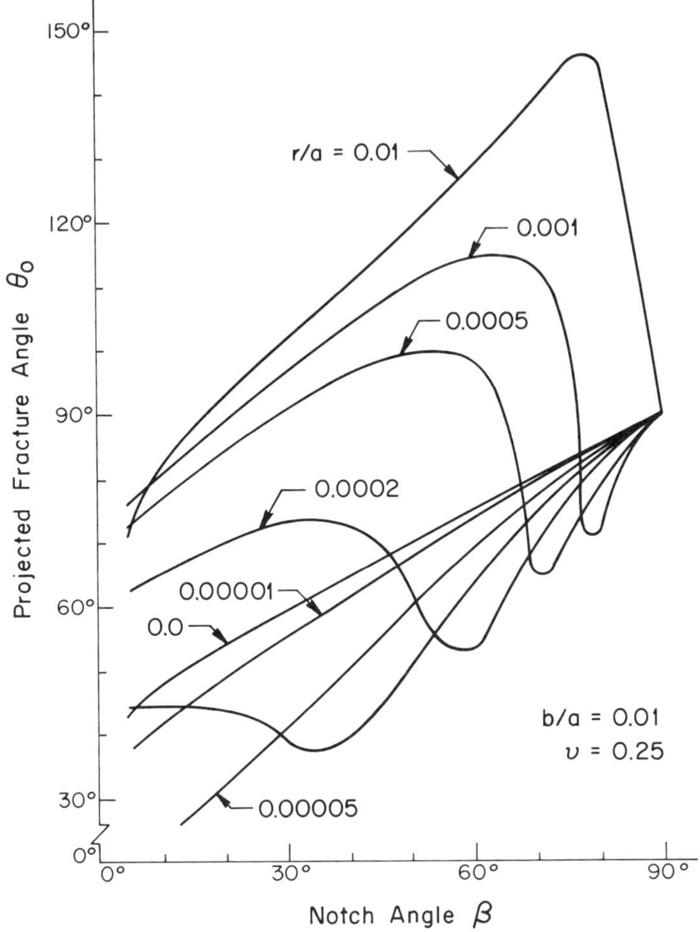

Figure 35. Variation of fracture angle with notch angle for $b/a = 0.01$ and compressive loading.

V Two external notches

Having discussed the details of the use of the strain energy density theory for the internal notch, the problem of two interacting external notches will be treated in this section to demonstrate certain features that are not intuitively obvious.

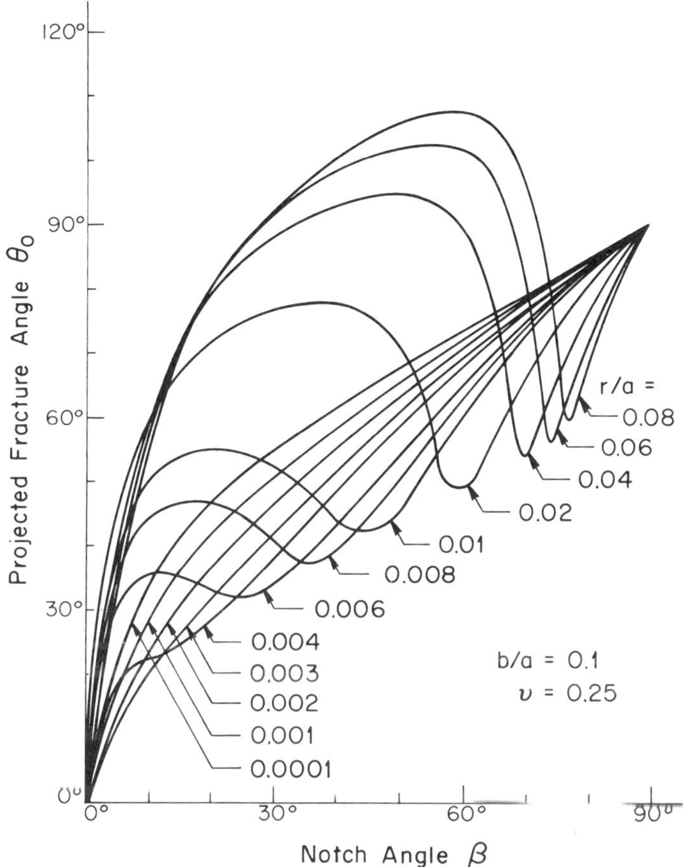

Figure 36. Variation of fracture angle with notch angle for $b/a = 0.1$ and compressive loading.

The elastic solution for plane extension of two external hyperbolic notches has been obtained by Griffith [23] and Neuber [10]. Two cases are considered, simple normal extension and pure shear, shown respectively in Figures 41a and 41b. The thickness is taken to be h, and assumed large enough to have plane strain prevail. P is the net tensile force applied a large distance from the waist between the notches; Q is the net shear force also applied a large distance, d, from the waist between the notches, with moments $Q \cdot d$ applied to maintain static equilibrium. These two cases may be synthesized to obtain a resultant

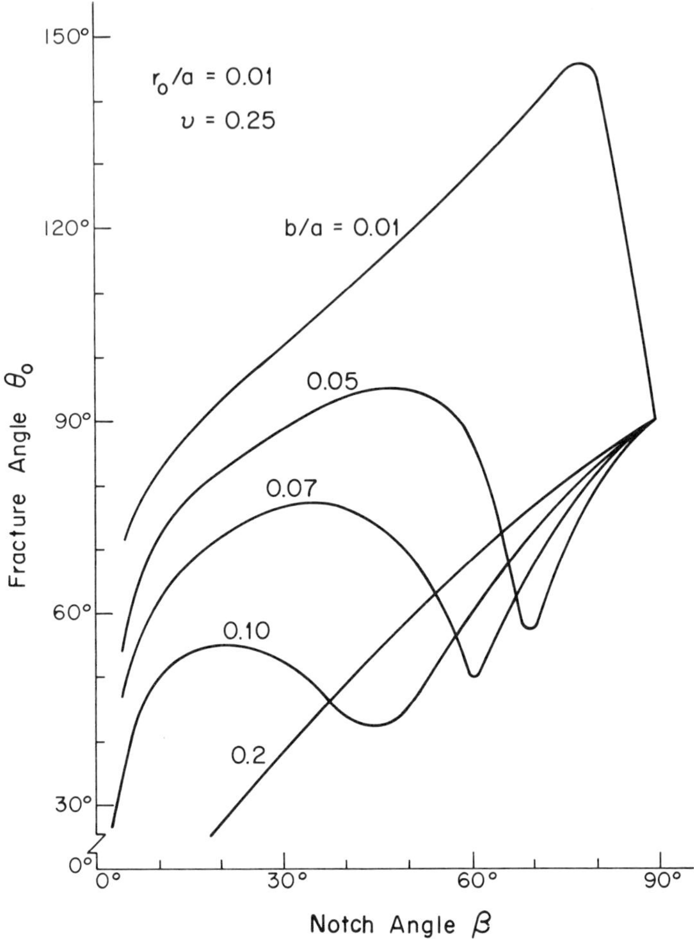

Figure 37. Variation of fracture angle and notch angle for fixed core dimension: r_0/a = 0.01 and compressive loading.

force R applied at an angle β to the x-axis, where $P = R \sin \beta$, and $Q = R \cos \beta$. Figure 42 illustrates the results of this combination of forces and couples to the single resultant force R. In addition, the geometry of the external notch, assuming the form of a plane hyperbola, is shown, with waist length $2a$, normal angle ϕ, and asymptotic slope(s) $\pm b/a$.

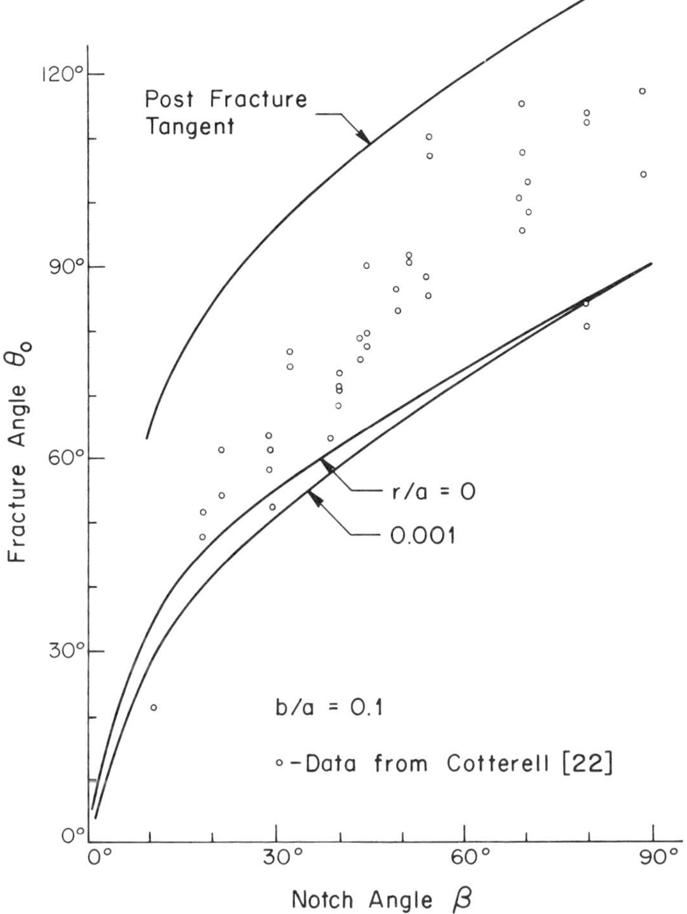

Figure 38. Variation of fracture angle with notch angle for $b/a = 0.1$ and compressive loading data [22].

The solutions to the two basic cases are available in elliptic coordinates (Griffith [23] and Neuber [10]), but a more convenient form is that of rectangular cartesian coordinates, and the transformed solutions for the stresses follow in terms of the complex variable $z = x + iy$. In tension, the stress combinations are

$$\sigma_{11} + \sigma_{22} = -2A_2 \operatorname{Im}\left[\frac{c}{\sqrt{z^2 - c^2}}\right] \tag{24a}$$

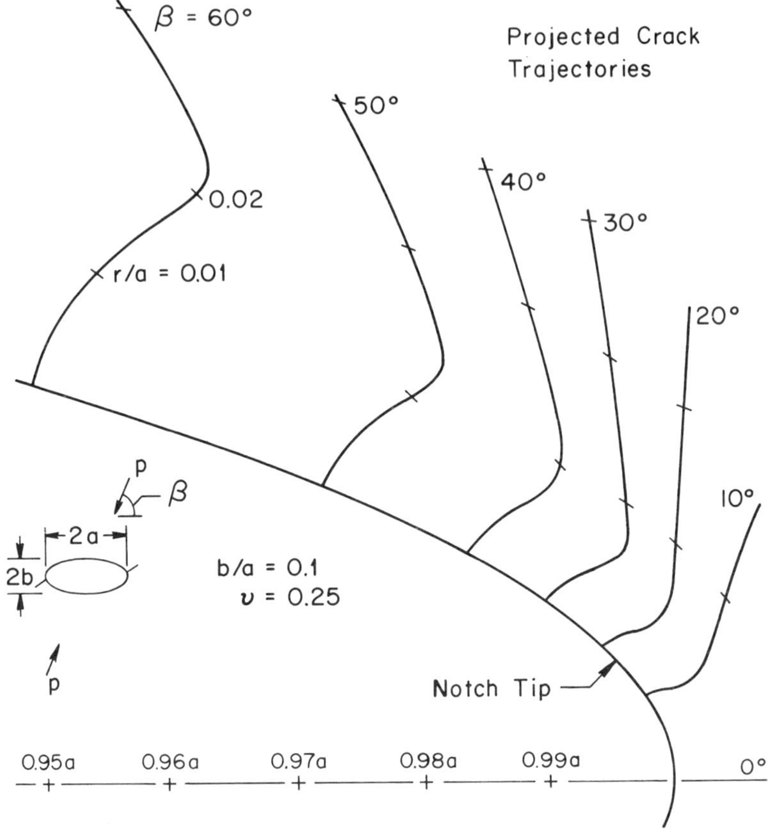

Figure 39. Predicted fracture trajectories for $b/a = 0.1$ and small angles of compressive loading.

$$\sigma_{22} - \sigma_{11} + 2i\sigma_{12} = iA_2 \frac{c}{\sqrt{z^2 - c^2}}\left[1 + \frac{a^2 - b^2 - z\bar{z}}{z^2 - c^2}\right] \tag{24b}$$

whereas, in shear, the stresses can be found from

$$\sigma_{11} + \sigma_{22} = 2A_1 \operatorname{Re}\left[\frac{c}{\sqrt{z^2 - c^2}}\right] \tag{25a}$$

$$\sigma_{22} - \sigma_{11} + 2i\sigma_{12} = -A_1 \frac{c}{\sqrt{z^2 - c^2}}\left[1 + \frac{z\bar{z} - a^2 + b^2}{z^2 - c^2}\right] \tag{25b}$$

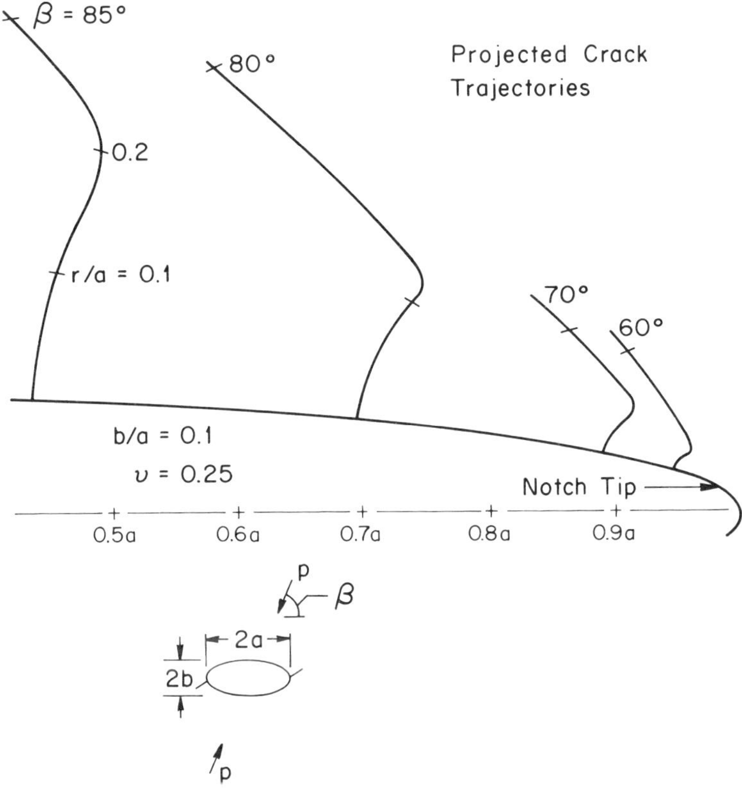

Figure 40. Predicted fracture trajectories for $b/a = 0.1$ and large angles of compressive loading.

The symbols Re and Im stand for the real and imaginary parts of a complex function. The constants A_1 and A_2 in equations (24) and (25) are given by

$$A_1 = \frac{R \sin \beta}{ch}\left[\pi - 2 \tan^{-1}\left(\frac{b}{a}\right) - 2ab/c^2\right]^{-1} \tag{26a}$$

$$A_2 = \frac{R \cos \beta}{ch}\left[\pi - 2 \tan^{-1}\left(\frac{b}{a}\right) + 2ab/c^2\right]^{-1} \tag{26b}$$

In equations (26), R is the applied force, h the thickness of the solid, B

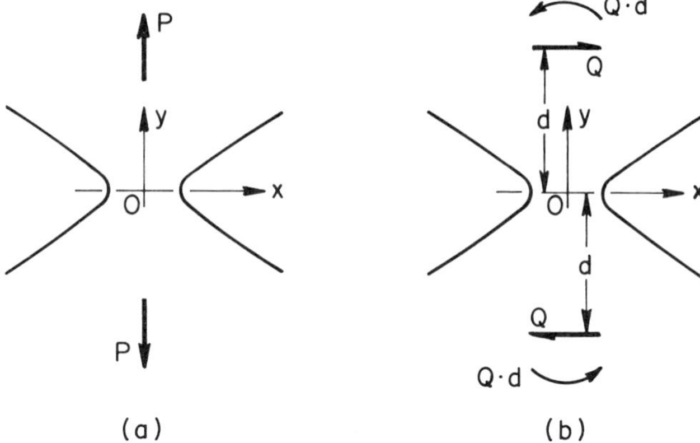

Figure 41. Extension and shearing of two external notches.

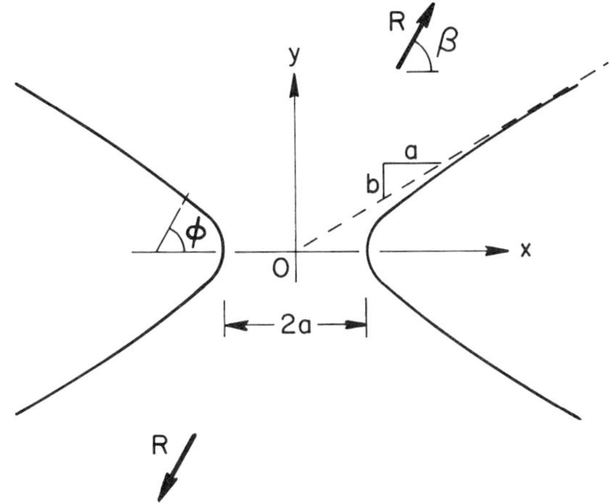

Figure 42. Geometry and loading of two hyperbolic notches.

the angle of inclination of R to the horizontal axis and $c^2 = a^2 + b^2$. Superposition of the tension and shear cases provides the stresses as a function of z. The branch cuts of the solution are such that the solution is correct only in the first and third quadrants. (The negative of the solutions is required for the remaining regions).

Surface layer energy. For non-vanishing ratios of b/a, (i.e., notches other than line crack, it is necessary to know the location of failure next to the notch boundary. As in the case of the internal notch, a formal boundary layer that separates the notch surface and interior regions is introduced. This boundary layer has a layer thickness, δ, of approximately continuum dimensions in thickness. The failure of this layer is to be analyzed using a surface layer energy quantity γ_e as given by equation (2) with σ_t given by

$$\sigma_t = \frac{2(-A_1 c \tan\phi + A_2)\sqrt{a^2 - b^2 \tan^2 \pi}}{ab(1+\tan^2\phi)} \tag{27}$$

where σ_t is the surface tangential stress along the left border (Figure 42). The location of failure on the notch boundary may be determined from the maximum of γ_e which is obtained by taking $\partial\gamma_e/\partial\phi = 0$. The resulting expression to be solved for the normal angle ϕ is

$$\tan^3\phi + \frac{A_1}{A_2}\left(2+\frac{a^2}{b^2}\right)\tan^2\phi - \left(\frac{2a^2}{b^2}+1\right)\tan\phi - \frac{A_1}{A_2}\frac{a^2}{b^2} = 0 \tag{28}$$

The coordinates (x_0, y_0) locating the point of failure can be found from

$$x_0 = -a^2/\sqrt{a^2 - b^2\tan^2\phi}$$

$$y_0 = b^2 \tan\phi/\sqrt{a^2 - b^2\tan^2\phi}$$

For the case of the line crack $(b/a = 0)$, the coordinates reduce to $(x_0, y_0) = (-a, 0)$, and the angle ϕ is not defined.

Strain energy density criterion. As discussed earlier, failure is expected to initiate in the region where γ_e is a maximum, although the failure loads are expected to be governed not by γ_e, but by material response interior to the free surface. Specifically, application of the strain energy density failure criterion requires the use of a local polar coordinate system; the complex variable z is assigned the form

$$z = x_0 + iy_0 + re^{i\theta}$$

i.e., the origin of the polar coordinate system is located where γ_e is maximum. The above expression for z is substituted into the stresses, and the stresses into the strain energy density function in equation (17).

Taking r/a to be constant, θ is varied until the strain energy density

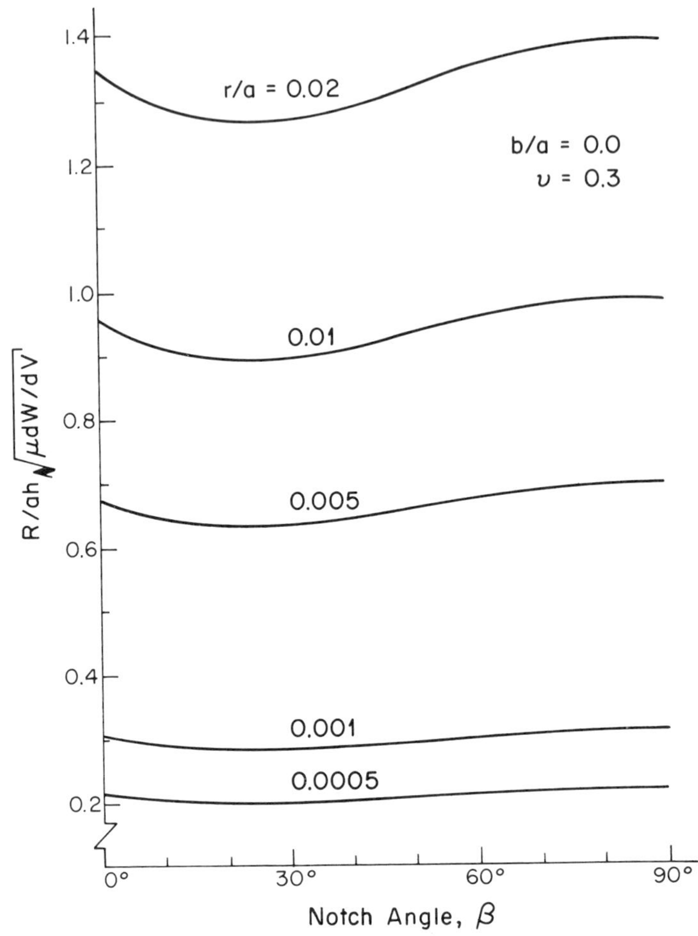

Figure 43. Load to failure for external line crack.

assumes its minimum value; failure is postulated to occur at this location, at a load corresponding to setting $\mathrm{d}W/\mathrm{d}V$ equal to some critical constant, $(\mathrm{d}W/\mathrm{d}V)_c$. As r is changed, a family of curves, as functions of β, is generated, of which one in particular is expected to describe a particular material, i.e., $(\mathrm{d}W/\mathrm{d}V)_c$ and one r/a are to be constants, resulting in a two parameter failure criteria. For any value of r/a, with $\mathrm{d}W/\mathrm{d}V = (\mathrm{d}W/\mathrm{d}V)_c$, the right hand side of equation (17) is of the form $(R/ah)^2$ times a non-dimensional function of the variables r/a, β, θ and

b/a. Then a required load to failure may be expressed as a non-dimensional force, $R/ah\sqrt{\mu \, \mathrm{d}W/\mathrm{d}V}$. It is emphasized that on the subsequent discussion, no particular material is considered, so that results are presented for a range of values of r/a and b/a.

Poisson's ratio has been set at 0.3 for convenience, and its variation will modify the numerical results slightly, but the tenor of the discussion is unaffected.

A characteristic of the internal notch is that the weakest position is in

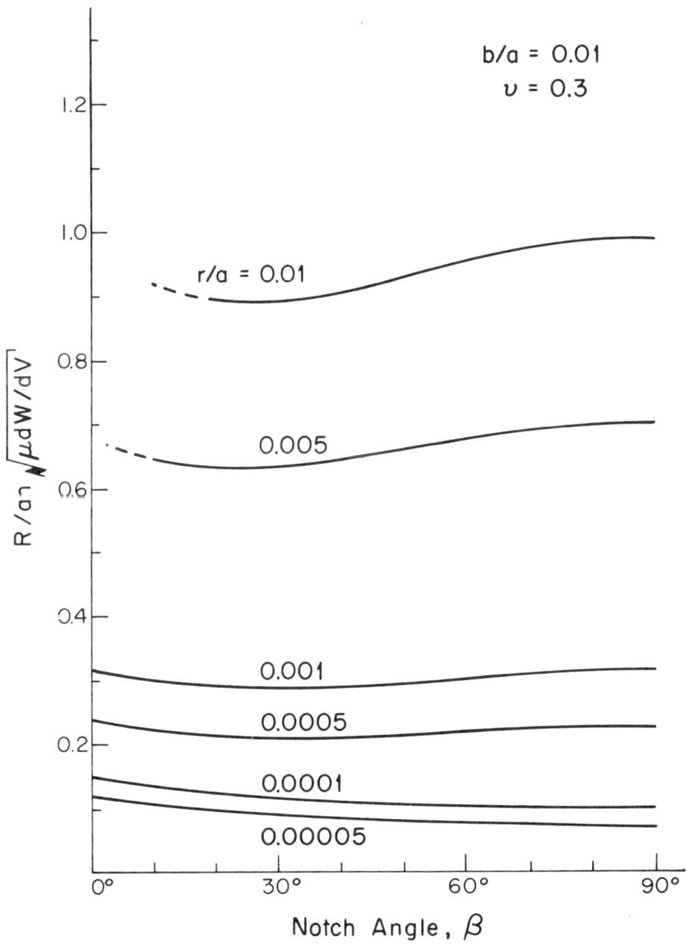

Figure 44. Load to failure for external notch with $b/a = 0.01$.

the normal loading configuration as shown earlier; in general, this is not the case for the external notch. Figure 43 indicates that for a wide range of r/a, (in fact, down to 10^{-6}), the minimum load occurs for $20° < \beta < 30°$, and further, that there is relatively little variation in the loads to failure for $0 \leqslant \beta \leqslant 90°$. In part, this may be attributed to the fact that the shearing contribution is more influential, and of a different nature, than in the internal notch case, i.e., net force loading, rather than uniform stress loading. As b/a is increased from zero to 0.01, Figure 44, the

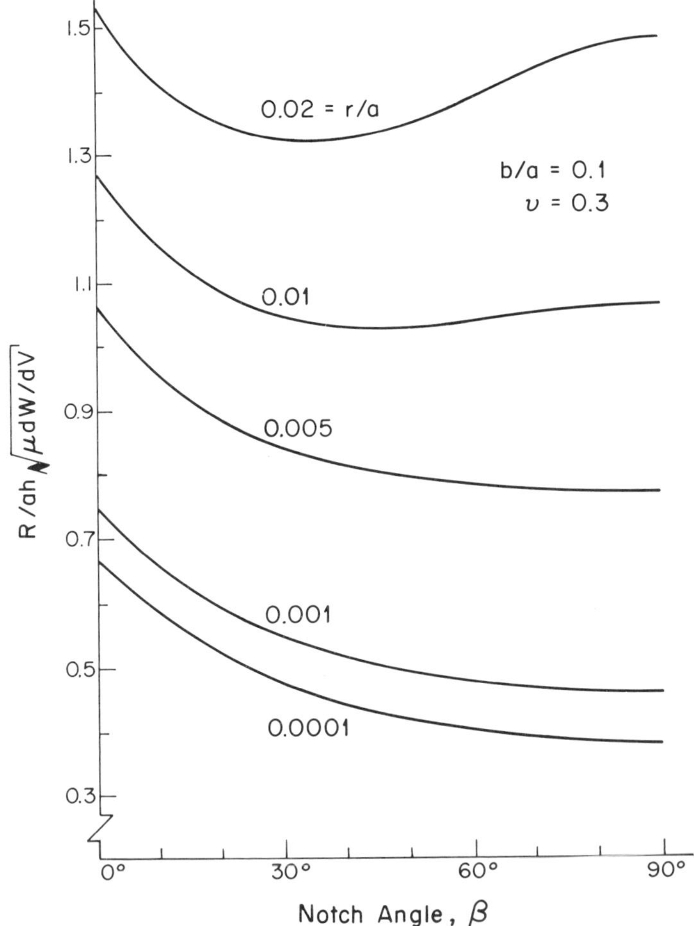

Figure 45. Load to failure for external notch with $b/a = 0.1$.

minimum is seen not to occur until $r/a \simeq 0.0005$. The minimum at distances $r/a \leqslant 0.0001$ does not reappear until $b/a \simeq 0.3$, Figure 47. Figures 45 and 46 show the variations of expected failure loads for $b/a = 0.1$, and 0.2. As may be observed, say in Figure 45, if r/a is chosen to be 0.005, the behavior differs substantially from that of $r/a = 0.01$. It is therefore imperative that for a given material, a value of r/a be found; this will allow, then, the description of several geometries, as described in Figure 48. In this figure, r/a is fixed at 0.001, and b/a is permitted to

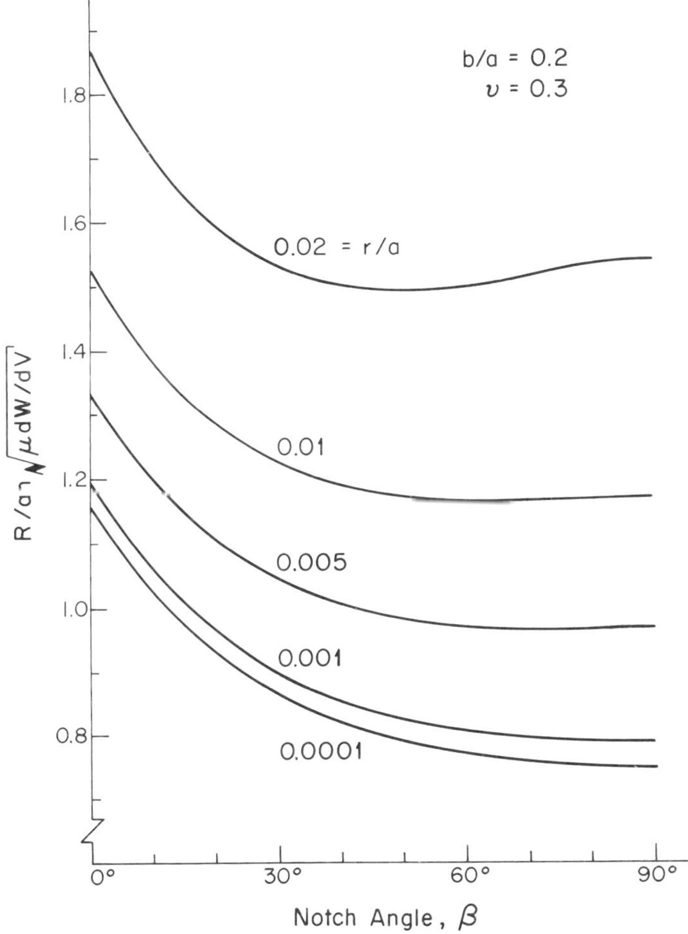

Figure 46. Load to failure for external notch with $b/a = 0.2$.

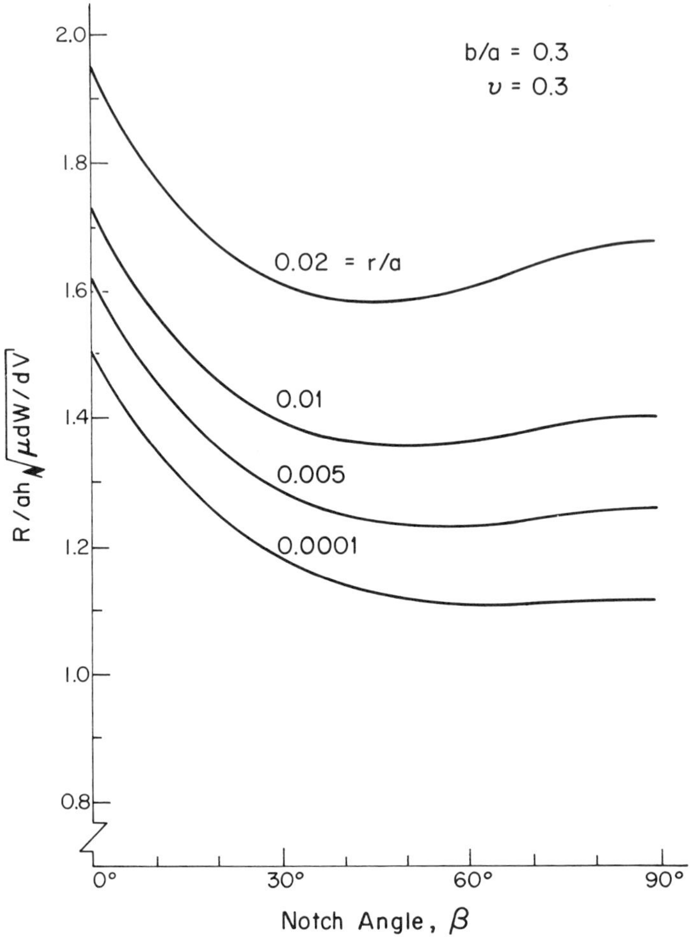

Figure 47. Load to failure for external notch with $b/a = 0.3$.

vary. Clearly, the blunter the notch, the greater the requisite failure loads with minima occurring at both symmetric and inclined loading configurations. The dashed lines in Figure 48 indicate where $\beta \leqslant \tan^{-1}(b/a)$; even so, at $\beta = 0$, pure shear prevails, and a finite load to failure still occurs.

Predicted fracture angles. Preliminary observations on double-edge cracked specimens under inclined loading suggest that the longer range

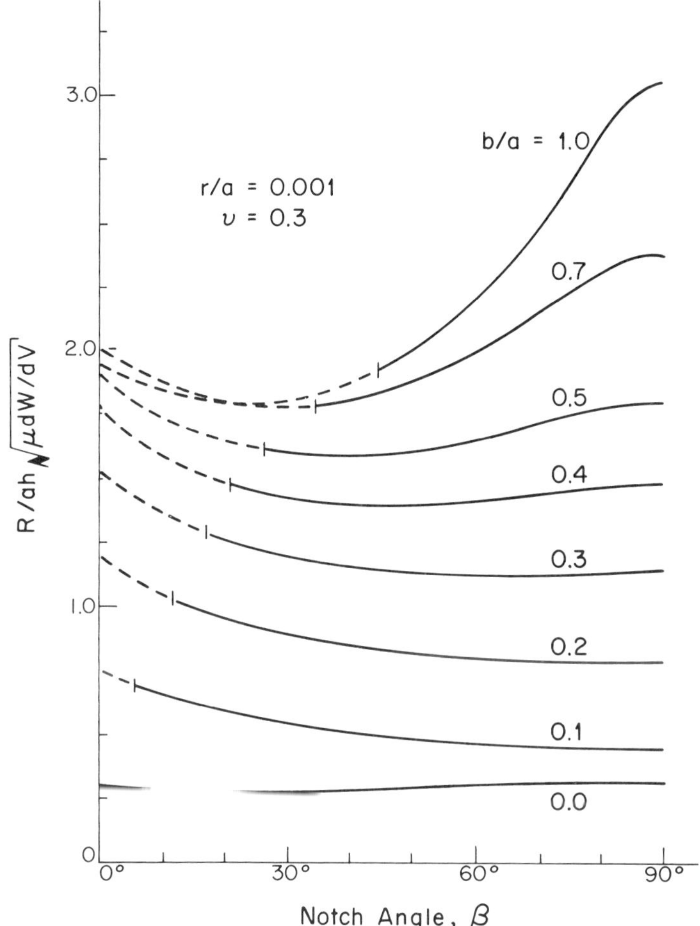

Figure 48. Load to failure for external notch with $r/a = 0.001$.

predicted fracture trajectories, that were quite satisfactory in the internal crack case, are no longer reliable. Equivalently, the external notch fracture trajectory is very dependent upon the growing-crack geometry. This especially is true for initially sharp cracks; for more blunted cracks or notches, initial failure results in a sudden jump in the crack, and this path is expected to be predicted with satisfactory accuracy.

Figures 49–51 indicate the stationary angles acquired in the search for minima of $\mathrm{d}W/\mathrm{d}V$ at fixed r/a. For $b/a = 0$, the angle is virtually

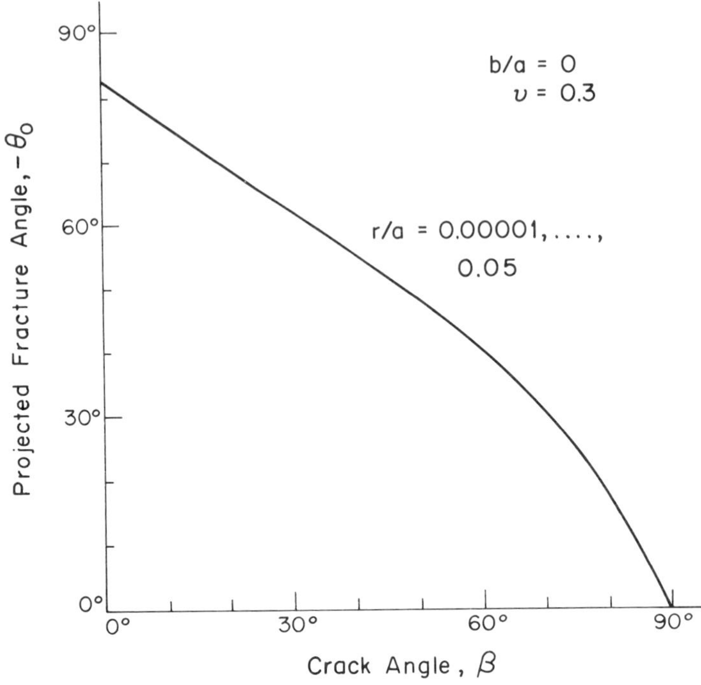

Figure 49. Fracture angles for external line cracks.

unchanged for a wide range of r/a. However, for blunted cracks, the angles at fixed β are seen to first decrease, then increase with increasing r/a; $r/a = 0$ corresponds to the surface normal angle. If the tangent angle is taken after θ begins to increase, it is observed to correspond to the line crack angle.

The summary of plots such as Figure 51 may be seen in Figure 52, where for $b/a = 0.3$, at a fixed angle β, the sequence of locations of minima for increasing r/a are plotted. One observes a bending from the surface normal towards the x-axis, then away into a tangent that parallels those of the line crack. This was observed by Cotterell [21], and predicted later by Kipp and Sih [15] for the internal notch. Additional results on fracture trajectories for the external notches can be found in Appendix B.

Some observations on the S-criterion. In the course of obtaining the minima of $\mathrm{d}W/\mathrm{d}V$, it was found that when r/a exceeded some value,

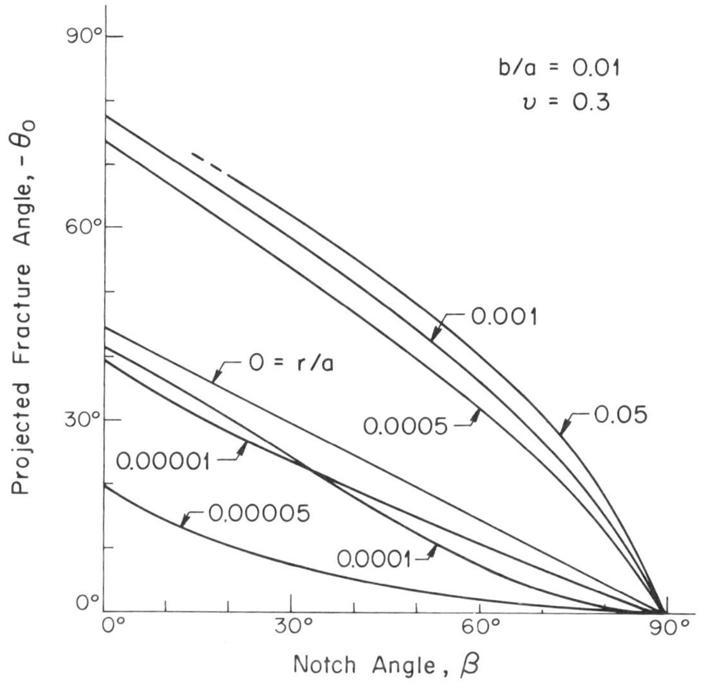

Figure 50. Fracture angles for two external notches with $b/a = 0.01$.

dependent on b/a, β, etc., a splitting of the locations of the minima occurred; that is, instead of acquiring a single minima, a maximum value for dW/dV appeared, flanked by two minima. Initially, it was thought that bifurcation was being predicted, but it now appears that the function is merely stating its own limitations. Such a phenomen was not observed for the internal notch, so it must indicate the limiting region where useful trajectory information can be found. If instead of fixing the origin at the surface of the notch, the origin is moved to the new minimum, at each stage, a floating origin is established that follows the path of predicted fracture. This scheme leaves the results of the strain energy density theory virtually unchanged; there are substantial changes in the results obtained from the maximum tangential stress theory. This occurs because of the directional sensitivity of the latter.

Most significant of the results previously discussed is the appearance of non-symmetrical weak orientations for notches. On the premise that failure is governed by interior behavior of the solid, it is clear that the

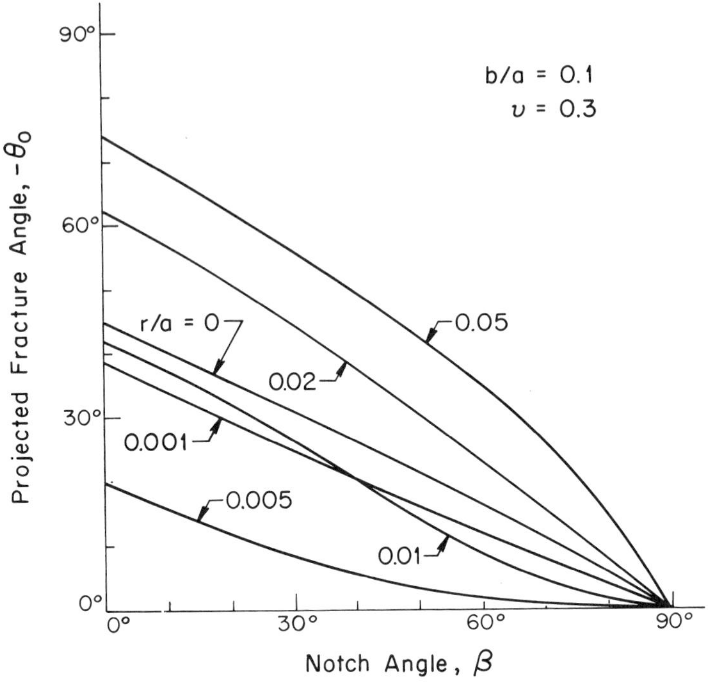

Figure 51. Fracture angles for two external notches with $b/a = 0.1$.

predicted failure loads are quite sensitive to the distance from the (free) surface, not only in magnitude, but also in character.

VI Concluding remarks

One of the most important purposes of the application of the theory at hand was to join the behavior of cracks to that of notches. This has essentially been accomplished through the consistent application of the strain energy density criterion to a thin layer near the notch boundary and material element in the interior regions of the solid. By specifying the failure to be dependent upon internal conditions being met before it can occur, the barrier that has separated notch and crack failure theories has been disposed. Another engaging result was the description of the phenomenon of post-fracture behavior. The local behavior has been shown to agree well with observable details of crack paths emanating from a notch tip.

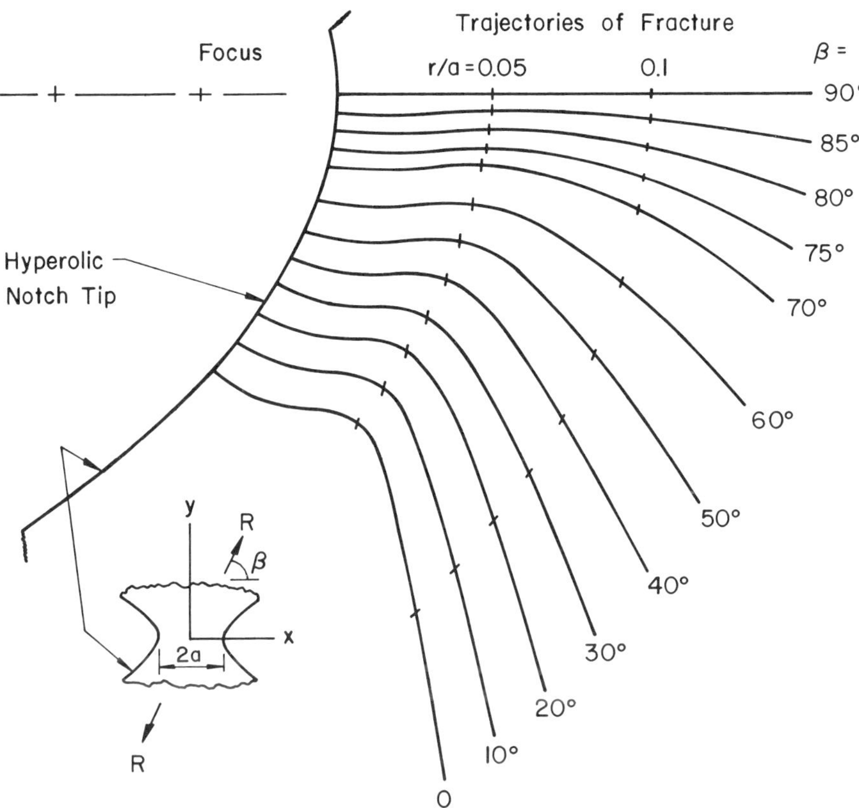

Figure 52. Local fracture trajectories for external notch with $b/a = 0.3$.

The theory upon which all the results were based is that of total strain energy density. Bridgeman [24] has observed that when hydrostatic pressure is applied to a specimen, the strain energy density may be raised to very high values, far exceeding those found under normal conditions, without ensuing failure. In such a case, the energy of volume change is much greater than the energy of distortion. But under tensile loading conditions, it is still held that the requirements are to locate the stationary values of the strain energy density as described, and determine whether the energy of volume change exceeds that of distortion or simply is sufficiently large to give the direction of fracture. The energy of distortion remains unchanged by presence of uniform triaxial stress states, and as long as other stress states are present as well, apparently stands as a good yielding failure theory. In the presence of a crack or notch, if there is yielding, it occurs in directions primarily other than that in which the crack eventually forms and propagates. A considerable amount of effort has been spent by Sih and his co-workers [25–27] in recent years to extend the strain energy density concept for characterizing the ductile fracture behavior of metal alloys. Basically, the material undergoes deformation that is no longer negligible prior to unstable fracture. During this initial stage, the crack is constrained by plasticity and may grow slowly in a stable fashion. Such a phenomenon has been referred to as necking in a tensile bar and the thinning of a plate specimen. Analytical modeling of ductile fracture is inherently three-dimensional in nature and is anti-symmetrical such as the cup-and-cone failure. In most cases, even though the initial load is applied normal to the crack plane, the last ligament of failure may be at an angle with the specimen boundary which necessitates the use of a fracture criterion that can treat mixed mode crack extension. One of the basic views taken in the strain energy density approach is that the critical value of $\mathrm{d}W/\mathrm{d}V$ is assumed to be a material constant at the very beginning and hence the objective of the analysis is to predict the failure load and the crack shape and size at instability. Preliminary results [27] obtained thus for have been very encouraging.

Finally, it is also appropriate to observe that failure theories must be developed that will allow continuity from non-flawed to flawed members. At present, once a flaw is postulated to be the mechanism whereby failure is expected to occur, then no matter how small the flaw becomes, it still dominates. In the case of prevailing fracture theories based on the crack tip singularity, although the strength of the singularity depends on the crack length, if the crack length becomes small, there is no other criterion that can be coordinated to apply in conjunction with the postulated crack. Parameters are required that can place the size of the flaw in context with some other characteristic material length, and allow a more unified evaluation of the failure of solids. It becomes clear that the final criterion cannot be restricted by the detailed geometry at the tip of the flaw but must apply to all geometries and loading conditions.

VI Appendix A: Fracture trajectory data for elliptical notch in tension

The strain energy density theory assumes that fracture occurs along the path of minimum strain energy density, $(dW/dV)_{min}$. The following Tables A1 to A16 give computed values of $(dW/dV)_{min}$ around the front of an elliptical notch (Figure A1) for $b/a = 0.01, 0.1, 0.5, 0.9$ and $\beta = 5°$, $30°$, $60°$, $90°$. In all cases, the Poisson's ratio is fixed at $\nu = 0.3$. The various increments of r_0/a are referred to as θ_0 and the coordinates (x_0, y_0) as indicated in Figure A1 while the location of r_0 on the elliptical boundary is determined by the normal angle ϕ.

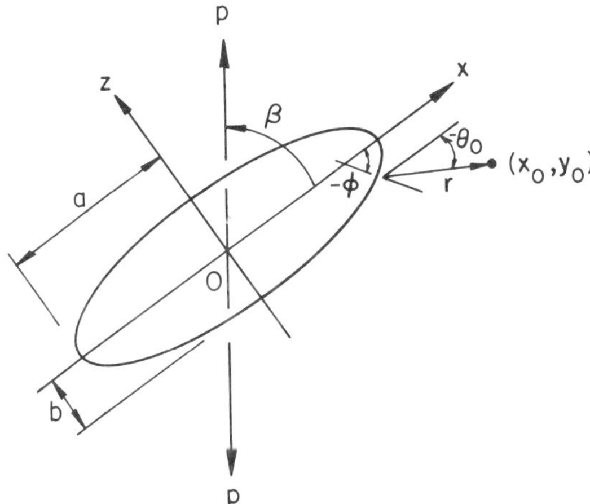

Figure A1. Radius vector measured from an elliptical notch in tension.

TABLE A1

Values of $(\mathrm{d}W/\mathrm{d}V)_{min}$ *for an elliptical notch with* $b/a = 0.01$, $\beta = 5°$ *and* $\phi = -45.76°$

r_0/a	θ_0	x_0	y_0	$(\mathrm{d}W/\mathrm{d}V)_{min}$
.000001	−45.1627	.999948	−.000103	15.87264
.000005	−42.7466	.999951	−.000106	14.55464
.000010	−39.8384	.999955	−.000109	13.11062
.000020	−34.0728	.999964	−.000114	10.77471
.000030	−28.6197	.999974	−.000117	9.01438
.000040	−23.5728	.999984	−.000119	7.69496
.000050	−19.3150	.999994	−.000119	6.72338
.000060	−16.7134	1.000005	−.000120	6.03024
.000070	−17.9634	1.000014	−.000124	5.55403
.000080	−23.6666	1.000021	−.000135	5.22329
.000090	−30.0103	1.000025	−.000148	4.97133
.000100	−35.3150	1.000029	−.000160	4.76272
.000200	−58.3150	1.000052	−.000273	3.52234
.000300	−65.7134	1.000071	−.000376	2.86588
.000400	−69.5377	1.000087	−.000477	2.44479
.000500	−71.8853	1.000103	−.000578	2.14938
.000600	−73.4791	1.000118	−.000678	1.92975
.000700	−74.6314	1.000133	−.000778	1.75956
.000800	−75.5377	1.000147	−.000877	1.62347
.000900	−76.2603	1.000161	−.000977	1.51195
.001000	−76.8170	1.000175	−.001076	1.41876
.002000	−79.7759	1.000302	−.002071	.94259

TABLE A2

Values of $(\mathrm{d}W/\mathrm{d}V)_{\min}$ *for an elliptical notch with* $b/a = 0.01$, $\beta = 30°$ *and* $\phi = -30.50°$

r_0/a	θ_0	x_0	y_0	$(\mathrm{d}W/\mathrm{d}V)_{\min}$
.000001	−29.8633	.999984	−.000059	963.51785
.000005	−27.3086	.999987	−.000061	841.99817
.000010	−24.1680	.999992	−.000063	719.80355
.000020	−18.3086	1.000002	−.000065	544.83854
.000030	−13.2735	1.000012	−.000066	430.45327
.000040	−9.5528	1.000022	−.000066	354.01002
.000050	−7.7930	1.000032	−.000066	302.21064
.000060	−8.7461	1.000042	−.000068	266.44527
.000070	−12.1211	1.000051	−.000074	240.6
.000080	−16.4493	1.000059	−.000082	221.14962
.000090	−20.6524	1.000067	−.000091	205.37243
.000100	−.24.3555	1.000074	−.000100	192.16559
.000200	−42.6211	1.000130	−.000194	119.82908
.000300	−49.1211	1.000179	−.000286	87.64288
.000400	−52.4493	1.000226	−.000376	69.17515
.000500	−54.4493	1.000273	−.000466	57.18138
.000600	−55.8106	1.000320	−.000555	48.76323
.000700	−56.8555	1.000365	−.000645	42.52843
.000800	−57.6055	1.000411	−.000734	37.72457
.000900	−58.1797	1.000457	−.000824	33.90928
.001000	−58.6524	1.000503	−.000913	30.80553
.002000	−60.9024	1.000955	−.001806	16.22480
.004000	−62.2383	1.001846	−.003598	8.49521
.006000	−62.8086	1.002724	−.005396	5.82958
.008000	−63.1211	1.003600	−.007195	4.47258
.010000	−63.4024	1.004460	−.009001	3.64791
.020000	−64.2383	1.008675	−.018071	1.96240
.040000	−65.1797	1.016774	−.036364	1.08521
1.200000	−64.5743	1.515192	−1.083830	.17573
1.400000	−63.1680	1.631909	−1.249327	.17432
1.600000	−61.8633	1.754505	−1.410979	.17363
1.800000	−60.6993	1.880891	−1.569772	.17332
2.000000	−59.6758	2.009766	−1.726424	.17321

TABLE A3

Values of $(dW/dV)_{min}$ *for an elliptical notch with* $b/a = 0.01$, $\beta = 60°$ *and* $\phi = -15.17°$

r_0/a	θ_0	x_0	y_0	$(dW/dV)_{min}$
.000001	−14.7552	.999997	−.000027	4418.38370
.000005	−13.1185	1.000001	−.000028	3734.96539
.000010	−11.1185	1.000006	−.000029	3089.11501
.000020	−7.5248	1.000016	−.000030	2233.68168
.000030	−4.6849	1.000026	−.000030	1713.36098
.000040	−2.7552	1.000036	−.000029	1376.36325
.000050	−1.8373	1.000046	−.000029	1146.84575
.000060	−1.8841	1.000056	−.000029	983.60391
.000070	−2.7201	1.000066	−.000030	862.85756
.000080	−4.1654	1.000076	−.000033	770.29914
.000090	−5.9076	1.000086	−.000036	697.06508
.000100	−7.8373	1.000095	−.000041	637.52829
.000200	−21.7552	1.000182	−.000101	350.65731
.000300	−27.7904	1.000262	−.000167	243.16198
.000400	−30.9427	1.000339	−.000233	186.12432
.000500	−32.8373	1.000416	−.000298	150.71539
.000600	−34.1185	1.000493	−.000364	126.58788
.000700	−34.9779	1.000570	−.000428	109.09144
.000800	−35.6654	1.000646	−.000494	95.82362
.000900	−36.1517	1.000723	−.000558	85.41760
.001000	−36.5677	1.000799	−.000623	77.03826
.002000	−38.2748	1.001566	−.001266	38.74878
.004000	−38.8490	1.003112	−.002536	19.29628
.006000	−38.8841	1.004667	−.003794	12.79320
.008000	−38.7904	1.006232	−.005039	9.54750
.010000	−38.6673	1.007804	−.006275	7.60546
.020000	−38.1654	1.015721	−.012386	3.75009
.040000	−37.3685	1.031786	−.024305	1.85882
.060000	−36.8841	1.047987	−.036039	1.24356
.080000	−36.4740	1.064326	−.047584	.94220
.100000	−36.1185	1.080776	−.058973	.76475
.200000	−35.0716	1.163683	−.114947	.42429
.300000	−34.2748	1.247900	−.168976	.31977
.400000	−33.6185	1.333093	−.221491	.27132
.500000	−32.9779	1.419437	−.272185	.24429
.600000	−32.4623	1.506243	−.322074	.22747
.700000	−31.9427	1.594000	−.370377	.21624
.800000	−31.4779	1.682270	−.417763	.20833
.900000	−31.0365	1.771152	−.464053	.20254
1.000000	−30.5970	1.860765	−.509024	.19818
1.200000	−29.9076	2.040193	−.598350	.19211
1.400000	−29.2904	2.221008	−.684958	.18820
1.600000	−28.7201	2.403161	−.768877	.18551
1.800000	−28.2220	2.586015	−.851229	.18359
2.000000	−27.8373	2.768551	−.933951	.18216

TABLE A4

Values of $(\mathrm{d}W/\mathrm{d}V)_{min}$ *for an elliptical notch with* $b/a = 0.01$, $\beta = 90°$ *and* $\phi = 0°$

r_0/a	θ_0	x_0	y_0	$(\mathrm{d}W/\mathrm{d}V)_{min}$
.000001	0.0000	1.000001	0.000000	6738.62426
.000005	0.0000	1.000005	0.000000	5628.00305
.000010	0.0000	1.000010	0.000000	4601.81923
.000020	0.0000	1.000020	0.000000	3278.06703
.000030	0.0000	1.000030	0.000000	2489.99771
.000040	0.0000	1.000040	0.000000	1982.79550
.000050	0.0000	1.000050	0.000000	1636.07736
.000060	0.0000	1.000060	0.000000	1387.45539
.000070	0.0000	1.000070	0.000000	1202.14329
.000080	0.0000	1.000080	0.000000	1059.55811
.000090	0.0000	1.000090	0.000000	946.90667
.000100	0.0000	1.000100	0.000000	855.90084
.000200	0.0000	1.000200	0.000000	440.39222
.000300	0.0000	1.000300	0.000000	299.47631
.000400	0.0000	1.000400	0.000000	227.70552
.000500	0.0000	1.000500	0.000000	183.90381
.000600	0.0000	1.000600	0.000000	154.28610
.000700	0.0000	1.000700	0.000000	132.88574
.000800	0.0000	1.000800	0.000000	116.68479
.000900	0.0000	1.000900	0.000000	103.98740
.001000	0.0000	1.001000	0.000000	93.76523
.002000	0.0000	1.002000	0.000000	47.02046
.004000	0.0000	1.004000	0.000000	23.24292
.006000	0.0000	1.006000	0.000000	15.30837
.008000	0.0000	1.008000	0.000000	11.35983
.010000	0.0000	1.010000	0.000000	9.00446
.020000	0.0000	1.020000	0.000000	4.36051
.040000	0.0000	1.040000	0.000000	2.11768
.060000	0.0000	1.060000	0.000000	1.40054
.080000	0.0000	1.080000	0.000000	1.05358
.100000	0.0000	1.100000	0.000000	.85117
.200000	0.0000	1.200000	0.000000	.46803
.300000	0.0000	1.300000	0.000000	.35149
.400000	0.0000	1.400000	0.000000	.29704
.500000	0.0000	1.500000	0.000000	.26613
.600000	0.0000	1.600000	0.000000	.24649
.700000	0.0000	1.700000	0.000000	.23303
.800000	0.0000	1.800000	0.000000	.22331
.900000	0.0000	1.900000	0.000000	.21602
1.000000	0.0000	2.000000	0.000000	.21036
1.200000	0.0000	2.200000	0.000000	.20224
1.400000	0.0000	2.400000	0.000000	.19675
1.600000	0.0000	2.600000	0.000000	.19283
1.800000	0.0000	2.800000	0.000000	.18993
2.000000	0.0000	3.000000	0.000000	.18770

TABLE A5

Values of $(\mathrm{d}W/\mathrm{d}V)_{\min}$ *for an elliptical notch with* $b/a = 0.10$, $\beta = 5°$ *and* $\phi = -66.91°$

r_0/a	θ_0	x_0	y_0	$(\mathrm{d}W/\mathrm{d}V)_{\min}$
.000100	−66.6511	.973619	−.022927	.46369
.000200	−66.4284	.973659	−.023018	.46175
.000400	−65.8952	.973742	−.023200	.45791
.000600	−65.4284	.973829	−.023381	.45411
.000800	−64.9401	.973918	−.023560	.45035
.001000	−64.4108	.974011	−.023737	.44663
.002000	−61.9655	.974519	−.024600	.42868
.003000	−59.4987	.975102	−.025420	.41172
.004000	−57.0651	.975754	−.026192	.39570
.005000	−54.5808	.976477	−.026910	.38057
.006000	−52.0886	.977266	−.027569	.36628
.007000	−49.5868	.978118	−.028164	.35281
.008000	−47.0300	.979032	−.028689	.34011
.009000	−44.4636	.980002	−.029139	.32816
.010000	−41.8620	.981027	−.029508	.31692
.011000	−39.1862	.982105	−.029785	.30638
.012000	−36.4284	.983234	−.029961	.29650
.013000	−33.5339	.984415	−.030017	.28725
.040000	−38.3112	1.005811	−.046522	.22920
.050000	−54.8151	1.002390	−.063700	.22913
.060000	−62.2761	1.001492	−.075947	.22823
.070000	−67.0651	1.000857	−.087301	.22713
.080000	−70.4987	1.000285	−.098246	.22598
.090000	−73.1120	.999724	−.108954	.22483
.100000	−75.1589	.999193	−.119499	.22369
.200000	−84.9714	.991110	−.222065	.21393
.300000	−89.3151	.977165	−.322814	.20646
.400000	−92.3405	.957244	−.422501	.20041
.500000	−94.8620	.931201	−.521036	.19538
.600000	−97.0651	.899781	−.618279	.19113
.700000	−98.9870	.864232	−.714242	.18753
.800000	−100.7214	.824752	−.808870	.18449
2.000000	−101.4089	.577961	−1.983316	.17339

TABLE A6

Values of $(dW/dV)_{min}$ *for an elliptical notch with* $b/a = 0.10$, $\beta = 30°$ *and* $\phi = -34.91°$

r_0/a	θ_0	x_0	y_0	$(dW/dV)_{min}$
.000100	−34.2277	.997656	−.007019	11.28986
.000200	−33.5968	.997740	−.007073	10.95166
.000400	−32.3156	.997911	−.007176	10.31849
.000600	−30.9913	.998088	−.007272	9.73796
.000800	−29.7257	.998268	−.007359	9.20471
.001000	−28.4679	.998452	−.007439	8.71408
.002000	−22.3449	.999423	−.007723	6.77341
.003000	−16.8663	1.000444	−.007833	5.45187
.004000	−12.3507	1.001481	−.007818	4.54020
.005000	−9.2257	1.002508	−.007764	3.90920
.006000	−8.5382	1.003507	−.007853	3.47161
.007000	−10.9913	1.004445	−.008297	3.16250
.008000	−15.5031	1.005282	−.009101	2.93262
.009000	−20.2218	1.006018	−.010073	2.75027
.010000	−24.4327	1.006678	−.011099	2.59835
.011000	−28.0226	1.007284	−.012131	2.46771
.012000	−31.0695	1.007852	−.013156	2.35301
.013000	−33.6554	1.008394	−.014167	2.25085
.014000	−35.9132	1.008912	−.015174	2.15889
.015000	−37.8663	1.009415	−.016170	2.07543
.016000	−39.6320	1.009896	−.017168	1.99920
.017000	−41.1632	1.010371	−.018152	1.92918
.018000	−42.5382	1.010836	−.019132	1.86459
.019000	−43.7726	1.011293	−.020107	1.80475
.020000	−44.8663	1.011748	−.021072	1.74914
.030000	−52.1632	1.015976	−.030655	1.35002
.040000	−56.0577	1.019907	−.040147	1.11169
.050000	−58.5031	1.023696	−.049596	.95225
.060000	−60.2218	1.027372	−.059040	.83769
.070000	−61.5382	1.030933	−.068502	.75119
.080000	−62.5851	1.034408	−.077978	.68345
.090000	−63.4327	1.037825	−.087459	.62888
.100000	−64.1281	1.041209	−.096940	.58393
.200000	−68.0577	1.072308	−.192475	.36474
.300000	−69.8195	1.101067	−.288546	.28467
.900000	−69.7843	1.308573	−.851521	.18426
1.000000	−69.1163	1.354045	−.941269	.18104
1.200000	−67.6906	1.453104	−1.117139	.17712
1.400000	−66.2804	1.560739	−1.288698	.17511
1.600000	−64.9288	1.675563	−1.456214	.17408
1.800000	−63.7726	1.793056	−1.621647	.17356
2.000000	−62.6906	1.915165	−1.784046	.17332

TABLE A7

Values of $(\mathrm{d}W/\mathrm{d}V)_{\min}$ *for an elliptical notch with* $b/a = 0.10$, $\beta = 60°$ *and* $\phi = -16.65°$

r_0/a	θ_0	x_0	y_0	$(\mathrm{d}W/\mathrm{d}V)_{\min}$
.000100	−16.2068	.999649	−.003018	48.76473
.000200	−15.7615	.999745	−.003044	46.69727
.000400	−14.8553	.999940	−.003092	42.94595
.000600	−14.0193	1.000135	−.003135	39.64069
.000800	−13.1522	1.000332	−.003172	36.71549
.001000	−12.3240	1.000530	−.003203	34.11594
.002000	−8.4647	1.001531	−.003238	24.65663
.003000	−5.4647	1.002539	−.003275	18.90012
.004000	−3.5584	1.003545	−.003238	15.17963
.005000	−2.8553	1.004547	−.003239	12.65497
.006000	−3.2772	1.005543	−.003333	10.86572
.007000	−4.6834	1.006530	−.003561	9.54457
.008000	−6.6834	1.007499	−.003921	8.53097
.009000	−8.9295	1.008444	−.004387	7.72652
.010000	−11.1541	1.009364	−.004924	7.06976
.011000	−13.3240	1.010257	−.005525	6.52122
.012000	−15.2772	1.011129	−.006152	6.05470
.013000	−17.1053	1.011978	−.006813	5.65211
.014000	−18.6678	1.012816	−.007471	5.30055
.015000	−20.1365	1.013636	−.008154	4.99054
.016000	−21.4295	1.014447	−.008835	4.71488
.017000	−22.6053	1.015247	−.009524	4.46803
.018000	−23.6522	1.016041	−.010211	4.24561
.019000	−24.6053	1.016828	−.010901	4.04411
.020000	−25.4647	1.017610	−.011589	3.86069
.030000	−30.8943	1.025296	−.018393	2.65068
.040000	−33.5115	1.032904	−.025074	2.01402
.050000	−34.9295	1.040546	−.031618	1.62308
.060000	−35.7615	1.048240	−.038054	1.35970
.070000	−36.3240	1.055951	−.044454	1.17082
.080000	−36.6522	1.063735	−.050746	1.02915
.090000	−36.8553	1.071567	−.056971	.91922
.100000	−37.0115	1.079404	−.063187	.83162
.200000	−36.9022	1.159485	−.123080	.44816
.300000	−36.2068	1.241620	−.180200	.33039
.400000	−35.5584	1.324962	−.235603	.27634
.500000	−34.9022	1.409618	−.289078	.24653
.600000	−34.2772	1.495347	−.340908	.22824
.700000	−33.7088	1.581861	−.391470	.21616
.800000	−33.2068	1.668912	−.441120	.20778
.900000	−32.7303	1.756656	−.489606	.20171
1.000000	−32.3240	1.844591	−.537696	.19719
1.200000	−31.5232	2.022467	−.630403	.19102
1.400000	−30.8553	2.201405	−.721010	.18712
1.600000	−30.2420	2.381802	−.808835	.18450
1.800000	−29.7088	2.562953	−.895055	.18265
2.000000	−29.2772	2.744082	−.981059	.18130

TABLE A8

Values of $(\mathrm{d}W/\mathrm{d}V)_{\min}$ *for an elliptical notch with* $b/a = 0.10$, $\beta = 90°$ *and* $\phi = 0°$

r_0/a	θ_0	x_0	y_0	$(\mathrm{d}W/\mathrm{d}V)_{\min}$
.000100	0.0000	1.000100	0.000000	73.49397
.000200	0.0000	1.000200	0.000000	70.07738
.000400	0.0000	1.000400	0.000000	63.94391
.000600	0.0000	1.000600	0.000000	58.61149
.000800	0.0000	1.000800	0.000000	53.94934
.001000	0.0000	1.001000	0.000000	49.85173
.002000	0.0000	1.002000	0.000000	35.29844
.003000	0.0000	1.003000	0.000000	26.68192
.004000	0.0000	1.004000	0.000000	21.16006
.005000	0.0000	1.005000	0.000000	17.39769
.006000	0.0000	1.006000	0.000000	14.70638
.007000	0.0000	1.007000	0.000000	12.70399
.008000	0.0000	1.008000	0.000000	11.16531
.009000	0.0000	1.009000	0.000000	9.95082
.010000	0.0000	1.010000	0.000000	8.97039
.011000	0.0000	1.011000	0.000000	8.16367
.012000	0.0000	1.012000	0.000000	7.48899
.013000	0.0000	1,013000	0.000000	6.91676
.014000	0.0000	1.014000	0.000000	6.42549
.015000	0.0000	1.015000	0.000000	5.99922
.016000	0.0000	1.016000	0.000000	5.62588
.017000	0.0000	1.017000	0.000000	5.29621
.018000	0.0000	1.018000	0.000000	5.00295
.019000	0.0000	1.019000	0.000000	4.74038
.020000	0.0000	1.020000	0.000000	4.50391
.030000	0.0000	1.030000	0.000000	3.00154
.040000	0.0000	1.040000	0.000000	2.24666
.050000	0.0000	1.050000	0.000000	1.79333
.060000	0.0000	1.060000	0.000000	1.49195
.070000	0.0000	1.070000	0.000000	1.27784
.080000	0.0000	1.080000	0.000000	1.11843
.090000	0.0000	1.090000	0.000000	.99549
.100000	0.0000	1.100000	0.000000	.89807
.200000	0.0000	1.200000	0.000000	.47876
.300000	0.0000	1.300000	0.000000	.35289
.400000	0.0000	1.400000	0.000000	.29529
.500000	0.0000	1.500000	0.000000	.26324
.600000	0.0000	1.600000	0.000000	.24322
.700000	0.0000	1.700000	0.000000	.22972
.800000	0.0000	1.800000	0.000000	.22011
.900000	0.0000	1.900000	0.000000	.21298
1.000000	0.0000	2.000000	0.000000	.20752
1.200000	0.0000	2.200000	0.000000	.19977
1.400000	0.0000	2.400000	0.000000	.19461
1.600000	0.0000	2.600000	0.000000	.19097
1.800000	0.0000	2.800000	0.000000	.18830
2.000000	0.0000	3.000000	0.000000	.18627

TABLE A9

Values of $(\mathrm{d}W/\mathrm{d}V)_{min}$ *for an elliptical notch with* $b/a = 0.50$, $\beta = 5°$ *and* $\phi = -82.54°$

r_0/a	θ_0	x_0	y_0	$(\mathrm{d}W/\mathrm{d}V)_{min}$
.000100	−82.5375	.253434	−.483777	.71979
.000500	−82.4906	.253487	−.484174	.71888
.001000	−82.4906	.253552	−.484669	.71775
.005000	−82.3500	.254087	−.488634	.70880
.010000	−82.2094	.254777	−.493586	.69784
.015000	−82.0219	.255503	−.498533	.68713
.020000	−81.8813	.256246	−.503478	.67666
.025000	−81.6645	.257046	−.508414	.66641
.030000	−81.4906	.257860	−.513348	.65640
.035000	−81.3500	.258685	−.518280	.64661
.040000	−81.1625	.259567	−.523203	.63704
.045000	−80.9906	.260468	−.528123	.62769
.050000	−80.8500	.261372	−.533042	.61854
.055000	−80.6938	.262315	−.537954	.60960
.060000	−80.4906	.263334	−.542854	.60085
.065000	−80.3500	.264317	−.547758	.59230
.070000	−80.1625	.265381	−.552649	.58395
.075000	−79.9691	.266485	−.557532	.57577
.080000	−79.8461	.267525	−.562425	.56778
.085000	−79.6820	.268646	−.567303	.55997
.090000	−79.4906	.269837	−.572168	.55233
.095000	−79.3500	.270978	−.577042	.54486
.100000	−79.1625	.272224	−.581894	.53755
.150000	−77.5375	.285791	−.630144	.47280
.200000	−75.9691	.301910	−.677711	.42090
.250000	−74.6625	.319547	−.724774	.37919
.300000	−73.6820	.337712	−.771593	.34558
.350000	−73.2563	.354253	−.818839	.31840
.400000	−73.5375	.366776	−.867280	.29633
.450000	−74.6000	.372922	−.917521	.27830
.500000	−76.2211	.372509	−.969289	.26342
.550000	−78.1625	.366246	−1.021981	.25100
.600000	−80.1274	.356297	−1.074793	.24051
.650000	−81.9047	.344954	−1.127201	.23158
.700000	−83.4906	.332777	−1.179165	.22393
.750000	−84.8149	.321202	−1.230609	.21734
.800000	−85.8813	.310880	−1.281612	.21165
.900000	−87.4086	.294113	−1.382758	.20247
1.000000	−88.2797	.283442	−1.483227	.19554
1.200000	−88.9906	.274560	−1.683492	.18631
1.400000	−89.0922	.275602	−1.883502	.18090
1.600000	−88.8969	.284224	−2.083381	.17768
1.800000	−88.6156	.296908	−2.283153	.17574
2.000000	−88.3500	.311009	−2.482849	.17457

TABLE A10

Values of $(\mathrm{d}W/\mathrm{d}V)_{min}$ *for an elliptical notch with* $b/a = 0.50$, $\beta = 30°$ *and* $\phi = -50.45°$

r_0/a	θ_0	x_0	y_0	$(\mathrm{d}W/\mathrm{d}V)_{min}$
.000100	−50.3998	.855512	−.259021	1.42941
.000500	−50.3529	.855767	−.259329	1.42412
.001000	−50.2592	.856087	−.259713	1.41753
.005000	−49.5033	.858695	−.262746	1.36645
.010000	−.48.5717	.862065	−.266442	1.30631
.015000	−47.6264	.865558	−.270026	1.25001
.020000	−46.7240	.869158	−.273505	1.19726
.025000	−45.7904	.872880	−.276864	1.14779
.030000	−44.8529	.876716	−.280103	1.10138
.035000	−43.9662	.880639	−.283242	1.05779
.040000	−43.0717	.884668	−.286261	1.01683
.045000	−42.1654	.888803	−.289152	.97833
.050000	−41.2592	.893035	−.291918	.94211
.055000	−40.3998	.897333	−.294591	.90802
.060000	−39.5033	.901743	−.297112	.87592
.065000	−38.6498	.906212	−.299541	.84569
.070000	−37.8139	.910749	−.301861	.81720
.075000	−37.0014	.915345	−.304082	.79035
.080000	−36.1889	.920014	−.306180	.76504
.085000	−35.3998	.924734	−.308183	.74117
.090000	−34.6498	.929486	−.310115	.71866
.095000	−33.8783	.934319	−.311900	.69743
.100000	−33.1654	.939158	−.313650	.67741
.150000	−23.1303	.987730	−.329666	.53012
.200000	−30.3178	1.028096	−.359903	.44745
.250000	−38.2240	1.051848	−.413629	.39661
.300000	−45.5561	1.065511	−.473125	.36021
.350000	−51.1889	1.074812	−.531670	.33175
.400000	−55.5033	1.081991	−.588608	.30857
.450000	−58.8529	1.088205	−.644073	.28932
.500000	−61.4858	1.094137	−.698293	.27314
.550000	−63.5561	1.100375	−.751398	.25944
.600000	−65.1889	1.107225	−.803562	.24779
.650000	−66.4623	1.115027	−.854863	.23785
.700000	−67.5033	1.123289	−.905675	.22934
.750000	−68.2592	1.133254	−.955596	.22205
.800000	−68.8529	1.144058	−1.005070	.21578
.900000	−69.5561	1.169810	−1.102257	.20574
1.000000	−69.8490	1.199943	−1.197732	.19824
1.200000	−69.6967	1.271836	−1.384387	.18833
1.400000	−69.0717	1.355527	−1.566584	.18257
1.600000	−68.3061	1.446885	−1.745619	.17912
1.800000	−67.5033	1.544181	−1.921967	.17702
2.000000	−66.7592	1.644641	−2.096653	.17572

TABLE A11

Values of $(\mathrm{d}W/\mathrm{d}V)_{\min}$ *for an elliptical notch with* $b/a = 0.50$, $\beta = 60°$ *and* $\phi = -23.05°$

r_0/a	θ_0	x_0	y_0	$(\mathrm{d}W/\mathrm{d}V)_{\min}$
.000100	−23.0042	.978198	−.104092	3.27395
.000500	−22.9573	.978567	−.104248	3.25073
.001000	−22.8636	.979028	−.104441	3.22206
.005000	−22.1956	.982736	−.105941	3.00570
.010000	−21.3284	.987422	−.107690	2.76470
.015000	−20.4886	.992158	−.109303	2.55193
.020000	−19.6409	.996943	−.110775	2.36336
.025000	−18.8284	1.001769	−.112121	2.19565
.030000	−18.0042	1.006637	−.113325	2.04598
.035000	−17.2542	1.011531	−.114434	1.91200
.040000	−16.4827	1.016463	−.115402	1.79171
.045000	−15.7932	1.021408	−.116300	1.68341
.050000	−15.1136	1.026377	−.117089	1.58566
.055000	−14.4886	1.031357	−.117813	1.49720
.060000	−13.8636	1.036359	−.118429	1.41697
.065000	−13.3636	1.041346	−.119076	1.34404
.070000	−12.8636	1.046350	−.119637	1.27761
.075000	−12.4534	1.051342	−.120226	1.21697
.080000	−12.1077	1.056327	−.120833	1.16152
.085000	−11.8284	1.061302	−.121476	1.11071
.090000	−11.6057	1.066266	−.122158	1.06407
.095000	−11.4573	1.071213	−.122923	1.02117
.100000	−11.4104	1.076130	−.123836	.98164
.150000	−14.6761	1.123212	−.142056	.71307
.200000	−20.7932	1.165080	−.175052	.56745
.250000	−25.8284	1.203132	−.212972	.47522
.300000	−29.3636	1.239564	−.251158	.41151
.350000	−31.7229	1.275817	−.288087	.36528
.400000	−33.3011	1.312425	−.323668	.33061
.450000	−34.3479	1.349638	−.357950	.30397
.500000	−35.0042	1.387662	−.390871	.28310
.550000	−35.4222	1.426304	−.422831	.26649
.600000	−35.6761	1.465503	−.453974	.25309
.650000	−35.7932	1.505343	−.484213	.24215
.700000	−35.8284	1.545648	−.513804	.23313
.750000	−35.8284	1.586187	−.543072	.22561
.800000	−35.7698	1.627204	−.571677	.21931
.900000	−35.5706	1.710166	−.627588	.20944
1.000000	−35.3011	1.794233	−.681925	.20221
1.200000	−34.7229	1.964406	−.787583	.19270
1.400000	−34.1604	2.136562	−.890170	.18702
1.600000	−33.6409	2.310148	−.990430	.18345
1.800000	−33.1761	2.484694	−1.089037	.18111
2.000000	−32.7347	2.660474	−1.185551	.17952

TABLE A12

Values of $(dW/dV)_{min}$ *for an elliptical notch with* $b/a = 0.50$, $\beta = 90°$ *and* $\phi = 0°$

r_0/a	θ_0	x_0	y_0	$(dW/dV)_{min}$
.000100	0.0000	1.000100	0.000000	4.36581
.000500	0.0000	1.000500	0.000000	4.32936
.001000	0.0000	1.001000	0.000000	4.28443
.005000	0.0000	1.005000	0.000000	3.94905
.010000	0.0000	1.010000	0.000000	3.58306
.015000	0.0000	1.015000	0.000000	3.26675
.020000	0.0000	1.020000	0.000000	2.99184
.025000	0.0000	1.025000	0.000000	2.75167
.030000	0.0000	1.030000	0.000000	2.54081
.035000	0.0000	1.035000	0.000000	2.35483
.040000	0.0000	1.040000	0.000000	2.19006
.045000	0.0000	1.045000	0.000000	2.04348
.050000	0.0000	1.050000	0.000000	1.91256
.055000	0.0000	1.055000	0.000000	1.79519
.060000	0.0000	1.060000	0.000000	1.68960
.065000	0.0000	1.065000	0.000000	1.59427
.070000	0.0000	1.070000	0.000000	1.50794
.075000	0.0000	1.075000	0.000000	1.42951
.080000	0.0000	1.080000	0.000000	1.35806
.085000	0.0000	1.085000	0.000000	1.29277
.090000	0.0000	1.090000	0.000000	1.23297
.095000	0.0000	1.095000	0.000000	1.17804
.100000	0.0000	1.100000	0.000000	1.12747
.150000	0.0000	1.150000	0.000000	.78437
.200000	0.0000	1.200000	0.000000	.60221
.250000	0.0000	1.250000	0.000000	.49249
.300000	0.0000	1.300000	0.000000	.42054
.350000	0.0000	1.350000	0.000000	.37044
.400000	0.0000	1.400000	0.000000	.33402
.450000	0.0000	1.450000	0.000000	.30664
.500000	0.0000	1.500000	0.000000	.28554
.550000	0.0000	1.550000	0.000000	.26892
.600000	0.0000	1.600000	0.000000	.25561
.650000	0.0000	1.650000	0.000000	.24480
.700000	0.0000	1.700000	0.000000	.23591
.750000	0.0000	1.750000	0.000000	.22852
.800000	0.0000	1.800000	0.000000	.22231
.900000	0.0000	1.900000	0.000000	.21259
1.000000	0.0000	2.000000	0.000000	.20543
1.200000	0.0000	2.200000	0.000000	.19588
1.400000	0.0000	2.400000	0.000000	.19004
1.600000	0.0000	2.600000	0.000000	.18625
1.800000	0.0000	2.800000	0.000000	.18369
2.000000	0.0000	3.000000	0.000000	.18188

TABLE A13

Values of $(dW/dV)_{min}$ *for an elliptical notch* $b/a = 0.90$, $\beta = 5°$ *and* $\phi = -84.72°$

r_0/a	θ_0	x_0	y_0	$(dW/dV)_{min}$
.000100	−84.7239	.102081	−.895399	1.37454
.000500	−84.7239	.102118	−.895797	1.37178
.001000	−84.7239	.102164	−.896295	1.36834
.005000	−84.6770	.102536	−.900278	1.34128
.010000	−84.6770	.103000	−.905256	1.30855
.020000	−84.6770	.103927	−.915213	1.24856
.040000	−84.5950	.105840	−.935121	1.13511
.060000	−84.5364	.107785	−.955027	1.03821
.080000	−84.4661	.109787	−.974926	.95366
.100000	−84.4075	.111817	−.994823	.87962
.120000	−84.3489	.113888	−1.014716	.81455
.140000	−84.2786	.116029	−1.034602	.75716
.160000	−84.2434	.118120	−1.054492	.70637
.180000	−84.1555	.120401	−1.074384	.66129
.200000	−84.1614	.122417	−1.094262	.62114
.220000	−84.0911	.124720	−1.114130	.58526
.240000	−84.0911	.126779	−1.134024	.55311
.260000	−84.0676	.128944	−1.153907	.52422
.280000	−84.0676	.131011	−1.173800	.49818
.300000	−84.0676	.133078	−1.193693	.47465
.320000	−84.0676	.135145	−1.213586	.45333
.340000	−84.0911	.137074	−1.233493	.43397
.360000	−84.1252	.138913	−1.253409	.41634
.380000	−84.1614	.140723	−1.273328	.40025
.400000	−84.2032	.142437	−1.293257	.38553
.420000	−84.2258	.144327	−1.313168	.37204
.440000	−84.3137	.145668	−1.333134	.35966
.460000	−84.3489	.147368	−1.353064	.34826
.480000	−84.4426	.148556	−1.373043	.33775
.500000	−84.5012	.149984	−1.392998	.32805
.550000	−84.6301	.153544	−1.442886	.30682
.600000	−84.8020	.156431	−1.492832	.28918
.650000	−84.9270	.159548	−1.542753	.27440
.700000	−85.0364	.162638	−1.592674	.26191
.750000	−85.0950	.166200	−1.642553	.25129
.800000	−85.1770	.169334	−1.692467	.24220
.850000	−85.1770	.173538	−1.742290	.23438
.900000	−85.2434	.176702	−1.792200	.22761
1.000000	−85.2786	.184383	−1.891906	.21661
1.200000	−85.3489	.199378	−2.891348	.20164
1.400000	−85.3137	.216452	−2.290619	.19242
1.600000	−85.3137	.232792	−2.489950	.18653
1.800000	−85.2786	.250232	−2.689191	.18265
2.000000	−85.2786	.266695	−2.888513	.18005

TABLE A14

Values of $(\mathrm{d}W/\mathrm{d}V)_{\min}$ *for an elliptical notch with* $b/a = 0.90$, $\beta = 30°$ *and* $\phi = -58.66°$

r_0/a	θ_0	x_0	y_0	$(\mathrm{d}W/\mathrm{d}V)_{\min}$
.000100	−58.6599	.560453	−.745485	1.47876
.000500	−58.6599	.560661	−.745826	1.47553
.001000	−58.6130	.560922	−.746253	1.47150
.005000	−58.5662	.563088	−.749666	1.43987
.010000	−58.4724	.565630	−.753923	1.40173
.020000	−58.3318	.570901	−.762421	1.32981
.040000	−58.0037	.581595	−.779323	1.20170
.060000	−57.6462	.592510	−.796085	1.09160
.080000	−57.3787	.603528	−.812780	.99654
.100000	−57.0662	.614768	−.829329	.91412
.120000	−56.7849	.626135	−.845794	.84234
.140000	−56.5193	.637633	−.862169	.77958
.160000	−56.2498	.649293	−.878434	.72448
.180000	−56.0271	.660985	−.894674	.67592
.200000	−55.8396	.672703	−.910893	.63298
.220000	−55.6990	.684380	−.927139	.59485
.240000	−55.5662	.696110	−.943347	.56090
.260000	−55.5310	.707550	−.959752	.53055
.280000	−55.5310	.718870	−.976241	.50335
.300000	−55.5662	.730037	−.992833	.47888
.320000	−55.6462	.740977	−1.009582	.45681
.340000	−55.7869	.751574	−1.026583	.43685
.360000	−55.9744	.761844	−1.043763	.41875
.380000	−56.2498	.771519	−1.061357	.40228
.400000	−56.5193	.781083	−1.079028	.38726
.420000	−56.7849	.790470	−1.096780	.37353
.440000	−57.0662	.799616	−1.114691	.36095
.460000	−57.4021	.808221	−1.132937	.34939
.480000	−57.7165	.816773	−1.151199	.33876
.500000	−58.0271	.825160	−1.169549	.32895
.550000	−58.7380	.845824	−1.215541	.30754
.600000	−59.3787	.866018	−1.261731	.28978
.650000	−59.8748	.886631	−1.307604	.27492
.700000	−60.2849	.907382	−1.353350	.26238
.750000	−60.6130	.928430	−1.398893	.25172
.800000	−60.8630	.949920	−1.444166	.24260
.850000	−61.0662	.971630	−1.489301	.23476
.900000	−61.2146	.993778	−1.534186	.22798
1.000000	−61.4021	1.039061	−1.623400	.21697
1.200000	−61.5310	1.132421	−1.800290	.20198
1.400000	−61.5193	1.228009	−1.975968	.19276
1.600000	−61.4373	1.325394	−2.150670	.18686
1.800000	−61.3435	1.423604	−2.324918	.18298
2.000000	−61.2146	1.523462	−2.498258	.18036

TABLE A15

Values of $(\mathrm{d}W/\mathrm{d}V)_{\min}$ *for an elliptical notch with* $b/a = 0.90$, $\beta = 60°$ *and* $\phi = -28.73°$

r_0/a	θ_0	x_0	y_0	$(\mathrm{d}W/\mathrm{d}V)_{\min}$
.000100	−28.7285	.896899	−.398220	1.70102
.000500	−28.7285	.897250	−.398412	1.69671
.001000	−28.6817	.897688	−.398652	1.69134
.005000	−28.6348	.901200	−.400568	1.64931
.010000	−28.5410	.905596	−.402950	1.59892
.020000	−28.4004	.914404	−.407685	1.50480
.040000	−28.0430	.932115	−.416977	1.34005
.060000	−27.7676	.949902	−.426125	1.20156
.080000	−27.4473	.967806	−.435047	1.08439
.100000	−27.1348	.985805	−.443781	.98465
.120000	−26.8848	1.003841	−.452436	.89926
.140000	−26.6348	1.021955	−.460934	.82573
.160000	−26.4121	1.040110	−.469344	.76209
.180000	−28.2129	1.058300	−.477680	.70672
.200000	−26.0957	1.076423	−.486147	.65831
.220000	−25.9785	1.094582	−.494540	.61579
.240000	−25.9434	1.112626	−.503168	.57829
.260000	−25.9785	1.130540	−.512061	.54507
.280000	−26.0371	1.148394	−.521079	.51551
.300000	−26.1660	1.166067	−.530464	.48913
.320000	−26.3184	1.183641	−.540047	.46548
.340000	−26.5410	1.200980	−.550097	.44421
.360000	−26.7910	1.218167	−.560438	.42502
.380000	−27.0410	1.235270	−.570931	.40765
.400000	−27.3535	1.252087	−.581964	.39188
.420000	−27.6348	1.268898	−.592982	.37751
.440000	−27.9434	1.285512	−.604356	.36440
.460000	−28.2481	1.302028	−.615886	.35239
.480000	−28.5059	1.318620	−.627252	.34138
.500000	−28.7910	1.335002	−.638980	.33125
.550000	−29.4004	1.375977	−.668173	.30923
.600000	−29.9082	1.416906	−.697339	.29108
.650000	−30.2832	1.458114	−.725951	.27595
.700000	−30.5996	1.499333	−.754497	.26324
.750000	−30.8067	1.540986	−.782279	.25248
.800000	−31.0059	1.582503	−.810273	.24330
.850000	−31.1348	1.624372	−.837667	.23542
.900000	−31.2285	1.666407	−.864780	.22863
1.000000	−31.3535	1.750784	−.918489	.21761
1.200000	−31.4004	1.921068	−1.023391	.20266
1.400000	−31.3535	2.092373	−1.126616	.19346
1.600000	−31.2481	2.264698	−1.228163	.18757
1.800000	−31.1348	2.437527	−1.328868	.18367
2.000000	−31.0410	2.610408	−1.429476	.18103

TABLE A16

Values of $(dW/dV)_{min}$ *for an elliptical notch with* $b/a = 0.90$, $\beta = 90°$ *and* $\phi = 0°$

r_0/a	θ_o	x_0	y_0	$(dW/dV)_{min}$
.000100	0.0000	1.000100	0.000000	1.81575
.000500	0.0000	1.000500	0.000000	1.81085
.001000	0.0000	1.001000	0.000000	1.80475
.005000	0.0000	1.005000	0.000000	1.75704
.010000	0.0000	1.010000	0.000000	1.69999
.020000	0.0000	1.020000	0.000000	1.59391
.040000	0.0000	1.040000	0.000000	1.40976
.060000	0.0000	1.060000	0.000000	1.25654
.080000	0.0000	1.080000	0.000000	1.12812
.100000	0.0000	1.100000	0.000000	1.01971
.120000	0.0000	1.120000	0.000000	.92759
.140000	0.0000	1.140000	0.000000	.84880
.160000	0.0000	1.160000	0.000000	.78101
.180000	0.0000	1.180000	0.000000	.72234
.200000	0.0000	1.200000	0.000000	.67128
.220000	0.0000	1.220000	0.000000	.62663
.240000	0.0000	1.240000	0.000000	.58738
.260000	0.0000	1.260000	0.000000	.55272
.280000	0.0000	1.280000	0.000000	.52198
.300000	0.0000	1.300000	0.000000	.49460
.320000	0.0000	1.320000	0.000000	.47012
.340000	0.0000	1.340000	0.000000	.44815
.360000	0.0000	1.360000	0.000000	.42837
.380000	0.0000	1.380000	0.000000	.41049
.400000	0.0000	1.400000	0.000000	.39430
.420000	0.0000	1.420000	0.000000	.37957
.440000	0.0000	1.440000	0.000000	.36616
.460000	0.0000	1.460000	0.000000	.35390
.480000	0.0000	1.480000	0.000000	.34267
.500000	0.0000	1.500000	0.000000	.33236
.550000	0.0000	1.550000	0.000000	.31001
.600000	0.0000	1.600000	0.000000	.29164
.650000	0.0000	1.650000	0.000000	.27639
.700000	0.0000	1.700000	0.000000	.26360
.750000	0.0000	1.750000	0.000000	.25280
.800000	0.0000	1.800000	0.000000	.24360
.850000	0.0000	1.850000	0.000000	.23571
.900000	0.0000	1.900000	0.000000	.22892
1.000000	0.0000	2.000000	0.000000	.21791
1.200000	0.0000	2.200000	0.000000	.20300
1.400000	0.0000	2.400000	0.000000	.19382
1.600000	0.0000	2.600000	0.000000	.18794
1.800000	0.0000	2.800000	0.000000	.18404
2.000000	0.0000	3.000000	0.000000	.18138

VII Appendix B: Fracture trajectory data for two hyperbolic external notches in tension

Referring to Figure B1, the minima of the strain energy density function, $(dW/dV)_{min}$, are computed with reference to the notch located on the negative x-axis. As in the case of the elliptical notch, the various values chosen for the parameters under consideration are $b/a = 0.01$, 0.1, 0.5, 0.9 and $\beta = 5°$, 30°, 60°, 90° with the Poisson's ratio $\nu = 0.3$. The trajectory of the crack path emanating from the notch boundary (as given by ϕ) is determined by the various increments r_0/a in Tables B1 to B16.

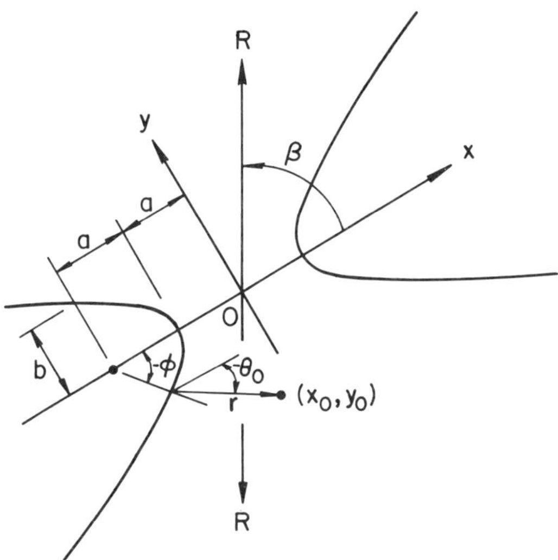

Figure B1. Radius vector measured from a hyperbolic notch in tension.

TABLE B1

Values of $(\mathrm{d}W/\mathrm{d}V)_{min}$ *for two hyperbolic notches with* $b/a = 0.01$, $\beta = 5°$ *and* $\phi = -42.53°$

r_0/a	θ_0	x_0	y_0	$(\mathrm{d}W/\mathrm{d}V)_{min}$
.000100	−39.3061	−.999965	−.000155	.13508
.000500	−70.6069	−.999876	−.000563	.22801
.001000	−74.7319	−.999779	−.001056	.30662
.005000	−74.7319	−.999779	−.001056	.30662
.010000	−74.7319	−.999779	−.001056	.30662
.020000	−74.7319	−.999779	−.001056	.30662
.040000	−74.7319	−.999779	−.001056	.30662
.060000	−74.7319	−.999779	−.001056	.30662
.080000	−74.7319	−.999779	−.001056	.30662
.100000	−74.7319	−.999779	−.001056	.30662
.120000	−74.7319	−.999779	−.001056	.30662
.140000	−74.7319	−.999779	−.001056	.30662
.160000	−74.7319	−.999779	−.001056	.30662
.180000	−74.7319	−.999779	−.001056	.30662
.200000	−74.7319	−.999779	−.001056	.30662
.220000	−74.7319	−.999779	−.001056	.30662
.240000	−74.7319	−.999779	−.001056	.30662
.260000	−74.7319	−.999779	−.001056	.30662
.280000	−74.7319	−.999779	−.001056	.30662
.300000	−74.7319	−.999779	−.001056	.30662
.320000	−74.7319	−.999779	−.001056	.30662
.340000	−74.7319	−.999779	−.001056	.30662
.360000	−74.7319	−.999779	−.001056	.30662
.380000	−74.7319	−.999779	−.001056	.30662
.400000	−74.7319	−.999779	−.001056	.30662
.420000	−74.7319	−.999779	−.001056	.30662
.440000	−74.7319	−.999779	−.001056	.30662
.460000	−74.7319	−.999779	−.001056	.30662
.480000	−74.7319	−.999779	−.001056	.30662
.500000	−74.7319	−.999779	−.001056	.30662
.550000	−74.7319	−.999779	−.001056	.30662
.600000	−74.7319	−.999779	−.001056	.30662
.650000	−77.3061	−.857210	−.634205	8.26769
.700000	−.76.9350	−.841803	−.681972	8.70721
.750000	−76.5600	−.825722	−.729552	9.15548
.800000	−76.1538	−.808589	−.776845	9.61487
.850000	−75.6558	−.789457	−.823593	10.08774
.900000	−75.2124	−.770329	−.870282	10.57651
.950000	−74.6850	−.749124	−.916356	11.08364
1.000000	−74.1538	−.726986	−.962090	11.61178
1.200000	−71.7085	−.623420	−1.139458	13.99427
1.400000	−68.9350	−.496846	−1.306535	16.98939
1.600000	−66.1538	−.353189	−1.463506	20.92958
1.800000	−63.7085	−.202753	−1.613885	26.28225

TABLE B2

Values of $(\mathrm{d}W/\mathrm{d}V)_{min}$ *for two hyperbolic notches with* $b/a = 0.01$, $\beta = 30°$ *and* $\phi = -30.16°$

r_0/a	θ_0	x_0	y_0	$(\mathrm{d}W/\mathrm{d}V)_{min}$
.000100	−24.4693	−.999926	−.000100	.11411
.000500	−54.1724	−.999724	−.000463	.21158
.001000	−58.2349	−.999490	−.000908	.29056
.005000	−61.5240	−.997633	−.004453	.63393
.010000	−61.9693	−.995317	−.008885	.89318
.020000	−62.1568	−.990676	−.017743	1.25949
.040000	−62.3130	−.981431	−.035478	1.77497
.060000	−62.3599	−.972182	−.053211	2.16752
.080000	−62.4341	−.962995	−.070976	2.49597
.100000	−62.4693	−.953794	−.088734	2.78323
.120000	−62.5161	−.944637	−.106515	3.04111
.140000	−62.5630	−.935509	−.124311	3.27667
.160000	−62.6099	−.926409	−.142121	3.49454
.180000	−62.6099	−.917209	−.159879	3.69794
.200000	−62.6587	−.908159	−.177715	3.88925
.220000	−62.7115	−.899153	−.195574	4.07028
.240000	−62.7818	−.890245	−.213483	4.24243
.260000	−62.7818	−.881098	−.231269	4.40684
.280000	−62.8404	−.872205	−.249185	4.56444
.300000	−62.8462	−.863103	−.266994	4.71601
.320000	−62.9341	−.854412	−.285013	4.86221
.340000	−62.9693	−.845498	−.302917	5.00359
.360000	−62.9693	−.836408	−.320733	5.14065
.380000	−63.0279	−.827665	−.338724	5.27380
.400000	−63.0630	−.818813	−.356660	5.40344
.420000	−63.1099	−.810059	−.374646	5.52989
.440000	−63.1099	−.801013	−.392483	5.65345
.460000	−63.1568	−.792303	−.410491	5.77441
.480000	−63.1431	−.783170	−.428284	5.89301
.500000	−63.2134	−.774683	−.446404	6.00947
.550000	−63.2134	−.752149	−.491038	6.29263
.600000	−63.2661	−.730109	−.535922	6.56661
.650000	−63.2661	−.707616	−.580577	6.83382
.700000	−63.2134	−.684549	−.624942	7.09637
.750000	−63.1958	−.661810	−.669473	7.35610
.800000	−63.0630	−.637609	−.713262	7.61468
.850000	−62.9693	−.613719	−.757207	7.87363
.900000	−62.7818	−.588374	−.800402	8.13439
.950000	−62.5630	−.562283	−.843200	8.39829
1.000000	−62.2349	−.534169	−.884923	8.66668
1.200000	−60.6099	−.411113	−1.045616	9.81175
1.400000	−57.9693	−.257493	−1.186927	11.14485
1.600000	−54.2349	−.064875	−1.298330	12.80661
1.800000	−49.2661	−.174566	−1.364006	15.06402

TABLE B3

Values of $(\mathrm{d}W/\mathrm{d}V)_{min}$ *for two hyperbolic notches with* $b/a = 0.01$, $\beta = 60°$ *and* $\phi = -15.16°$

r_0/a	θ_0	x_0	y_0	$(\mathrm{d}W/\mathrm{d}V)_{min}$
.000100	−7.5649	−.999905	−.000040	.10800
.000500	−32.8423	−.999584	−.000298	.22114
.001000	−36.8423	−.999203	−.000627	.30823
.005000	−40.0649	−.996177	−.003245	.68032
.010000	−40.5024	−.992400	−.006522	.95950
.020000	−40.7485	−.984852	−.013082	1.35269
.040000	−40.9360	−.969786	−.026236	1.90318
.060000	−41.0649	−.954766	−.039442	2.31950
.080000	−41.1899	−.939801	−.052712	2.66530
.100000	−41.2837	−.924858	−.066006	2.96540
.120000	−41.3618	−.909937	−.079324	3.23258
.140000	−41.4712	−.895103	−.092741	3.47449
.160000	−41.5649	−.880291	−.106182	3.69611
.180000	−41.6606	−.865527	−.119676	3.90093
.200000	−41.7837	−.850870	−.133291	4.09153
.220000	−41.9009	−.836257	−.146953	4.26982
.240000	−41.9712	−.821568	−.160529	4.43734
.260000	−42.0649	−.806983	−.174220	4.59529
.280000	−42.1977	−.792571	−.188101	4.74466
.300000	−42.3032	−.778126	−.201943	4.88628
.320000	−42.4087	−.763731	−.215840	5.02084
.340000	−42.4731	−.749222	−.229610	5.14892
.360000	−42.5903	−.734968	−.243658	5.27105
.380000	−42.7134	−.720796	−.257793	5.38766
.400000	−42.8305	−.706657	.271960	5.49916
.420000	−42.9360	−.692515	−.286123	5.60588
.440000	−43.0649	−.678548	−.300471	5.70813
.460000	−43.1899	−.664623	−.314860	5.80621
.480000	−43.3149	−.650758	−.329311	5.90037
.500000	−43.4009	−.636721	−.343576	5.99084
.550000	−43.7485	−.602694	−.380349	6.20224
.600000	−44.0649	−.568872	−.417311	6.39487
.650000	−44.4555	−.536038	−.455258	6.57128
.700000	−44.8423	−.503668	−.493637	6.73371
.750000	−45.2368	−.471870	−.532544	6.88421
.800000	−45.6606	−.440878	−.572197	7.02464
.850000	−46.1450	−.411093	−.612958	7.15680
.900000	−46.6782	−.382518	−.654788	7.28235
.950000	−47.1899	−.354412	−.696957	7.40289
1.000000	−47.7134	−.327164	−.739815	7.51992
1.200000	−49.9009	−.227069	−.917944	7.97950

TABLE B4

Values of $(dW/dV)_{min}$ *for two hyperbolic notches with* $b/a = 0.01$, $\beta = 90°$ *and* $\phi = 0°$

r_0/a	θ_0	x_0	y_0	$(dW/dV)_{min}$
.000100	0.0000	−.999900	0.000000	.10764
.000500	0.0000	−.999500	0.000000	.23059
.001000	0.0000	−.999000	0.000000	.32093
.005000	0.0000	−.995000	0.000000	.70502
.010000	0.0000	−.990000	0.000000	.99343
.020000	0.0000	−.980000	0.000000	1.39967
.040000	0.0000	−.960000	0.000000	1.96820
.060000	0.0000	−.940000	0.000000	2.39771
.080000	0.0000	−.920000	0.000000	2.75405
.100000	0.0000	−.900000	0.000000	3.06285
.120000	0.0000	−.880000	0.000000	3.33734
.140000	0.0000	−.860000	0.000000	3.58541
.160000	0.0000	−.840000	0.000000	3.81221
.180000	0.0000	−.820000	0.000000	4.02136
.200000	0.0000	−.800000	0.000000	4.21547
.220000	0.0000	−.780000	0.000000	4.39654
.240000	0.0000	−.760000	0.000000	4.56613
.260000	0.0000	−.740000	0.000000	4.72546
.280000	0.0000	−.720000	0.000000	4.87554
.300000	0.0000	−.700000	0.000000	5.01721
.320000	0.0000	−.680000	0.000000	5.15116
.340000	0.0000	−.660000	0.000000	5.27797
.360000	0.0000	−.640000	0.000000	5.39815
.380000	0.0000	−.620000	0.000000	5.51213
.400000	0.0000	−.600000	0.000000	5.62029
.420000	0.0000	−.580000	0.000000	5.72296
.440000	0.0000	−.560000	0.000000	5.82043

TABLE B5

Values of $(dW/dV)_{min}$ *for two hyperbolic notches with* $b/a = 0.10$, $\beta = 5°$ *and* $\phi = -42.55°$

r_0/a	θ_0	x_0	y_0	$(dW/dV)_{min}$
.000100	−41.9182	−1.004166	−.009285	.62069
.000500	−39.4104	−1.003854	−.009536	.65297
.001000	−36.3284	−1.003435	−.009811	.69387
.005000	−17.4221	−.999470	−.010716	1.00599
.010000	−39.8948	−.996568	−.015632	1.20796
.020000	−59.3635	−.994048	−.026427	1.46393
.040000	−69.1057	−.989974	−.046588	1.86765
.060000	−72.4573	−.986155	−.066428	2.20104
.080000	−74.1760	−.982426	−.086187	2.49276
.100000	−75.1760	−.978655	−.105890	2.75715
.120000	−75.8596	−.974924	−.125583	2.99879
.140000	−76.3635	−.971234	−.145272	3.22552
.160000	−76.7073	−.967452	−.164932	3.43965
.180000	−77.0041	−.963762	−.184608	3.64356
.200000	−77.2229	−.960008	−.204266	3.83904
.220000	−77.3635	−.956112	−.223890	4.02746
.240000	−77.5041	−.952312	−.243533	4.20995
.260000	−77.5041	−.952312	−.243533	4.20995
.280000	−77.5041	−.952312	−.243533	4.20995
.300000	−77.5041	−.952312	−.243533	4.20995
.320000	−77.5041	−.952312	−.243533	4.20995
.340000	−77.5041	−.952312	−.243533	4.20995
.360000	−77.5041	−.952312	−.243533	4.20995
.380000	−77.5041	−.952312	−.243533	4.20995
.400000	77.5041	−.952312	−.243533	4.20995
.420000	−77.5041	−.952312	−.243533	4.20995
.440000	−77.5041	−.952312	−.243533	4.20995
.460000	−77.5041	−.952312	−.243533	4.20995
.480000	−77.5041	−.952312	−.243533	4.20995
.500000	−77.5041	−.952312	−.243533	4.20995
.550000	−77.4221	−.884468	−.546019	6.69746
.600000	−77.2229	−.871545	−.594361	7.08320
.650000	−76.9221	−.857161	−.642360	7.47301
.700000	−76.6291	−.842363	−.690244	7.86916
.750000	−76.2698	−.826227	−.737787	8.27382
.800000	−75.8948	−.809277	−.785098	8.68907
.850000	−75.4573	−.790803	−.831985	9.11705
.900000	−75.0041	−.771366	−.878569	9.55991
.950000	−74.5041	−.750430	−.924686	10.01991
1.000000	−73.9534	−.727820	−.970256	10.49943
1.200000	−71.5901	−.625264	−1.147804	12.66724
1.400000	−68.9221	−.500749	−1.315548	15.39917
1.600000	−66.2229	−.359152	−1.473412	18.99640
1.800000	−63.9182	−.212863	−1.625920	23.87631

TABLE B6

Values of $(dW/dV)_{min}$ *for two hyperbolic notches with* $b/a = 0.10$, $\beta = 30°$ *and* $\phi = -31.44°$

r_0/a	θ_0	x_0	y_0	$(dW/dV)_{min}$
.000100	−30.7530	−1.001788	−.006176	.47525
.000500	−28.1807	−1.001433	−.006361	.50782
.001000	−24.9932	−1.000967	−.006547	.54860
.005000	−8.4952	−.996929	−.006863	.84264
.010000	−26.4386	−.992920	−.010577	1.05098
.020000	−44.6182	−.987638	−.020172	1.32703
.040000	−54.2979	−.978531	−.038607	1.74504
.060000	−57.5948	−.969720	−.056781	2.07866
.080000	−59.3448	−.961084	−.074945	2.36382
.100000	−60.3917	−.952467	−.093067	2.61634
.120000	−61.0636	−.943813	−.111144	2.84497
.140000	−61.5948	−.935275	−.129269	3.05515
.160000	−61.9932	−.926742	−.147387	3.25053
.180000	−62.3448	−.918327	−.165561	3.43374
.200000	−62.5655	−.909727	−.183632	3.60671
.220000	−62.8057	−.901332	−.201806	3.77096
.240000	−62.9932	−.892891	−.219953	3.92766
.260000	−63.1807	−.884568	−.238158	4.07777
.280000	−63.3096	−.876107	−.256290	4.22209
.300000	−63.4386	−.867727	−.274461	4.36127
.320000	−63.5636	−.859408	−.292662	4.49588
.340000	−63.6417	−.850919	−.310777	4.62641
.360000	−63.7511	−.842656	−.329002	4.75328
.380000	−63.8448	−.834368	−.347214	4.87686
.400000	−63.9386	−.826140	−.365454	4.99748
.420000	−63.9932	−.817713	−.383596	5.11542
.440000	−64.0636	−.809429	−.401808	5.23097
.460000	−64.1221	−.801105	−.419999	5.34435
.480000	−64.1807	−.792818	−.438207	5.45579
.500000	−64.2159	−.784383	−.456344	5.56548
.550000	−64.2979	−.763343	−.501708	5.83327
.600000	−64.3448	−.742101	−.546974	6.09383
.650000	−64.3448	−.720454	−.592045	6.34931
.700000	−64.3096	−.698419	−.636930	6.60160
.750000	−64.2511	−.676052	−.681654	6.85237
.800000	−64.1221	−.652711	−.725906	7.10313
.850000	−63.9405	−.628465	−.769712	7.35528
.900000	−63.7511	−.603829	−.813317	7.61015
.950000	−63.5011	−.578002	−.856320	7.86901
1.000000	−63.1807	−.550696	−.898559	8.13314
1.200000	−61.3917	−.427291	−1.059621	9.26856
1.400000	−58.6534	−.273574	−1.201775	10.60480
1.600000	−54.8702	−.081185	−1.314685	12.28725
1.800000	−50.0636	.153614	−1.386287	14.58992

TABLE B7

Values of $(dW/dV)_{min}$ *for two hyperbolic notches with* $b/a = 0.10$, $\beta = 60°$ *and* $\phi = -16.64°$

r_0/a	θ_0	x_0	y_0	$(dW/dV)_{min}$
.000100	−16.1949	−1.000351	−.003018	.40316
.000500	−14.3824	−.999963	−.003114	.43791
.001000	−12.2300	−.999470	−.003202	.48086
.005000	−1.9546	−.995450	−.003161	.78447
.010000	−8.7652	−.990564	−.004514	1.04403
.020000	−23.6792	−.982131	−.011022	1.39430
.040000	−33.3238	−.967024	−.024965	1.89503
.060000	−36.7183	−.952352	−.038863	2.28126
.080000	−38.4996	−.937838	−.052791	2.60501
.100000	−39.5933	−.923388	−.066724	2.88744
.120000	−40.3589	−.909007	−.080699	3.13976
.140000	−40.9175	−.894655	−.094686	3.36878
.160000	−41.3296	−.880299	−.108653	3.57902
.180000	−41.6968	−.866045	−.122724	3.77365
.200000	−42.0425	−.851917	−.136927	3.95501
.220000	−42.3121	−.837759	−.151087	4.12490
.240000	−42.5464	−.823632	−.165275	4.28473
.260000	−42.7964	−.809666	−.179633	4.43560
.280000	−42.9996	−.795666	−.193948	4.57846
.300000	−43.1421	−.781549	−.208133	4.71408
.320000	−43.3589	−.767786	−.222691	4.84310
.340000	−43.5113	−.753866	−.237079	4.96608
.360000	−43.6968	−.740165	−.251693	5.08351
.380000	−43.8550	−.726431	−.266268	5.19580
.400000	−44.0074	−.712747	−.280890	5.30334
.420000	−44.1421	−.699049	−.295495	5.40644
.440000	−44.3238	−.685670	−.310423	5.50542
.460000	−44.4527	−.672086	−.325137	5.60053
.480000	−44.5933	−.658635	−.339984	5.69204
.500000	−44.7613	−.645424	−.355067	5.78015
.550000	−45.1597	−.612624	−.392981	5.98693
.600000	−45.5464	−.580248	−.431281	6.17668
.650000	−45.9488	−.548501	−.470157	6.35184
.700000	−46.3296	−.517091	−.509317	6.51460
.750000	−46.7964	−.487003	−.549685	6.66691
.800000	−47.2300	−.457202	−.590259	6.81058
.850000	−47.6968	−.428351	−.631645	6.94726
.900000	−48.1597	−.400096	−.673497	7.07853
.950000	−48.6402	−.372701	−.716036	7.20585
1.000000	−49.0933	−.345618	−.758767	7.33056
1.200000	−50.7964	−.241954	−.932876	7.82685
1.400000	−51.5933	−.130712	−1.100059	8.36261
1.600000	−50.5933	.015266	−1.239245	8.98091
1.800000	−47.0933	.225005	−1.321424	9.73236

TABLE B8

Values of $(\mathrm{d}W/\mathrm{d}V)_{min}$ *for two hyperbolic notches with* $b/a = 0.10$, $\beta = 90°$ *and* $\phi = 0°$

r_0/a	θ_0	x_0	y_0	$(\mathrm{d}W/\mathrm{d}V)_{min}$
.000100	0.0000	−.999900	0.000000	.38444
.000500	0.0000	−.999500	0.000000	.42057
.001000	0.0000	−.999000	0.000000	.46498
.005000	0.0000	−.995000	0.000000	.77814
.010000	0.0000	−.990000	0.000000	1.07336
.020000	0.0000	−.980000	0.000000	1.48997
.040000	0.0000	−.960000	0.000000	2.05514
.060000	0.0000	−.940000	0.000000	2.47716
.080000	0.0000	−.920000	0.000000	2.82700
.100000	0.0000	−.900000	0.000000	3.13051
.120000	0.0000	−.880000	0.000000	3.40067
.140000	0.0000	−.860000	0.000000	3.64515
.160000	0.0000	−.840000	0.000000	3.86892
.180000	0.0000	−.820000	0.000000	4.07547
.200000	0.0000	−.800000	0.000000	4.26734
.220000	0.0000	−.780000	0.000000	4.44645
.240000	0.0000	−.760000	0.000000	4.61430
.260000	0.0000	−.740000	0.000000	4.77208
.280000	0.0000	−.720000	0.000000	4.92078
.300000	0.0000	−.700000	0.000000	5.06120
.320000	0.0000	−.680000	0.000000	5.19402
.340000	0.0000	−.660000	0.000000	5.31980
.360000	0.0000	−.640000	0.000000	5.43904
.380000	0.0000	−.620000	0.000000	5.55217
.400000	0.0000	−.600000	0.000000	5.65954
.420000	0.0000	−.580000	0.000000	5.76148
.440000	0.0000	−.560000	0.000000	5.85828
.460000	0.0000	−.540000	0.000000	5.95019

TABLE B9

Values of $(\mathrm{d}W/\mathrm{d}V)_{\min}$ *for two hyperbolic notches with* $b/a = 0.50$, $\beta = 5°$ *and* $\phi = -38.25°$

r_0/a	θ_0	x_0	y_0	$(\mathrm{d}_W/\mathrm{d}V)_{\min}$
.000100	−38.2032	−1.088016	−.214508	1.80586
.000500	−38.1212	−1.087701	−.214755	1.80998
.001000	−37.9395	−1.087306	−.215061	1.81513
.005000	−36.8751	−1.084095	−.217447	1.85647
.010000	−35.5001	−1.079953	−.220254	1.90852
.020000	−32.7520	−1.071274	−.225267	2.01368
.040000	−27.5469	−1.052629	−.232946	2.22683
.060000	−22.7032	−1.032743	−.237604	2.43927
.080000	−18.6563	−1.012298	−.240038	2.64278
.100000	−16.1212	−.992027	−.242214	2.82658
.120000	−16.8751	−.973262	−.249281	2.98089
.140000	−22.1094	−.958389	−.267139	3.10511
.160000	−28.7032	−.947755	−.291290	3.20992
.180000	−34.4063	−.939585	−.316157	3.30493
.200000	−39.0274	−.932725	−.340385	3.39482
.220000	−42.7696	−.926595	−.363838	3.48180
.240000	−45.8399	−.920895	−.386622	3.56704
.260000	−48.3751	−.915389	−.408799	3.65123
.280000	−50.5274	−.910096	−.430587	3.73479
.300000	−52.3751	−.904947	−.452054	3.81804
.320000	−53.9688	−.899862	−.473230	3.90119
.340000	−55.3594	−.894829	−.494176	3.98442
.360000	−56.5626	−.889725	−.514862	4.06787
.380000	−57.6173	−.884577	−.535353	4.15164
.400000	58.5626	.879468	.555731	4.23584
.420000	−59.3770	−.874152	−.575873	4.32055
.440000	−60.1563	−.869135	−.596097	4.40587
.460000	−60.8048	−.863712	−.616009	4.49185
.480000	−61.3770	−.858153	−.635786	4.57858
.500000	−61.9337	−.852848	−.655648	4.66612
.550000	−63.0274	−.838634	−.704620	4.88897
.600000	−63.8399	−.823566	−.752986	5.11834
.650000	−64.4532	−.807783	−.800898	5.35524
.700000	−64.8751	−.790879	−.848216	5.60072
.750000	−65.1563	−.772986	−.895040	5.85584
.800000	−65.3067	−.753886	−.941292	6.12177
.850000	−65.3770	−.733946	−.987155	6.39973
.900000	−65.3946	−.713365	−1.032724	6.69105
.950000	−65.3067	−.691222	−1.077576	6.99719
1.000000	−65.1563	−.667950	−1.121904	7.31972
1.200000	−64.0626	−.563227	−1.293573	8.81106
1.400000	−62.5626	−.443003	−1.456967	10.73810
1.600000	−61.1212	−.315360	−1.615475	13.29985
1.800000	−60.2032	−.193628	−1.776474	16.73911

TABLE B10

Values of $(\mathrm{d}W/\mathrm{d}V)_{\min}$ *for two hyperbolic notches with* $b/a = 0.50$, $\beta = 30°$ *and* $\phi = -32.84°$

r_0/a	θ_0	x_0	y_0	$(\mathrm{d}W/\mathrm{d}V)_{\min}$
.000100	−32.7917	−1.056440	−.170527	1.59510
.000500	−32.6979	−1.056103	−.170742	1.59926
.001000	−32.5573	−1.055681	−.171011	1.60446
.005000	−31.3932	−1.052256	−.173077	1.64610
.010000	−29.9479	−1.047859	−.175465	1.69829
.020000	−27.0807	−1.038717	−.179577	1.80299
.040000	−21.6159	−1.019337	−.185208	2.01200
.060000	−16.6510	−.999040	−.187665	2.21603
.080000	−12.5807	−.978445	−.187898	2.40780
.100000	−10.0182	−.958049	−.187869	2.57907
.120000	−10.3932	−.938493	−.192121	2.72365
.140000	−14.6159	−.921055	−.205800	2.84222
.160000	−20.5280	−.906684	−.226579	2.94284
.180000	−26.0534	−.894815	−.249530	3.03331
.200000	−30.6510	−.884467	−.272434	3.11791
.220000	−34.4635	−.875137	−.294966	3.19878
.240000	−37.5807	−.866325	−.316843	3.27709
.260000	−40.2409	−.858057	−.338433	3.35355
.280000	−42.5104	−.850121	−.359675	3.42862
.300000	−44.4635	−.842415	−.380609	3.50262
.320000	−46.1471	−.834825	−.401231	3.57578
.340000	−47.6510	−.827485	−.421751	3.64828
.360000	−48.9635	−.820170	−.442017	3.72027
.380000	−50.1354	−.812954	−.462146	3.79188
.400000	−51.1510	−.805616	−.481993	3.86321
.420000	−52.0885	−.798458	−.501836	3.93435
.440000	−52.9167	−.791215	−.521486	4.00539
.460000	−53.6979	−.784185	−.541189	4.07639
.480000	−54.3405	−.776700	−.560470	4.14744
.500000	−54.9948	−.769699	−.580022	4.21859
.550000	−56.2702	−.751122	−.627888	4.39732
.600000	−57.2917	−.732307	−.675332	4.57799
.650000	−58.0534	−.712590	−.722024	4.76144
.700000	−58.6510	−.692350	−.768283	4.94847
.750000	−59.0885	−.671240	−.813944	5.13988
.800000	−59.3405	−.648576	−.858643	5.33649
.850000	−59.5221	−.625400	−.903024	5.53911
.900000	−59.5807	−.600833	−.946581	5.74863
.950000	−59.5573	−.575181	−.989502	5.96597
1.000000	−59.4284	−.547909	−1.031466	6.19214
1.200000	−58.1823	−.423862	−1.190148	7.20881
1.400000	−55.9479	−.272599	−1.330413	8.48076
1.600000	−53.0417	−.094550	−1.448989	10.16349
1.800000	−49.8776	.103437	−1.546878	12.52316

TABLE B11

Values of $(dW/dV)_{min}$ *for two hyperbolic notches with* $b/a = 0.50$, $\beta = 60°$ *and* $\phi = -22.49°$

r_0/a	θ_0	x_0	y_0	$(dW/dV)_{min}$
.000100	−22.4479	−1.022058	−.105858	1.65130
.000500	−22.3659	−1.021688	−.106010	1.65660
.001000	−22.2370	−1.021225	−.106198	1.66323
.005000	−21.1784	−1.017488	−.107626	1.71617
.010000	−19.8971	−1.012747	−.109223	1.78209
.020000	−17.3542	−1.003061	−.111786	1.91290
.040000	−12.5104	−.983100	−.114485	2.16865
.060000	−8.3073	−.962780	−.114489	2.41276
.080000	−4.8542	−.942437	−.112590	2.64032
.100000	−2.4479	−.922242	−.110091	2.84712
.120000	−1.4479	−.902189	−.108852	3.03078
.140000	−1.9968	−.882235	−.110698	3.19153
.160000	−4.2721	−.862595	−.117739	3.33213
.180000	−7.6042	−.843733	−.129639	3.45682
.200000	−11.2721	−.826008	−.144914	3.56977
.220000	−14.8659	−.809514	−.162263	3.67411
.240000	−18.1198	−.794052	−.180461	3.77196
.260000	−20.9968	−.779414	−.198982	3.86472
.280000	−23.6042	−.765577	−.217936	3.95334
.300000	−25.8971	−.752277	−.236847	4.03849
.320000	−27.9264	−.739415	−.255688	4.12065
.340000	−29.7917	−.727086	−.274748	4.20020
.360000	−31.4479	−.715029	−.293640	4.27743
.380000	−32.9479	−.703268	−.312493	4.35261
.400000	−34.3073	−.691740	−.331273	4.42593
.420000	−35.5514	−.680441	−.350022	4.49757
.440000	−36.6979	−.669360	−.368762	4.56768
.460000	−37.7448	−.658408	−.387407	4.63640
.480000	−38.6979	−.647533	−.405923	4.70387
.500000	−39.6042	−.636917	−.424560	4.77018
.550000	−41.6042	−.610888	−.471009	4.93162
.600000	−43.2721	−.585287	−.517099	5.08798
.650000	−44.6979	−.560114	−.563010	5.24050
.700000	−45.9323	−.535295	−.608783	5.39029
.750000	−46.9479	−.510153	−.653870	5.53834
.800000	−47.8620	−.485416	−.699045	5.68554
.850000	−48.5729	−.459734	−.743149	5.83271
.900000	−49.2370	−.434512	−.787495	5.98064
.950000	−49.7448	−.408267	−.830835	6.13003
1.000000	−50.1198	−.380966	−.873207	6.28158
1.200000	−50.6979	−.262060	−1.034401	6.92274
1.400000	−49.6042	−.114860	−1.172040	7.65474
1.600000	−46.8073	.072976	−1.272309	8.54211
1.800000	−41.9792	.315948	−1.309769	9.71316

TABLE B12

Values of $(\mathrm{d}W/\mathrm{d}V)_{\min}$ *for two hyperbolic notches with* $b/a = 0.50$, $\beta = 90°$ *and* $\phi = 0°$

r_0/a	θ_0	x_0	y_0	$(\mathrm{d}W/\mathrm{d}V)_{\min}$
.000100	0.0000	−.999900	0.000000	1.80314
.000500	0.0000	−.999500	0.000000	1.81013
.001000	0.0000	−.999000	0.000000	1.81887
.005000	0.0000	−.995000	0.000000	1.88844
.010000	0.0000	−.990000	0.000000	1.97454
.020000	0.0000	−.980000	0.000000	2.14366
.040000	0.0000	−.960000	0.000000	2.46828
.060000	0.0000	−.940000	0.000000	2.77352
.080000	0.0000	−.920000	0.000000	3.05918
.100000	0.0000	−.900000	0.000000	3.32590
.120000	0.0000	−.880000	0.000000	3.57478
.140000	0.0000	−.860000	0.000000	3.80711
.160000	0.0000	−.840000	0.000000	4.02421
.180000	0.0000	−.820000	0.000000	4.22737
.200000	0.0000	−.800000	0.000000	4.41779
.220000	0.0000	−.780000	0.000000	4.59654
.240000	0.0000	−.760000	0.000000	4.76462
.260000	0.0000	−.740000	0.000000	4.92288
.280000	0.0000	−.720000	0.000000	5.07212
.300000	0.0000	−.700000	0.000000	5.21300
.320000	0.0000	−.680000	0.000000	5.34615
.340000	0.0000	−.660000	0.000000	5.47210
.360000	0.0000	−.640000	0.000000	5.59133
.380000	0.0000	−.620000	0.000000	5.70426
.400000	0.0000	−.600000	0.000000	5.81128
.420000	0.0000	−.580000	0.000000	5.91272
.440000	0.0000	−.560000	0.000000	6.00887
.460000	0.0000	−.540000	0.000000	6.10002
.480000	0.0000	−.520000	0.000000	6.18640
.500000	0.0000	−.500000	0.000000	6.26823
.550000	0.0000	−.450000	0.000000	6.45415
.600000	0.0000	−.400000	0.000000	6.61538
.650000	0.0000	−.350000	0.000000	6.75394
.700000	0.0000	−.300000	0.000000	6.87139
.750000	0.0000	−.250000	0.000000	6.96898
.800000	0.0000	−.200000	0.000000	7.04766

TABLE B13

Values of $(dW/dV)_{min}$ *for two hyperbolic notches with* $b/a = 0.90$, $\beta = 5°$ *and* $\phi = -31.27°$

r_0/a	θ_0	x_0	y_0	$(dW/dV)_{min}$
.000100	−31.2659	−1.193979	−.587325	1.91084
.000500	−31.2190	−1.193637	−.587533	1.91253
.001000	−31.1721	−1.193209	−.587791	1.91464
.005000	−30.7854	−1.189769	−.589833	1.93159
.010000	−30.3225	−1.185433	−.592322	1.95286
.020000	−29.3440	−1.176631	−.597075	1.99569
.040000	−27.4690	−1.158574	−.605724	2.08243
.060000	−25.6721	−1.139987	−.613267	2.17049
.080000	−23.9378	−1.120946	−.619733	2.25964
.100000	−22.2659	−1.101521	−.625164	2.34959
.120000	−20.7034	−1.081814	−.629697	2.43996
.140000	−19.2659	−1.061905	−.633467	2.53028
.160000	−18.0081	−1.041903	−.636738	2.61997
.180000	−16.9495	−1.021883	−.639749	2.70837
.200000	−16.1721	−1.001979	−.642978	2.79480
.220000	−15.7854	−.982361	−.647121	2.87852
.240000	−15.8557	−.963196	−.652845	2.95891
.260000	−16.4456	−.944702	−.660881	3.03552
.280000	−17.6721	−.927278	−.672273	3.10822
.300000	−19.3870	−.911075	−.686858	3.17724
.320000	−21.5159	−.896364	−.704637	3.24314
.340000	−23.7854	−.882943	−.724400	3.30662
.360000	−26.0784	−.870715	−.745530	3.36837
.380000	−28.2190	−.859229	−.766954	3.42898
.400000	30.3049	.848724	.789114	3.48893
.420000	−32.1721	−.838555	−.810909	3.54858
.440000	−33.8909	−.828820	−.832623	3.60822
.460000	−35.4690	−.819427	−.854194	3.66807
.480000	−36.8909	−.810170	−.875414	3.72832
.500000	−38.2190	−.801239	−.896608	3.78911
.550000	−41.0081	−.779025	−.948165	3.94423
.600000	−43.2190	−.756820	−.998147	4.10505
.650000	−45.0081	−.734510	−1.046958	4.27278
.700000	−46.4807	−.712046	−1.094874	4.44850
.750000	−47.6370	−.688695	−1.141441	4.63324
.800000	−48.6096	−.665116	−1.187451	4.82803
.850000	−49.3909	−.640804	−1.232566	5.03393
.900000	−50.0081	−.615653	−1.276795	5.25207
.950000	−50.5159	−.589994	−1.320484	5.48367
1.000000	−50.8909	−.563265	−1.363220	5.73005
1.200000	−51.8206	−.452313	−1.530568	6.89351
1.400000	−52.2190	−.336362	−1.693775	8.42807
1.600000	−52.7678	−.225991	−1.861178	10.46414
1.800000	−53.9378	−.134470	−2.042354	13.11930

TABLE B14

Values of $(\mathrm{d}W/\mathrm{d}V)_{\min}$ *for two hyperbolic notches with* $b/a = 0.90$, $\beta = 30°$ *and* $\phi = -28.81°$

r_0/a	θ_0	x_0	y_0	$(\mathrm{d}W/\mathrm{d}V)_{\min}$
.000100	−28.8052	−1.150716	−.512613	1.82007
.000500	−28.7583	−1.150365	−.512805	1.82172
.001000	−28.7115	−1.149926	−.513045	1.82379
.005000	−28.3247	−1.146402	−.514937	1.84034
.010000	−27.7915	−1.141957	−.517227	1.86107
.020000	−26.8365	−1.132957	−.521593	1.90268
.040000	−24.8833	−1.114517	−.529395	1.98640
.060000	−23.0083	−1.095577	−.536016	2.07063
.080000	−21.2115	−1.076223	−.541509	2.15515
.100000	−19.4771	−1.056526	−.545907	2.23964
.120000	−17.7915	−1.036543	−.549231	2.32376
.140000	−16.3072	−1.016436	−.551875	2.40705
.160000	−15.0083	−.996261	−.553998	2.48904
.180000	−13.8618	−.976046	−.555689	2.56919
.200000	−13.0083	−.955936	−.557583	2.64693
.220000	−12.4888	−.936009	−.560139	2.72173
.240000	−12.4302	−.916429	−.564225	2.79312
.260000	−12.8833	−.897349	−.570536	2.86081
.280000	−13.9146	−.879020	−.579897	2.92474
.300000	−15.5474	−.861781	−.592975	2.98514
.320000	−17.5825	−.845753	−.609230	3.04251
.340000	−19.8443	−.830993	−.627982	3.09749
.360000	−22.1724	−.817425	−.648427	3.15068
.380000	−24.3950	−.804730	−.669514	3.20265
.400000	−26.4947	−.792813	−.691010	3.25381
.420000	−28.4888	−.781661	−.712899	3.30449
.440000	−30.2896	−.770869	−.734488	3.35497
.460000	−31.9146	−.760338	−.755745	3.40543
.480000	−33.4302	−.750216	−.777006	3.45603
.500000	−34.8052	−.740255	−.797959	3.50690
.550000	−37.7115	−.715698	−.848991	3.63593
.600000	−40.0552	−.691549	−.898680	3.76858
.650000	−41.9146	−.667111	−.946779	3.90582
.700000	−43.4302	−.642455	−.993794	4.04848
.750000	−44.6177	−.616947	−1.039344	4.19734
.800000	−45.5825	−.590899	−1.083972	4.35316
.850000	−46.3072	−.563630	−1.127160	4.51673
.900000	−46.8833	−.535666	−1.169532	4.68889
.950000	−47.3599	−.507282	−1.211406	4.87056
1.000000	−47.6646	−.477334	−1.251779	5.06276
1.200000	−47.9322	−.346791	−1.403387	5.96109
1.400000	−47.3052	−.201473	−1.541531	7.14471
1.600000	−46.3247	−.045891	−1.669789	8.76134
1.800000	−45.7583	.105032	−1.802091	11.00881

TABLE B15

Values of $(dW/dV)_{min}$ *for two hyperbolic notches with* $b/a = 0.90$, $\beta = 60°$ *and* $\phi = -23.41°$

r_0/a	θ_0	x_0	y_0	$(dW/dV)_{min}$
.000100	−23.4131	−1.085758	−.380888	2.16108
.000500	−23.3662	−1.085390	−.381047	2.16319
.001000	−23.3193	−1.084931	−.381245	2.16583
.005000	−22.9326	−1.081245	−.382797	2.18696
.010000	−22.3994	−1.076604	−.384659	2.21334
.020000	−21.4443	−1.067234	−.388161	2.26599
.040000	−19.4912	−1.048142	−.394195	2.37082
.060000	−17.6162	−1.028663	−.399007	2.47480
.080000	−15.7842	−1.008866	−.402610	2.57766
.100000	−14.0381	−.988836	−.405105	2.67912
.120000	−12.3662	−.968634	−.406548	2.77883
.140000	−10.7803	−.948320	−.407035	2.87642
.160000	−9.3662	−.927982	−.406888	2.97148
.180000	−8.0850	−.907638	−.406164	3.06364
.200000	−6.9678	−.887326	−.405111	3.15248
.220000	−6.0967	−.867094	−.404214	3.23764
.240000	−5.4912	−.846951	−.403815	3.31883
.260000	−5.2725	−.826949	−.404741	3.39580
.280000	−5.4131	−.807098	−.407263	3.46847
.300000	−6.0381	−.787514	−.412406	3.53687
.320000	−7.1904	−.768366	−.420902	3.60126
.340000	−8.8506	−.749898	−.433161	3.66203
.360000	−10.8506	−.732286	−.448618	3.71973
.380000	−13.0381	−.715646	−.466576	3.77498
.400000	−15.3193	−.700062	−.486528	3.82831
.420000	−17.5381	−.685372	−.507411	3.88021
.440000	−19.6631	−.671507	−.528904	3.93106
.460000	−21.6631	−.658339	−.550657	3.98117
.480000	−23.4912	−.645631	−.572181	4.03076
.500000	−25.2256	−.633531	−.593940	4.08003
.550000	−28.9678	−.604659	−.647224	4.20266
.600000	−32.0029	−.577037	−.698826	4.32577
.650000	−34.4697	−.549973	−.748730	4.45045
.700000	−36.5576	−.523569	−.797790	4.57753
.750000	−38.2256	−.496664	−.844918	4.70766
.800000	−39.5693	−.469166	−.890458	4.84144
.850000	−40.6631	−.441078	−.934717	4.97942
.900000	−41.5693	−.412511	−.978022	5.12216
.950000	−42.2256	−.382370	−1.019298	5.27024
1.000000	−42.7568	−.351608	−1.059737	5.42426
1.200000	−43.3193	−.212800	−1.204126	6.11459
1.400000	−42.1318	−.047605	−1.320023	6.97272
1.600000	−39.3193	.151953	−1.394676	8.10972
1.800000	−34.9678	.389205	−1.412457	9.76749

TABLE B16

Values of $(dW/dV)_{min}$ *for two hyperbolic notches with* $b/a = 0.90$, $\beta = 90°$ *and* $\phi = 0°$

r_0/a	θ_0	x_0	y_0	$(dW/dV)_{min}$
.000100	0.0000	−.999900	0.000000	2.87347
.000500	0.0000	−.999500	0.000000	2.87691
.001000	0.0000	−.999000	0.000000	2.88121
.005000	0.0000	−.995000	0.000000	2.91553
.010000	0.0000	−.990000	0.000000	2.95817
.020000	0.0000	−.980000	0.000000	3.04259
.040000	0.0000	−.960000	0.000000	3.20790
.060000	0.0000	−.940000	0.000000	3.36844
.080000	0.0000	−.920000	0.000000	3.52411
.100000	0.0000	−.900000	0.000000	3.67489
.120000	0.0000	−.880000	0.000000	3.82079
.140000	0.0000	−.860000	0.000000	3.96183
.160000	0.0000	−.840000	0.000000	4.09806
.180000	0.0000	−.820000	0.000000	4.22954
.200000	0.0000	−.800000	0.000000	4.35637
.220000	0.0000	−.780000	0.000000	4.47861
.240000	0.0000	−.760000	0.000000	4.59638
.260000	0.0000	−.740000	0.000000	4.70977
.280000	0.0000	−.720000	0.000000	4.81887
.300000	0.0000	−.700000	0.000000	4.92379
.320000	0.0000	−.680000	0.000000	5.02464
.340000	0.0000	−.660000	0.000000	5.12150
.360000	0.0000	−.640000	0.000000	5.21449
.380000	0.0000	−.620000	0.000000	5.30369
.400000	0.0000	−.600000	0.000000	5.38920
.420000	0.0000	−.580000	0.000000	5.47110
.440000	0.0000	−.560000	0.000000	5.54949
.460000	0.0000	−.540000	0.000000	5.62445
.480000	0.0000	−.520000	0.000000	5.69605
.500000	0.0000	−.500000	0.000000	5.76437
.550000	0.0000	−.450000	0.000000	5.92129
.600000	0.0000	−.400000	0.000000	6.05917
.650000	0.0000	−.350000	0.000000	6.17890
.700000	0.0000	−.300000	0.000000	6.28123
.750000	0.0000	−.250000	0.000000	6.36680
.800000	0.0000	−.200000	0.000000	6.43613
.850000	0.0000	−.150000	0.000000	6.48963

References

[1] Griffith, A. A., The theory of rupture, *Proceedings of First International Congress of Applied Mechanics*, Delft, pp. 55–93 (1924).

[2] Griffith, A. A., The phenomena of rupture and flow in solids, *Philosophical Transactions, Royal Society of London, Series A221*, pp. 163–198 (1921).

[3] Sih, G. C., A special theory of crack propagation: methods of analysis and solutions of crack problems, *Mechanics of Fracture I*, edited by G. C. Sih, Noordhoff International Publishing, Leyden, pp. 21–45 (1973).

[4] Sih, G. C., Surface layer energy and strain energy density for a blunted crack or notch, *Prospects of Fracture Mechanics*, edited by G. C. Sih, H. C. van Elst and D. Broek, Noordhoff International Publishing, Leyden, pp. 85–102 (1974).

[5] Sih, G. C. and Kipp, M. E., discussion on Fracture under complex stress—the angled crack problem, *International Journal of Fracture Mechanics*, 10, pp. 261–265 (1974).

[6] Williams, J. G. and Ewing, P. D., Fracture under complex stress—the angled crack problem, *International Journal of Fracture Mechanics*, 8, pp. 441–446 (1972).

[7] Matthaes, K., Betrachtungen zur theorie der werkstoff—festigkeit, *Zeitschrift für Metallkunde*, 43, pp. 11–19 and 90–95 (1952).

[8] Gillemot, L. F., Criterion of crack initiation and spreading, *Journal of Engineering Fracture Mechanics*, 8, pp. 239–253 (1976).

[9] Sih, G. C., A three-dimensional strain energy density factor theory of crack propagation: three dimensional crack problems, *Mechanics of Fracture II*, edited by G. C. Sih, Noordhoff International Publishing, Leyden, pp. 15–53 (1975).

[10] Neuber, H., *Theory of Notches*, Edwards, Michigan (1946).

[11] Muskhelishvili, N. I., *Some Basic Problems of the Mathematical Theory of Elasticity*, Noordhoff, Groningen (1953).

[12] McClintock, F. A. and Walsh, J. B., Friction on Griffith Cracks in Rocks under Pressure, *Proceedings of Fourth U.S. National Congress of Applied Mechanics*, pp. 1015–1021 (1962).

[13] Haigh, B. P., The strain energy function and the elastic limit, *British Association of Advancement of Sciences*, pp. 486–495 (1919).

[14] Sih, G. C., Application of strain energy density theory to fundamental fracture problems, *Proceedings of the 10th Annual* Meeting of the Society of Engineering Science, North Carolina State University, pp. 221–234 (1975).

[15] Kipp, M. E. and Sih, G. C., The strain energy density failure criterion applied to notched elastic solids, *International Journal of Solids and Structures*, 2, pp. 153–173 (1975).

[16] Hartranft, R. J. and Sih, G. C., Stress singularity for a crack with an arbitrarily curved front, *Journal of Engineering Fracture Mechanics*, (in press) (1977).

[17] Palaniswamy, K., *Crack propagation under general in-plane loading*, Ph.D. Dissertation, California Institute of Technology (1972).

[18] Erdogan, F. and Sih, G. C., On the crack extension in plates under plane loading and transverse shear, *Journal of Basic Engineering*, pp. 519–525 (1963).

[19] Kassir, M. K., Three dimensional notch problems: stress analysis of notches, *Mechanics of Fracture V*, edited by G. C. Sih, Noordhoff International Publishing, Alphen a/d Rijn (in press) (1978).

[20] McClintock, F. A., discussion on Crack extension in plates under plane loading and transverse shear, *Journal of Basic Engineering*, pp. 525–527 (1963).

[21] Cotterell, B., The paradox between the theories for tensile and compressive fracture, *International Journal of Fracture Mechanics*, 5, pp. 251–252 (1969).

[22] Cotterell, B., Brittle fracture in compression, *International Journal of Fracture Mechanics*, 8, pp. 195–208 (1972).

[23] Griffith, A. A., Stresses in a plate bounded by a hyperbolic cylinder, *Great Britain Aeronautical Research Council Technical Report*, 2, pp. 668–677 (1928).

[24] Bridgeman, P. W., *Studies in Large Plastic Flow and Fracture*, McGraw-Hill, New York (1952).

[25] Sih, G. C., Elastic-plastic fracture mechanics, *Prospects of Fracture Mechanics*, edited by G. C. Sih, H. C. van Elst and D. Broek, Noordhoff International Publishing, Leyden, pp. 613–621 (1974).

[26] Sih, G. C., Fracture toughness concept, ASTM STP 605, *American Society for Testing of Materials*, pp. 3–15 (1976).

[27] Sih, G. C., Mechanics of Ductile Fracture, *Proceedings of Fracture Mechanics and Technology*, edited by G. C. Sih and C. L. Chow, Noordhoff International Publishing, Alphen a/d Rijn, pp. 767–784 (1977).

H. Nisitani

1 Solutions of notch problems by body force method

1.1 Introduction

Various methods of calculating stress concentration factors or stress intensity factors have been developed for two-dimensional problems [1] dealing with notches or cracks. There are, however, only few methods of analysis available for three-dimensional problems. In the present chapter, a method of analysis applicable to two- and three-dimensional problems will be described and illustrated, and the numerical results will be presented in the form of tables and graphs. The method of solution makes use of stress field derived from point forces acting in an infinite plate or infinite body. The given boundary conditions are satisfied by applying body force (continuously embedded point forces) along the prospective sites of notches or cracks.

1.2 Principle of the body force method

The present method for solving two-dimensional problems is based on using the stress field of a point force in an infinite plate without any notch or crack [2]. The solutions are obtained by superposing the stress fields of a point force so as to satisfy a given boundary condition. By means of these fundamental stress fields, all problems can be solved in principle. The problem of a semi-infinite plate or a strip is more conveniently solved by using the stress field of a point force in a semi-infinite plate or in a strip.

In general, the boundary conditions are satisfied by applying the body force (continuously embedded point forces) to some imaginary boundaries in the elastic medium and adjusting its density so as to satisfy the specified conditions. The imaginary boundary stands for the prospective notch or crack site and should be free from stresses. As will be shown later, the density of body force in a simple problem can be obtained in closed form. In general, the density of body force has to be determined by a numerical procedure. The method is equivalent to determining the density of the body force.

In the present chapter, by taking as an example the tension of a semi-infinite plate with an elliptic hole or a crack, the method of solutions will be illustrated. In the three-dimensional case, the method is based on the stress field due to a point force acting in an infinite body [3].

Tension of an infinite plate with an elliptic hole. In this section, the plane stress conditions are assumed. Consider an infinite plate (unit thickness) with an elliptic hole subjected to a tensile stress σ_y^∞ at infinity, as shown in Figure 1.1. The x- and y-components of the displacement u, v at (ξ, η) on the elliptic hole are independent of the Poisson's ratio ν and are expressed as follows [4].

$$u = -\frac{\sigma_y^\infty}{E}\xi \tag{1.1a}$$

$$v = \frac{\sigma_y^\infty}{E}\left(1 + 2\frac{a}{b}\right)\eta \tag{1.1b}$$

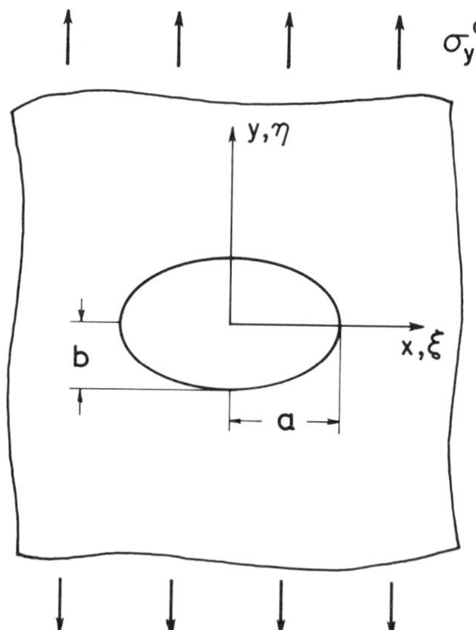

Figure 1.1. Tension of an infinite plate with an elliptic hole.

Next, consider an elliptic plate (unit thickness) having the same size as the elliptic hole. If the displacements at (ξ, η) on the rim of the elliptic plate coincide with those of equations (1.1), the strain components ε_x, ε_y, and γ_{xy} at an arbitrary point in the plate are constant and are given by

$$\varepsilon_x = -\frac{\sigma_y^\infty}{E} \tag{1.2a}$$

$$\varepsilon_y = \frac{\sigma_y^\infty}{E}\left(1 + 2\frac{a}{b}\right) \tag{1.2b}$$

$$\gamma_{xy} = 0 \tag{1.2c}$$

From Hooke's law, the stresses σ_x, σ_y, and τ_{xy} corresponding to the strains in equation (1.2) can be expressed as

$$\sigma_x = \frac{E}{1-\nu^2}(\varepsilon_x + \nu\varepsilon_y) = -\frac{\sigma_y^\infty}{1-\nu^2}\left\{1 - \nu\left(1 + 2\frac{a}{b}\right)\right\} \tag{1.3a}$$

$$\sigma_y = \frac{E}{1-\nu^2}(\varepsilon_y + \nu\varepsilon_x) = \frac{\sigma_y^\infty}{1-\nu^2}\left\{\left(1 + 2\frac{a}{b}\right) - \nu\right\} \tag{1.3b}$$

$$\tau_{xy} = G\gamma_{xy} = 0 \tag{1.3c}$$

The elliptic plate subjected to the surface stresses given by equations (1.3) is inserted into the elliptic hole in an infinite plate having an elliptic hole subjected to the stress σ_y^∞ at infinity. This implies that the solution of the infinite plate with an elliptic hole can be obtained by superposing the stress fields due to a point force acting in an infinite plate without any hole (Figure 1.2). In other words, the problem of a plate with a hole can be reduced to the problem of an infinite plate without any hole. This is analogous to the Eshelby problem of inclusions and inhomogeneities [5].

The densities of body force ρ_x and ρ_y in this case can be obtained from equations (1.3):

$$\rho_x = -\frac{\sigma_y^\infty}{1-\nu^2}\left\{1 - \nu\left(1 + 2\frac{a}{b}\right)\right\} \tag{1.4a}$$

$$\rho_y = \frac{\sigma_y^\infty}{1-\nu^2}\left\{\left(1 + 2\frac{a}{b}\right) - \nu\right\} \tag{1.4b}$$

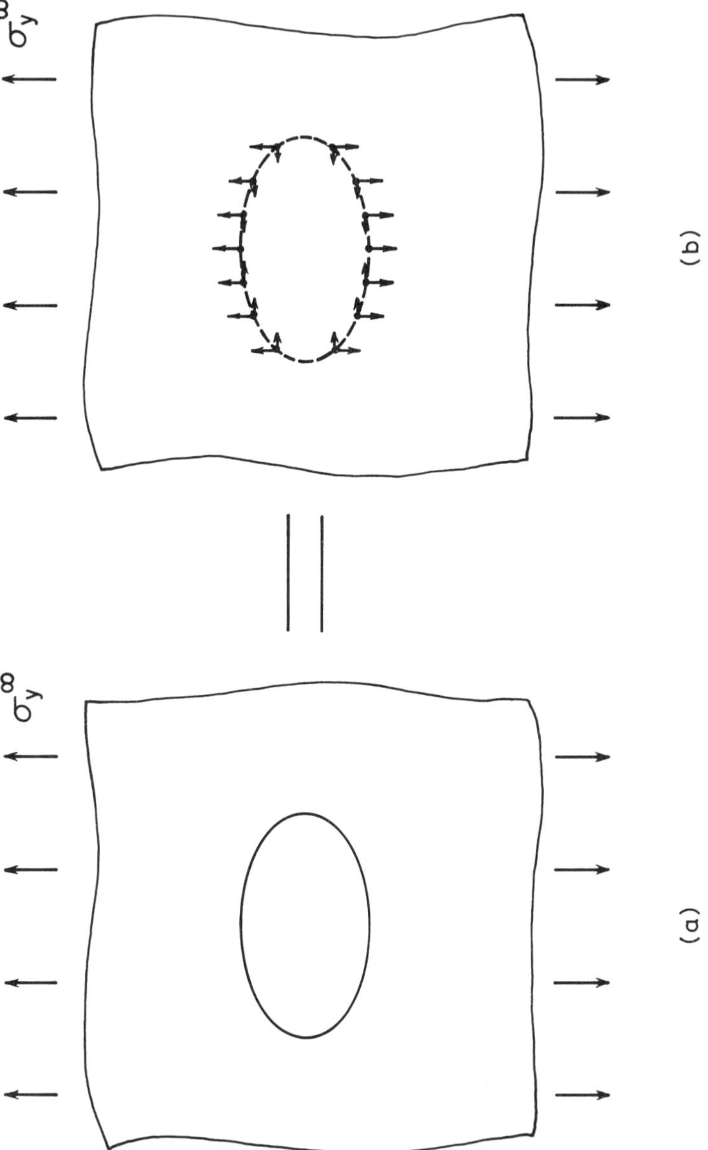

Figure 1.2. Illustration of body force method. (a) Infinite plate with an elliptic hole. (b) Infinite plate without any hole.

The densities for an infinite plate subjected to a tensile stress σ_x^∞ at infinity can be obtained in a similar manner as follows:

$$\rho_x = \frac{\sigma_x^\infty}{1-\nu^2}\left\{\left(1+2\frac{b}{a}\right)-\nu\right\} \tag{1.5a}$$

$$\rho_y = -\frac{\sigma_x^\infty}{1-\nu^2}\left\{1-\nu\left(1+2\frac{b}{a}\right)\right\} \tag{1.5b}$$

The definitions of the densities of body force ρ_x and ρ_y are given by

$$\rho_x = \frac{\mathrm{d}F_x}{\mathrm{d}\eta}, \qquad \rho_y = -\frac{\mathrm{d}F_y}{\mathrm{d}\xi} \tag{1.6}$$

where $\mathrm{d}F_x$ and $\mathrm{d}F_y$ denote the x- and y-components, respectively, of the embedded point forces acting on the element $\mathrm{d}S = \sqrt{(\mathrm{d}\xi)^2 + (\mathrm{d}\eta)^2}$ of an imaginary ellipse whose boundary is to be freed from stresses.

When the body force, whose densities are expressed by equations (1.4), acts on an imaginary ellipse in an infinite plate without any hole, the normal and shear stresses σ_n and τ_{nt} appearing at an infinitesimally small distance away from the ellipse boundary cancel the stresses σ_n^∞ and τ_{nt}^∞ appearing at the same location by the uniform stress $\sigma_y = \sigma_y^\infty$. This can be shown directly by calculation.

In the problem shown in Figure 1.1, the stress field owing to an elliptic hole can be deduced from the following complex potentials [4, 6]

$$\psi(z) = -\frac{a}{2}\frac{1}{w}\sigma_y^\infty \tag{1.7a}$$

$$\chi(z) = \left(-\frac{a^2}{2}\ln w + \frac{ab}{4}\frac{1}{w^2}\right)\sigma_y^\infty \tag{1.7b}$$

where w is the function of z determined from

$$z = \frac{c}{2}\left(Rw + \frac{1}{Rw}\right) \tag{1.8}$$

in which

$$R = \sqrt{\frac{a+b}{a-b}}, \qquad c = \sqrt{a^2-b^2}$$

The stresses σ_x, σ_y, and τ_{xy} can be calculated from

$$\sigma_x + \sigma_y = 4\,\mathrm{Re}\{\psi'(z)\} \tag{1.9a}$$

$$\sigma_y - \sigma_x + 2i\tau_{xy} = 2\{\bar{z}\psi''(z) + \chi''(z)\} \tag{1.9b}$$

When a concentrated force $X + iY$ acts at an arbitrary point $\zeta = \xi + i\eta$ in the x–y plane, the complex potentials are expressed as [7]

$$\psi(z) = -\frac{1+\nu}{8\pi}(X + iY)\ln(z - \zeta) \tag{1.10a}$$

$$\chi(z) = \frac{3-\nu}{8\pi}(X - iY)(z - \zeta) + \frac{1+\nu}{8\pi}(X + iY)\bar{\zeta}\ln(z - \zeta) \tag{1.10b}$$

For a body force acting on an imaginary ellipse, the densities are given by equations (1.4) and the complex potentials for the stress field are obtained by integrating the complex potentials for a concentrated force $\rho_x\,\mathrm{d}\eta + i\rho_y(-\mathrm{d}\xi)$ acting on the element $\mathrm{d}\zeta = \mathrm{d}\xi + i\,\mathrm{d}\eta$ of the ellipse. Hence, the complex potentials for this case become

$$\psi(z) = -\frac{1+\nu}{8\pi}\oint \ln(z - \zeta)(\rho_x\,\mathrm{d}\eta - i\rho_y\,\mathrm{d}\xi) \tag{1.11a}$$

$$\begin{aligned}\chi(z) = {} & \frac{3-\nu}{8\pi}\oint (z - \zeta)\ln(z - \zeta)(\rho_x\,\mathrm{d}\eta + i\rho_y\,\mathrm{d}\xi) \\ & + \frac{1+\nu}{8\pi}\oint \bar{\zeta}\ln(z - \zeta)(\rho_x\,\mathrm{d}\eta - i\rho_y\,\mathrm{d}\xi)\end{aligned} \tag{1.11b}$$

The ellipse in the ζ-plane is now mapped into the unit circle in the ω-plane by the relation

$$\zeta = \frac{c}{2}\left(R\omega + \frac{1}{R\omega}\right) \tag{1.12}$$

where

$$R = \sqrt{\frac{a+b}{a-b}}, \qquad c = \sqrt{a^2 - b^2}$$

The path of integration in equation (1.11) may be taken along the unit circle by making use of equation (1.12).

The various relations connected with the unit circle in the ω-plane are

$$\omega = e^{i\theta} \tag{1.13}$$

and

$$d\zeta = d\xi + i\,d\eta = \frac{c}{2}\left(R - \frac{1}{R\omega^2}\right)d\omega \tag{1.14a}$$

$$d\bar{\zeta} = d\xi - i\,d\eta = \frac{c}{2}\left(R - \frac{1}{R\bar{\omega}^2}\right)d\bar{\omega} = \frac{c}{2}\left(\frac{1}{R} - \frac{R}{\omega^2}\right)d\omega \tag{1.14b}$$

It follows from equation (1.14) that

$$\begin{aligned} d\xi &= \frac{c}{4}\left(R + \frac{1}{R}\right)\left(1 - \frac{1}{\omega^2}\right)d\omega \\ d\eta &= -i\frac{c}{4}\left(R - \frac{1}{R}\right)\left(1 + \frac{1}{\omega^2}\right)d\omega \end{aligned} \tag{1.15}$$

By using equation (1.8), $z - \zeta$ can be expressed in terms of w as

$$z - \zeta = \frac{c}{2}Rw\left(1 - \frac{\omega}{w}\right)\left(1 - \frac{1}{R^2 w\omega}\right) \tag{1.16}$$

Substituting equations (1.4), (1.12), (1.13), (1.15) and (1.16) into equation (1.11) and using Cauchy's integral formula, the complex potentials $\psi(z)$ and $\chi(z)$ in the present case can be written as

$$\psi(z) = -\frac{a}{2}\frac{1}{w}\sigma_y^\infty \tag{1.17a}$$

$$\chi(z) = \left(-\frac{a^2}{2}\ln w + \frac{ab}{4}\frac{1}{w^2}\right)\sigma_y^\infty \tag{1.17b}$$

which coincide with equations (1.7). That is, when the body force densities given by equations (1.4) acts on an ellipse in an infinite plate (without any hole) subjected to a tensile stress σ_y^∞ at infinity, the state of stress exterior to the ellipse is equal to that of an infinite plate with an elliptic hole subjected to σ_y^∞ at infinity.

Although the densities of body force acting on an imaginary ellipse is constant in the problem shown by Figure 1.1, the densities of body force in general may vary along the imaginary boundary. Therefore, the determination of the densities of body force usually results in calculation involving a numerical procedure.

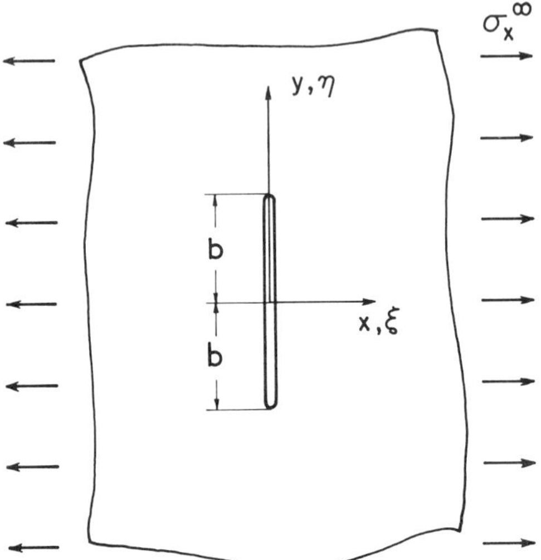

Figure 1.3. Tension of an infinite plate with a crack.

Tension of an infinite plate with a crack. Next, consider an infinite plate (unit thickness) with a crack subjected to tension, as shown in Figure 1.3. A crack can be considered as an extremely slender ellipse ($b/a \to \infty$). As $b/a \to \infty$, the points at which body forces are applied become closer and closer to each other. The fundamental stress field is obtained from a pair of point forces, each having the same magnitude but opposite in direction. Referring to Figure 1.4, the strength of the pair of point forces is expressed by the product of the magnitude of the force and the distance between the forces. The stress field is obtained by differentiating the stress field for a point force.

When a point force X acting in the x-direction is applied to a point (ξ, η) in an infinite plate, the stress σ_x^X at (x, y) is obtained from equations (1.9) and (1.10), and is given by*

$$\sigma_x^X = -\frac{(x-\xi)\{3(x-\xi)^2+(y-\eta)^2\}}{4\pi\{(x-\xi)^2+(y-\eta)^2\}^2} X \tag{1.18}$$

* Since ν does not affect the end results, it has been set to zero.

Now, consider the stress at $(0, y)$ generated by a body force with densities given by equations (1.5). The force acts on a very slender ellipse $(b/a \to \infty)$ in an infinite plate subjected to a tensile stress σ_x^∞ (Figure 1.5). Since ρ_y is negligible when $\nu = 0$, $d\sigma_x$ at (x, y) owing to a body force acting on the element $d\eta$ can be obtained from Figure 1.4 and equation (1.18). The result is

$$d\sigma_x = \lim_{\xi \to 0} \left[(\rho_x \, d\eta) \frac{\partial \sigma_x^X}{\partial \xi} \bigg|_{X=1} 2\xi \right] = \frac{\partial \sigma_x^X}{\partial \xi} \bigg|_{\substack{X=1 \\ \xi=0}} (\rho_x 2\xi) \bigg|_{\xi \to 0} d\eta \tag{1.19}$$

From equations (1.5) and (1.18)

$$\begin{aligned} (\rho_x 2\xi)\bigg|_{\xi \to 0} &= 4\sigma_x^\infty \sqrt{b^2 - \eta^2} \\ \frac{\partial \sigma_x^X}{\partial \xi}\bigg|_{\substack{x=1 \\ \xi=0}} &= \frac{\{-3x^4 + 6x^2(y-\eta)^2 + (y-\eta)^4\}}{4\pi\{x^2 + (y-\eta)^2\}^3} \end{aligned} \tag{1.20}$$

The stress field in equation (1.19) may be added to a uniform tensile stress field σ_x^∞ resulting in the stress $\sigma_x(y)$ acting at an arbitrary point $(0, y)$ on the y-axis, i.e.,

$$\sigma_x(y) = \lim_{x \to 0} \left[\int_{-b}^{b} \frac{\{-3x^4 + 6x^2(y-\eta)^2 + (y-\eta)^4\}}{4\pi\{x^2 + (y-\eta)^2\}^3} (4\sigma_x^\infty \sqrt{b^2 - \eta^2}) \, d\eta \right] + \sigma_x^\infty \tag{1.21}$$

(1) When $|y| > b$, equation (1.21) takes the form

$$\begin{aligned} \sigma_x(y) &= \frac{\sigma_x^\infty}{\pi} \int_{-b}^{b} \frac{\sqrt{b^2 - \eta^2}}{(y-\eta)^2} \, d\eta + \sigma_x^\infty \\ &= \frac{\sigma_x^\infty}{\pi} \left[\frac{\sqrt{b^2 - \eta^2}}{y - \eta} - \sin^{-1}\frac{\eta}{b} + \frac{|y|}{\sqrt{y^2 - b^2}} \sin^{-1}\left\{ \frac{y\eta - b^2}{b(y-\eta)} \right\} \right]_{-b}^{b} + \sigma_x^\infty \\ &= \frac{|y|}{\sqrt{y^2 - b^2}} \sigma_x^\infty \cong \frac{\sigma_x^\infty \sqrt{b}}{\sqrt{2r}} = \frac{k_1}{\sqrt{2r}}, \quad (r = |y| - b \ll b) \end{aligned} \tag{1.22}$$

(2) When $|y| < b$, the integral in equation (1.21) involves a singular term. Therefore, the integral has to be divided into two parts.

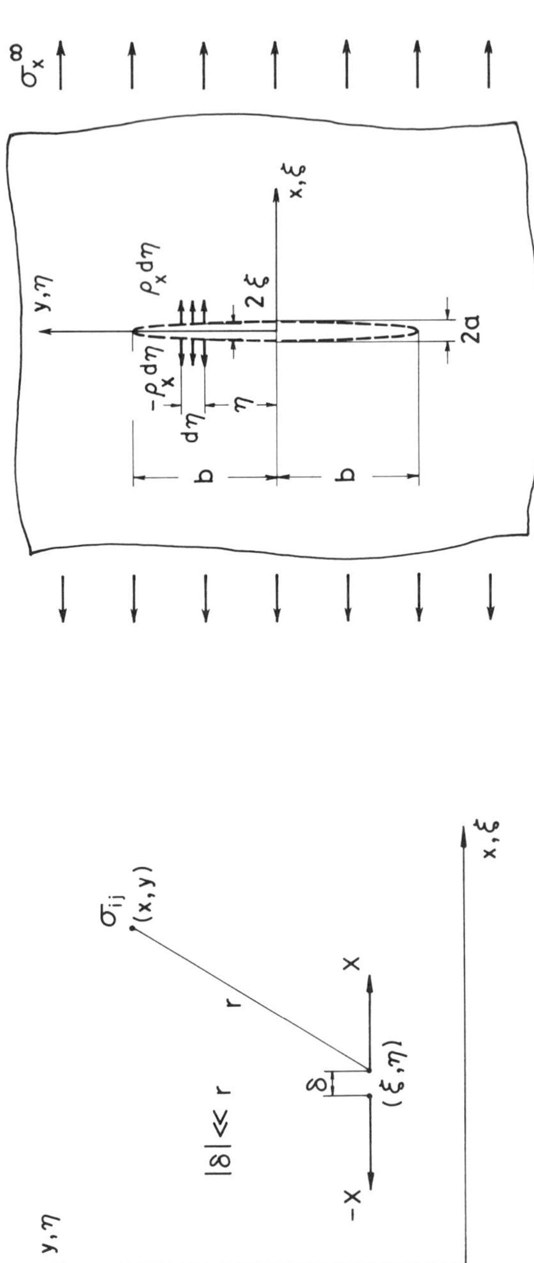

$$\sigma_{ij} = X\sigma_{ij}^{X}(\xi, \eta, x, y)\Big|_{X=1} - X\sigma_{ij}^{X}(\xi-\delta, \eta, x, y)\Big|_{X=1}$$

$$\cong X\frac{\partial \sigma_{ij}^{X}}{\partial \xi}\Big|_{X=1} \cdot \delta = \frac{\partial \sigma_{ij}^{X}}{\partial \xi}\Big|_{X=1} \cdot (X\delta)$$

Figure 1.4. Stress field due to a pair of point forces.

$$2\xi = 2\frac{a}{b}\sqrt{b^2 - \eta^2}$$

$$\rho_x = \left(1 + 2\frac{b}{a}\right)\sigma_x^{\infty} \quad [\text{Eq. (1.5)}]$$

Figure 1.5. Body force in a very slender ellipse.

$$\sigma_x(y) = \lim_{x\to 0} (I_1 + I_2) + \sigma_x^\infty \tag{1.23}$$

in which

$$I_1 = \int_{-b}^{y-\varepsilon} f(\eta)\,\mathrm{d}\eta + \int_{y+\varepsilon}^{b} f(\eta)\,\mathrm{d}\eta \tag{1.24a}$$

$$I_2 = \int_{y-\varepsilon}^{y+\varepsilon} f(\eta)\,\mathrm{d}\eta, \quad |\varepsilon| \ll b \tag{1.24b}$$

where

$$f(\eta) = \frac{-3x^4 + 6x^2(y-\eta)^2 + (y-\eta)^4}{4\pi\{x^2 + (y-\eta)^2\}^3}(4\sigma_x^\infty\sqrt{b^2-\eta^2})$$

In the case when $|\varepsilon| \ll b$, I_1 becomes

$$\begin{aligned}\lim_{x\to 0} I_1 &= \{F(y-\varepsilon) - F(-b)\} + \{F(b) - F(y+\varepsilon)\}\\ &\cong \frac{\sigma_x^\infty}{\pi}\left[\frac{2\sqrt{b^2-y^2}}{\varepsilon}\right] - \sigma_x^\infty\end{aligned} \tag{1.25}$$

in which

$$\begin{aligned}F(\eta) &= \frac{\sigma_x^\infty}{\pi}\int \frac{\sqrt{b^2-\eta^2}}{(y-\eta)^2}\,\mathrm{d}\eta\\ &= \frac{\sigma_x^\infty}{\pi}\left[\frac{\sqrt{b^2-\eta^2}}{y-\eta} - \sin^{-1}\left(\frac{\eta}{b}\right)\right.\\ &\quad \left. + \frac{y}{\sqrt{b^2-y^2}}\ln\left|\frac{(b^2-y\eta) + \sqrt{(b^2-y^2)(b^2-\eta^2)}}{b(y-\eta)}\right|\right]\end{aligned}$$

and I_2 can be approximated by

$$\begin{aligned}I_2 &\cong \frac{\sigma_x^\infty\sqrt{b^2-y^2}}{\pi}\int_{-\varepsilon}^{\varepsilon}\frac{-3x^4 + 6x^2t^2 + t^4}{(x^2+t^2)^3}\,\mathrm{d}t\\ &= -\frac{2\sigma_x^\infty\sqrt{b^2-y^2}}{\pi}\left[\frac{\varepsilon}{x^2+\varepsilon^2} + \frac{2x^2\varepsilon}{(x^2+\varepsilon^2)^2}\right]\end{aligned} \tag{1.26}$$

Inserting the results into equation (1.23) gives

$$\sigma_x(y) = \lim_{x\to 0}(I_1 + I_2) + \sigma_x^\infty = -\sigma_x^\infty + \sigma_x^\infty = 0 \tag{1.27}$$

Equations (1.22) and (1.27) show that the problem of an infinite plate with a crack can be reduced to the problem of an infinite plate without any crack but subjected to body force. The stresses at an arbitrary point are obtained by integrating the stress field for a pair of point forces, $(\rho_x 2\xi)|_{\xi\to 0}\,\mathrm{d}\eta$.

The density γ of a pair of body forces for a crack in an infinite plate subjected to tension is constant along the imaginary crack line:

$$\gamma = \frac{1}{4\sqrt{b^2-\eta^2}}\frac{\mathrm{d}R}{\mathrm{d}\eta} \tag{1.28}$$

Referring to Figure 1.3, $\gamma = \sigma_x^\infty$. In equation (1.28), $\mathrm{d}R$ is the strength of the pair of point forces acting on the element $\mathrm{d}\eta$. Since the density of a pair of body forces may vary along the imaginary crack line, the value of γ has to be determined by a numerical procedure. If the definition of equation (1.28) can be used, the variation of γ along the imaginary crack line is then small and the obtained solution is very accurate.

Tension of a semi-infinite plate with an elliptic hole. As an example, consider the case of tension of a semi-infinite plate with an elliptic hole (Figure 1.8). Use is made of the stress field for a point force acting in a semi-infinite plate. The density of body force in this case varies slightly along an imaginary boundary (Figure 1.6). Here, the stepped distribution (constant in each interval) of the body force is substituted for the continuously varying distribution. The values of the densities of body force are determined from the boundary conditions at each interval. That is, by dividing the elliptic arc AB in Figure 1.6 into MM equal intervals and writing the conditions to make the midpoint of the M-th interval ($M:1\sim MM$) free from stresses by using the densities of the N-th interval ρ_{xN} and ρ_{yN} (which are assumed to be constant in each interval). Then, the stresses at an arbitrary point can be expressed in the form of linear combination of ρ_{xN} and ρ_{yN}. The values of ρ_{xN} and ρ_{yN}, in the case of tension of an infinite plate with an elliptic hole (Figure 1.1), are constant in all intervals as shown in the previous section. If equations (1.6) are used as the densities of body force, the change in ρ_{xN} and ρ_{yN} along an

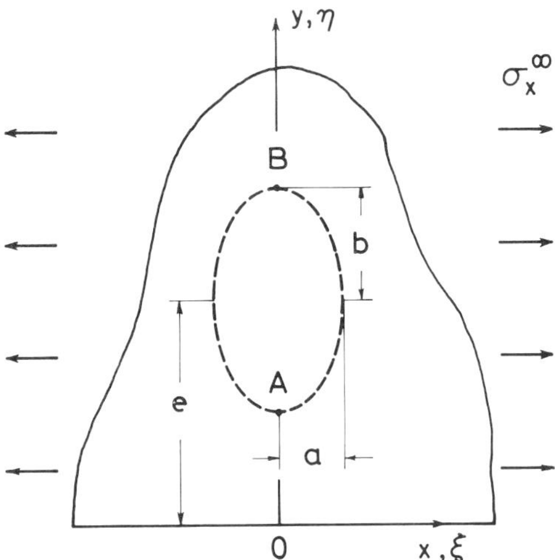

Figure 1.6. Imaginary ellipse in a semi-infinite plate.

imaginary elliptic boundary is comparatively small. It is this small change in ρ_{xN} and ρ_{yN} that leads to high accuracy of the proposed method.

The various steps for determining the values of ρ_{xN} and ρ_{yN} for obtaining the solution to notch problems may be outlined as follows:

Step (1): Divide the elliptic arc *AB* in Figure 1.6 into *MM* equal intervals involving φ which is a variable in the parametric equations of the ellipse:

$$\xi = a \cos \varphi, \qquad \eta = b \sin \varphi \tag{1.29}$$

The values of φ at both ends of the *N*-th interval are $(N-1)\pi/(2MM)$ and $N\pi/(2MM)$, respectively.

Step (2): Calculate the influence coefficients σ_{xM}^{XN}, σ_{yM}^{XN}, and τ_{xyM}^{XN} or σ_{xM}^{YN}, σ_{yM}^{YN}, and τ_{xyM}^{YN}, which are the stresses appearing at the mid-point of the *M*-th interval by the body force acting on the *N*-th interval (whose densities are $\rho_{xN} = 1$ or $\rho_{yN} = 1$).

Step (3): Write the conditions that make the mid-point of the *M*-th interval free from stresses by using the unknowns ρ_{xN} and ρ_{yN} and the influence coefficients σ_{xM}^{XN}, σ_{xM}^{YN}, etc.

The following set of $2MM$ linear equations are found:

$$\sum_{N=1}^{MM} \rho_{xN}(\sigma_{xM}^{XN} \cos\theta + \tau_{xyM}^{XN} \sin\theta) + \sum_{N=1}^{MM} \rho_{yN}(\sigma_{xM}^{YN} \cos\theta + \tau_{xyM}^{YN} \sin\theta) + \sigma_x^{\infty} \cos\theta = 0 \tag{1.30a}$$

$$\sum_{N=1}^{MM} \rho_{xN}(\sigma_{yM}^{XN} \sin\theta + \tau_{xyM}^{XN} \cos\theta) + \sum_{N=1}^{MM} \rho_{yN}(\sigma_{yM}^{YN} \sin\theta + \tau_{xyM}^{YN} \cos\theta) = 0 \ \ (M: 1 \sim MM) \tag{1.30b}$$

Step (4): The stresses at an arbitrary point can be expressed in terms of linear combination of ρ_{xN} and ρ_{yN}.

Step (5): Since the errors caused by the finite number of divisions MM are nearly proportional to $1/MM$, as shown in Figure 1.8, the true value of stress concentration factor K_t is obtained by extrapolation using the two values of K_t corresponding to two values of MM.

The influence coefficients in step (2) can be calculated from the stress fields for a point force X or Y acting at (ξ, η) in a semi-infinite plate. Since the relations $\mathrm{d}\eta = b \cos\varphi \, \mathrm{d}\varphi$ and $-\mathrm{d}\xi = a \sin\varphi \, \mathrm{d}\varphi$ prevail along the ellipse, the influence coefficients can be expressed as

$$\sigma_{xM}^{XN} = \int_N \sigma_x^X(\xi, \eta, x, y)|_{X=1} \, b \cos\varphi \, \mathrm{d}\varphi \tag{1.31a}$$

$$\sigma_{yM}^{XN} = \int_N \sigma_y^X(\xi, \eta, x, y)|_{X=1} \, b \cos\varphi \, \mathrm{d}\varphi \tag{1.31b}$$

$$\tau_{xyM}^{XN} = \int_N \tau_{xy}^X(\xi, \eta, x, y)|_{X=1} \, b \cos\varphi \, \mathrm{d}\varphi \tag{1.31c}$$

and

$$\sigma_{xM}^{YN} = \int_N \sigma_x^Y(\xi, \eta, x, y)|_{Y=1} \, a \sin\varphi \, \mathrm{d}\varphi \tag{1.32a}$$

$$\sigma_{yM}^{YN} = \int_N \sigma_y^Y(\xi, \eta, x, y)|_{Y=1} \, a \sin\varphi \, \mathrm{d}\varphi \tag{1.32b}$$

$$\tau_{xyM}^{YN} = \int_N \tau_{xy}^{Y}(\xi, \eta, x, y)|_{Y=1}\, a \sin\varphi\, \mathrm{d}\varphi \tag{1.32c}$$

where $\int_N$ stands for integration of the N-th interval and σ_x^X, σ_y^Y, etc. are given by equations (1.59) and (1.60). In the above equations, the variables x, y, etc. are given by

$$x = a \cos \varphi_M, \qquad y = e + b \sin \varphi_M \tag{1.33a}$$

where

$$\varphi_M = \frac{\pi(M - 0.5)}{2\, MM}$$

and

$$\xi = a \cos \varphi, \qquad \eta = e + b \sin \varphi \tag{1.33b}$$

The integrals in equations (1.31) and (1.32) are calculated by Simpson's or Gauss's integral formula. Since the integral involves a singular term for $M = N$, the principal values [2] of equations (1.31) and (1.32) must be taken. Furthermore, in order to find the stresses at a point infinitesimally close to the ellipse subjected to a body force, the stresses $\Delta\sigma_x^X$, $\Delta\sigma_x^Y$ etc. must be added. Consider an elliptic arc, whose length is $2\varepsilon(\ll b)$ and is subjected to the body force with a density ρ_x as shown in Figure 1.7. The stresses at the point C close to the elliptic arc may be found by considering the equilibrium of an infinitesimal rectangle $EFFE$. It can be shown that the force acting on the section EE is $-\frac{1}{2}\rho_x\varepsilon \cos\theta$ and the force acting on the section FF is $\frac{1}{2}\rho_x\varepsilon \cos\theta$. This means that the stresses at C on the section EE, i.e., $\sigma_{x'}$, $\sigma_{y'}$ and $\tau_{x'y'}$ are

$$\sigma_{x'} = 0 \tag{1.34a}$$

$$\sigma_{y'} = -\tfrac{1}{2}\rho_x \cos^2\theta \tag{1.34b}$$

$$\tau_{x'y'} = -\tfrac{1}{2}\rho_x \sin\theta \cos\theta \tag{1.34c}$$

Equation (1.34a) follows from the fact that the lengths of EE and FF do not change:

$$\varepsilon_{x'} = \frac{1}{E}(\sigma_{x'} - \nu\sigma_{y'}) = \frac{1}{E}\sigma_{x'} = 0 \tag{1.35}$$

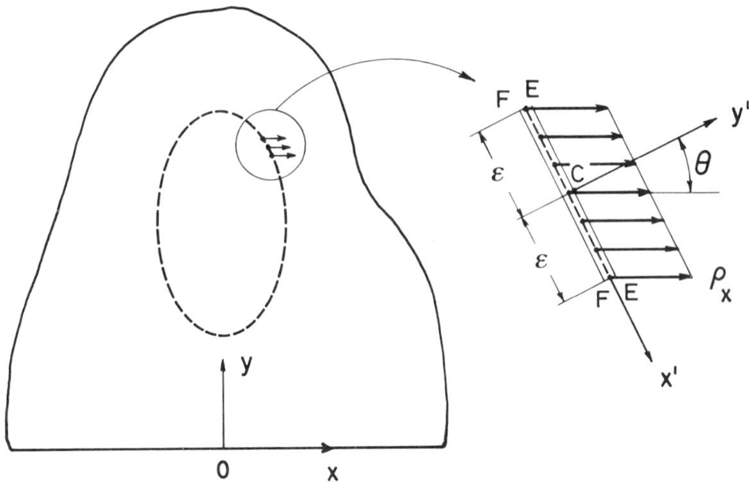

Figure 1.7. Stresses for $M = N$.

Equations (1.34) thus give the stresses $\Delta\sigma_x^X$, $\Delta\sigma_y^X$, and $\Delta\tau_{xy}^X$ at the point C due to a body force, acting on part 2ε. The result is

$$\Delta\sigma_x^X = -\frac{\rho_x}{16}(5 + 4\cos 2\theta - \cos 4\theta) \tag{1.36a}$$

$$\Delta\sigma_y^X = \frac{\rho_x}{16}(1 - \cos 4\theta) \tag{1.36b}$$

$$\Delta\tau_{xy}^X = -\frac{\rho_x}{16}(2\sin 2\theta - \sin 4\theta) \tag{1.36c}$$

In a similar manner, the stresses corresponding to ρ_y are given by

$$\Delta\sigma_x^Y = \frac{\rho_y}{16}(1 - \cos 4\theta) \tag{1.37a}$$

$$\Delta\sigma_y^Y = -\frac{\rho_y}{16}(5 - 4\cos 2\theta - \cos 4\theta) \tag{1.37b}$$

$$\Delta\tau_{xy}^Y = -\frac{\rho_y}{16}(2\sin 2\theta + \sin 4\theta) \tag{1.37c}$$

Note that $\Delta\sigma_x^X$ and $\Delta\sigma_x^Y$, etc. do not involve ε and are the terms not accounted for in calculating the principal values of the integrals. Therefore, when $M = N$, these terms must be added to equation (1.31) and (1.32).

Table 1.1 compares the results obtained from the present method with those of Isida [8]. The errors based on the finite number of divisions MM is nearly proportional to $1/MM$, as shown in Figure 1.8. The stress concentration factor in the limit as $MM \to \infty$ is the true value of K_t. The results in Table 1.1 were obtained by extrapolation using the values of K_t for $MM = 32$ and 48.

Tension of a semi-infinite plate with a crack. As another example, the case of tension of a semi-infinite plate with a crack (Figure 1.10) will be considered. The derivative of the stress field of a point force in a semi-infinite plate will be used. The density of a pair of body forces γ defined by equation (1.28) varies slightly along the imaginary crack line (Figure 1.9). Here, the continuously varying distribution is again replaced by the stepped distribution, constant in each interval. The value of γ in each interval is determined from the boundary condition. In particular, the line AB is divided into MM equal intervals such that γ_N of the N-th interval is obtained by satisfying the free stress condition at the mid-point of the M-th interval ($M: 1 \sim MM$). The stresses at an arbitrary point can then be expressed in terms of linear combination of γ_N.

The singularity at both ends A and B of a crack (Figure 1.9) is determined by γ_1 and γ_{MM} alone, respectively. As it is seen from

TABLE 1.1

Values of K_t of a semi-infinite plate with an elliptic hole

$\frac{b}{a}$	$\frac{b}{e}$	K_t	
		Nisitani	Isida [8]
0.5	0.2	5.084	5.085
	0.4	5.422	5.420
	0.6	6.269	–
	0.8	8.592	–

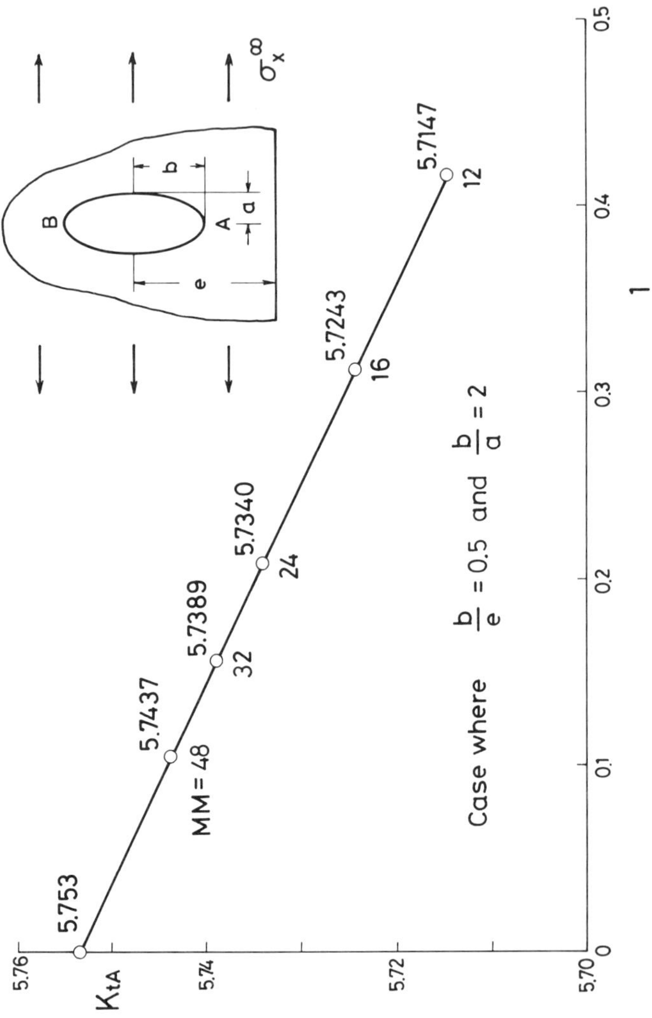

Figure 1.8. Method of extrapolation.

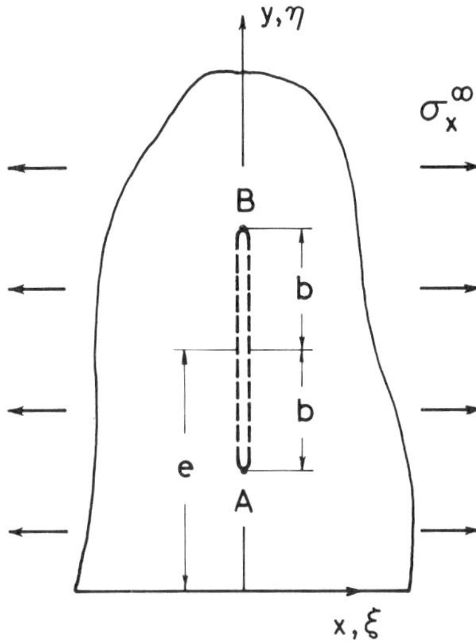

Figure 1.9. Imaginary crack line in a semi-infinite plate.

equation (1.22), when $\gamma_N = \text{constant} = \sigma_x^\infty$, k_1 is equal to $\gamma_N\sqrt{b}$. This implies that the stress intensity factors at points A and B can be written as

$$k_{1A} = \sigma_1\sqrt{b} \tag{1.38a}$$

$$k_{1B} = \sigma_{MM}\sqrt{b} \tag{1.38b}$$

It has been shown in the preceding section that γ_N in the problem of Figure 1.3 is constant in all intervals. Applying equation (1.28) to the density of a pair of body forces, the change in γ_N along an imaginary crack line is comparatively small, which tends to improve the accuracy of the solution.

The steps to determine the value of γ_N and to obtain a solution are as follows. These steps also apply to crack problems in general.

Step (1): Divide the line AB in Figure 1.9 into MM equal intervals. That is, the value of η at both ends of the N-th interval are $(e-b)+2b(N-1)/MM$ and $(e-b)+2bN/MM$, respectively.

Step (2): Calculate the influence coefficient $\sigma_{xM}^{\gamma_N}$ which is the stress appearing at the mid-point of the M-th interval for a pair of body forces of the N-th interval whose density γ_N is equal to 1.

Step (3): Apply the stress free condition to the mid-point of the M-th interval by using the unknown γ_N and the influence coefficient $\sigma_{xM}^{\gamma N}$. A set of MM linear equations is thus obtained:

$$\sum_{N=1}^{MM} \gamma_N \sigma_{xM}^{\gamma N} + \sigma_x^{\infty} = 0, \quad (M: 1 \sim MM) \tag{1.39}$$

Step (4): Determine the value of γ_N by solving equation (1.39).

Step (5): Stress intensity factors k_{1A} and k_{1B} are obtained from equation (1.38).

Step (6): The stresses at an arbitrary point can be expressed in the form of linear combination of γ_N.

Step (7): Since the errors due to the finiteness of MM are nearly proportional to $1/MM$, as shown in Figure 1.10, the value of stress intensity factor corresponding to $MM \to \infty$ is obtained by extrapolation of the two values of k_1 corresponding to two values of MM.

The influence coefficient $\sigma_{xM}^{\gamma N}$ in step (2) is calculated from the derivative of the stress field for a point force acting in a semi-infinite plate and is given by equations (1.59). From equations (1.28) and (1.59), and the information in Figure 1.4, the following result with $\nu = 0$ is obtained:

$$\begin{aligned}\sigma_{xM}^{\gamma N} &= \int_N \left.\frac{\partial \sigma_x^X}{\partial \xi}\right|_{\substack{X=1\\ x=\xi=0}} \{4\gamma_N \sqrt{b^2-(\eta-e)^2}\}|_{\gamma_N=1}\, \mathrm{d}\eta \\ &= \int_{\eta_{N-1}}^{\eta_N} \left[\frac{1}{4\pi(y-\eta)^2} + \frac{1}{4\pi}\left\{-\frac{1}{(y+\eta)^2} + \frac{12y}{(y+\eta)^3} - \frac{12y^2}{(y+\eta)^4}\right\}\right] \\ &\quad \times \{4\sqrt{b^2-(\eta-e)^2}\}\, \mathrm{d}\eta \end{aligned} \tag{1.40}$$

where η_{N-1} and η_N denote the values of η at both ends of the N-th interval. The integral in equation (1.40) can be obtained in closed form as follows:

$$\sigma_{xM}^{\gamma N} = F(\eta_N) - F(\eta_{N-1}) \tag{1.41}$$

in which $F(\eta)$ stands for

$$
\begin{aligned}
F(\eta) = {} & \frac{1}{\pi}\left[\frac{\sqrt{b^2-(\eta-e)^2}}{y-\eta} - \sin^{-1}\frac{\eta-e}{b} + \frac{y-e}{\sqrt{b^2-(y-e)^2}}\right. \\
& \left.\times \ln\left|\frac{b^2-(y-e)(\eta-e)+\sqrt{\{b^2-(y-e)^2\}\{b^2-(\eta-e)^2\}}}{b(y-\eta)}\right|\right] \\
& -\frac{1}{\pi}\left[-\frac{\sqrt{b^2-(\eta-e)^2}}{y+\eta} - \sin^{-1}\frac{\eta-e}{b}\right. \\
& \left. + \frac{y+e}{\sqrt{(y+e)^2-b^2}}\sin^{-1}\frac{b^2+(y+e)(\eta-e)}{b(y+\eta)}\right] \\
& + \frac{6y}{\pi}\left[-\frac{\sqrt{b^2-(\eta-e)^2}}{(y+\eta)^2}\left\{1-\frac{(y+e)(y+\eta)}{(y+e)^2-b^2}\right\} + \frac{b^2}{\{(y+e)^2-b^2\}^{\frac{3}{2}}}\right. \\
& \left.\times \sin^{-1}\frac{b^2+(y+e)(\eta-e)}{b(y+\eta)}\right] - \frac{2y^2}{\pi}\left[\frac{\sqrt{b^2-(\eta-e)^2}}{(y+\eta)^3}\right. \\
& \times\left\{-2+\frac{(y+e)(y+\eta)}{(y+e)^2-b^2}+\frac{\{2b^2+(y+e)^2\}(y+\eta)^2}{\{(y+e)^2-b^2\}^2}\right\} \\
& \left.+\frac{3b^2(y+e)}{\{(y+e)^2-b^2\}^{\frac{5}{2}}}\sin^{-1}\frac{b^2+(y+e)(\eta-e)}{b(y+\eta)}\right]
\end{aligned}
$$

Although the integral in equation (1.40) also involves a singular term in the case of $M = N$, no special consideration is necessary as illustrated in equations (1.23) to (1.27). Table 1.2 shows the comparison of the

TABLE 1.2

Values of F_I of a semi-infinite plate with a crack

$\frac{b}{e}$	F_I	
	Nisitani	Isida [9]
0.2	1.0112	1.0112
0.4	1.0528	1.0528
0.6	1.1491	1.1490
0.8	1.3879	1.3875

$$F_I = \frac{k_1}{\sigma_x^\infty\sqrt{b}}$$

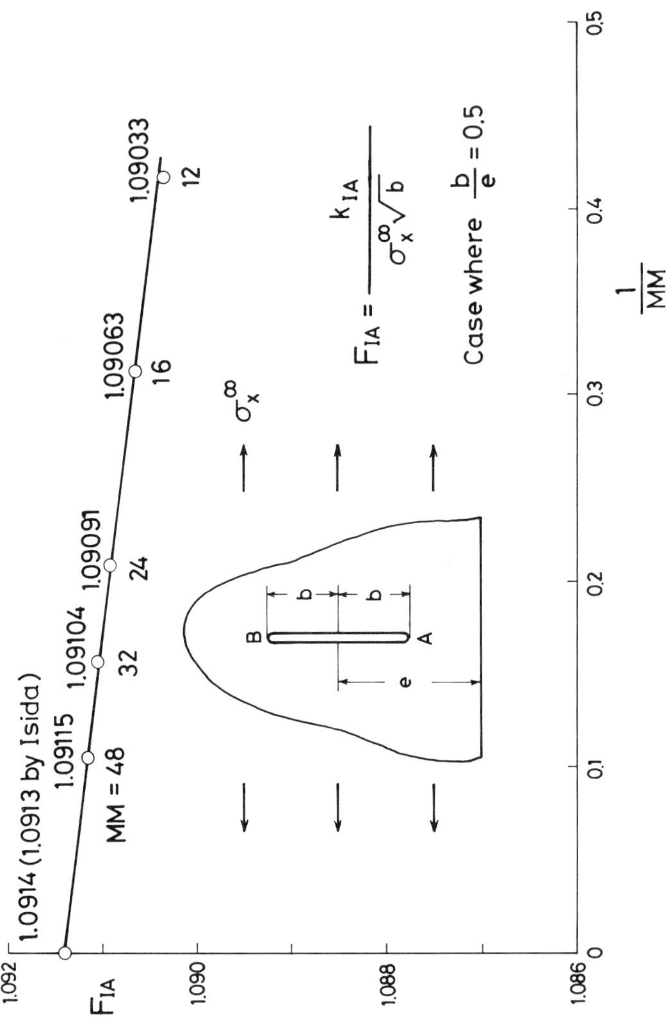

Figure 1.10. Method of extrapolation for a crack problem.

numerical results with those of Isida [9]. The errors due to the finiteness of MM are nearly proportional to $1/MM$ as mentioned earlier.

1.3 Influence coefficients and fundamental stress fields

The present method involves dividing a curved line (in a two-dimensional problem) or a curved surface (in a three-dimensional problem) into MM equal intervals to be free from stresses and then the density of the body force for the N-th interval ($N:1 \sim MM$) is determined from the stress free boundary condition. In the numerical procedure, the density of body force that vary continuously along a curved line or surface is replaced by a stepped density. The change of density along the notch boundary should be as small as possible. Refer to equations (1.6) and (1.28). The errors arising from the stepped density diminishes as $MM \to \infty$. In practice accurate solution can be obtained by extrapolation of the solutions for two values of MM. The errors due to the finiteness of MM are nearly proportional to $1/MM$ as shown in Figures 1.8 and 1.10.

The density of body force can be determined from the influence coefficient which represents the stress appearing at the mid-point of the M-th interval for a body force of unit density acting on the N-th interval. By using the values of influence coefficients, a set of simultaneous equations, whose unknown is the density of the N-th interval, is obtained from the condition which makes the mid-point of the M-th interval free from stresses. This set of equations can be solved for the density of body force. To calculate the influence coefficients, it is necessary to know the fundamental stress field. In what follows, the fundamental stress fields of various problems will be presented.

1.4 Fundamental stress field in problems of an infinite plate

Tension [2, 10]. The stress field for a point force acting in an infinite plate (Figure 1.11) will be used to solve the problem of an infinite plate with several holes or cracks subjected to tension. For plane stress conditions, the stresses are

$$\sigma_x^X = -Fl\{(3+\nu)l^2 + (1-\nu)m^2\}X \tag{1.42a}$$

$$\sigma_y^X = Fl\{(1-\nu)l^2 - (1+3\nu)m^2\}X \tag{1.42b}$$

$$\tau_{xy}^X = -Fm\{(3+\nu)l^2 + (1-\nu)m^2\}X \tag{1.42c}$$

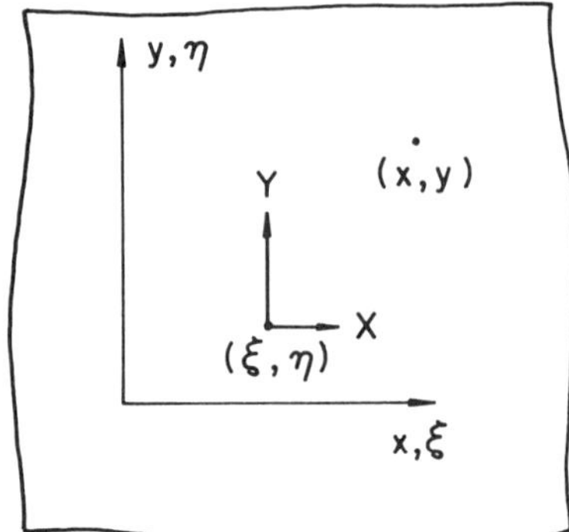

Figure 1.11. Fundamental stress field for the tension of an infinite plate.

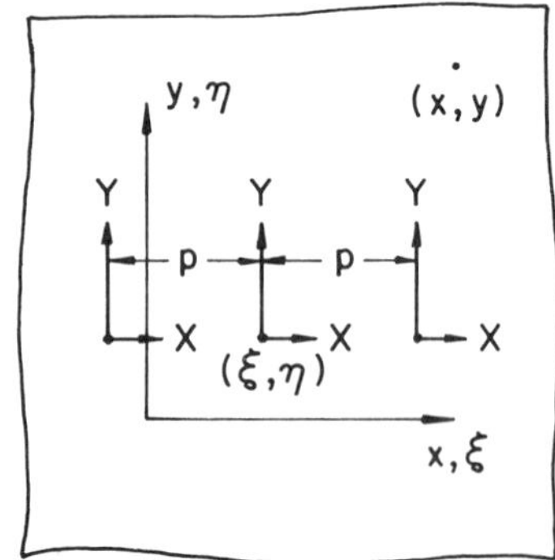

Figure 1.12. Stress field for a periodic array of in-plane forces in an infinite plate.

and

$$\sigma_x^Y = -Fm\{(1+3\nu)l^2-(1-\nu)m^2\}Y \tag{1.43a}$$

$$\sigma_y^Y = -Fm\{(1-\nu)l^2+(3+\nu)m^2\}Y \tag{1.43b}$$

$$\tau_{xy}^Y = -Fl\{(1-\nu)l^2+(3+\nu)m^2\}Y \tag{1.43c}$$

where

$$l=\frac{x-\xi}{y}, \qquad m=\frac{y-\eta}{y}, \qquad F=\frac{1}{4\pi y(l^2+m^2)^2} \tag{1.44}$$

For plane strain, ν in equations (1.42) and (1.43) must be replaced by $\nu/(1-\nu)$.

Consider an infinite plate with a periodic array of elliptical holes as shown in Figure A1.1. The fundamental stress field derived for this problem must also have this periodicity as illustrated in Figure 1.12. Since ν does not enter into this problem, it may be set to zero and becomes

$$\sigma_x^X = (3K_2-2m^2K_4)X \tag{1.45a}$$

$$\sigma_y^X = (-K_2+2m^2K_4)X \tag{1.45b}$$

$$\tau_{xy}^X = (-3mK_1+2m^3K_3)X \tag{1.45c}$$

and

$$\sigma_x^Y = (-mK_1 + 2m^3K_3)Y \tag{1.46a}$$

$$\sigma_y^Y = (-mK_1 - 2m^3K_3)Y \tag{1.46b}$$

$$\tau_{xy}^Y = (K_2 + 2m^2K_4)Y \tag{1.46c}$$

where

$$\begin{aligned}
K_1 &= \frac{\sinh \alpha m}{4pm(\cosh \alpha m - \cos \alpha l)} \\
K_2 &= -\frac{\sin \alpha l}{4p(\cosh \alpha m - \cos \alpha l)} \\
K_3 &= \frac{\sinh \alpha m(\cosh \alpha m - \cos \alpha l) - \alpha m(1 - \cosh \alpha m \cos \alpha l)}{8pm^3(\cosh \alpha m - \cos \alpha l)^2} \\
K_4 &= -\frac{\alpha \sinh \alpha m \sin \alpha l}{8pm(\cosh \alpha m - \cos \alpha l)^2}
\end{aligned} \tag{1.47}$$

in which α, l and m are given by

$$\alpha = \frac{2\pi y}{p}, \qquad l = \frac{x - \xi}{y}, \qquad m = \frac{y - \eta}{y}$$

The numerical solutions obtained from equations (1.45) and (1.46) are shown in Figures A1.1 and A1.2, and Table A1.5 of the Appendix.

Anti-plane shear [11]. The fundamental stress field for a point force acting along a straight line in an infinite body (Figure 1.13) is used to solve the anti-plane shear problem of an infinite plate with several holes or cracks.

The stresses for a point force are

$$\tau_{yz}^T = -\frac{m}{2\pi y(l^2 + m^2)} T \tag{1.48a}$$

$$\tau_{zx}^T = -\frac{l}{2\pi y(l^2 + m^2)} T \tag{1.48b}$$

where

$$l = \frac{x - \xi}{y}, \qquad m = \frac{y - \eta}{y} \tag{1.49}$$

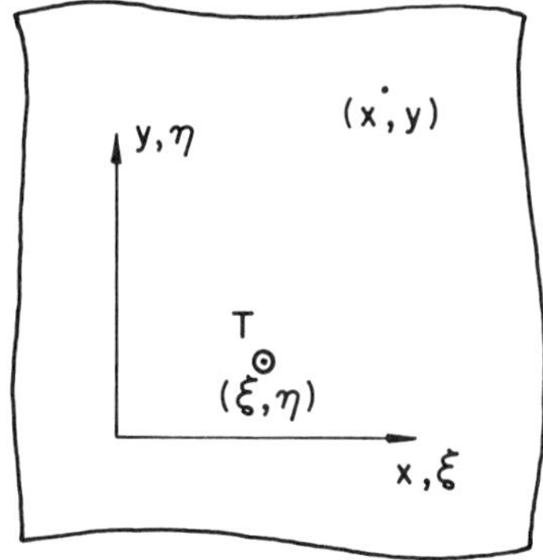

Figure 1.13. Fundamental stress field for the anti-plane shear of an infinite body.

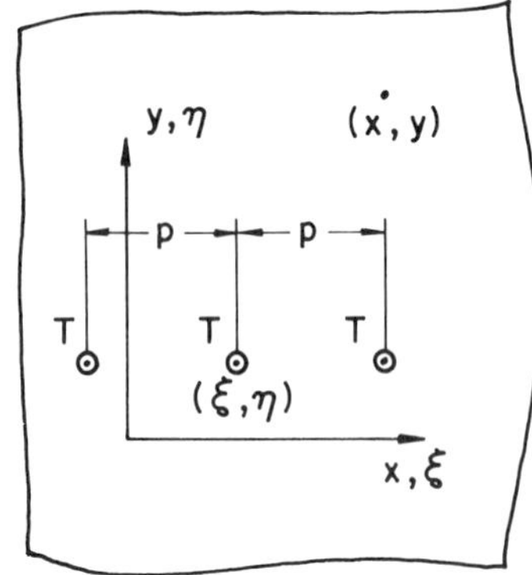

Figure 1.14. Stress field for a periodic array of anti-plane shear forces in an infinite body.

The solutions obtained from equation (1.48) are given in Figure A1.9 and Table A1.11 of the Appendix.

Figures A1.3 and A1.4 show a periodic array of elliptical holes. For these two problems, the fundamental anti-plane shear solution for a system of point forces spaced periodically in an infinite body is required. Referring to Figure 1.14, the solution is

$$\tau_{yz}^{T} = -\frac{\sinh \alpha m}{2p(\cosh \alpha m - \cos \alpha l)} T \tag{1.50a}$$

$$\tau_{zx}^{T} = -\frac{\sin \alpha l}{2p(\cosh \alpha m - \cos \alpha l)} T \tag{1.50b}$$

where

$$\alpha = \frac{2\pi y}{p}, \qquad l = \frac{x - \xi}{y}, \qquad m = \frac{y - \eta}{y} \tag{1.51}$$

The numerical results obtained by using equation (1.50) are shown in Figures A1.3 and A1.4.

Density of body force for anti-plane shear. The density of body force for the anti-plane shear problem of an infinite plate with holes can be expressed as

$$\rho_t = \frac{dF_z}{d\eta} \tag{1.52}$$

where dF_z denotes the force in the z-direction acting on the element $dS = \sqrt{(d\xi)^2 + (d\eta)^2}$ which has a unit length in the z-direction. In the case of an elliptic hole (Figure 1.15), the density of body force ρ_t, defined by equation (1.52), is constant along the entire periphery of an imaginary ellipse (Figure 1.16) and is given by

$$\rho_t = \left(1 + \frac{b}{a}\right)\tau_{zx}^{\infty} \tag{1.53}$$

It can be verified that the surface of an ellipse in an infinite plate subjected to shear stress τ_{zx}^{∞} at infinity can be freed from tractions by applying the body force having a density given by equation (1.53).

Referring to the stress notations in Figure 1.16b, the shear stress on the plane normal to n is

$$\tau_{zn} = \tau_{zx} \cos\theta + \tau_{zy} \sin\theta \tag{1.54}$$

On the ellipse, the following relations are given

$$\begin{aligned} \xi &= a \cos\varphi, \qquad \eta = b \sin\varphi \\ x &= a \cos\varphi_0, \qquad y = b \sin\varphi_0 \end{aligned} \tag{1.55}$$

such that

$$\tan\theta = \frac{a}{b} \tan\varphi_0$$

Making use of equation (1.53) for the density of body force on an ellipse in an infinite plate without any hole, the stresses appearing at (x, y) may be found from equation (1.48) as

$$\tau_{zn} = \oint (\tau_{zx}^T|_{T=1} \cos\theta + \tau_{zy}^T|_{T=1} \sin\theta)\rho_t \, d\eta - \tfrac{1}{2}\rho_t \cos\theta + \tau_{zx}^{\infty} \cos\theta \tag{1.56}$$

The term $-\frac{1}{2}\rho_t \cos\theta$ corresponds to equation (1.34) in the case of tensile

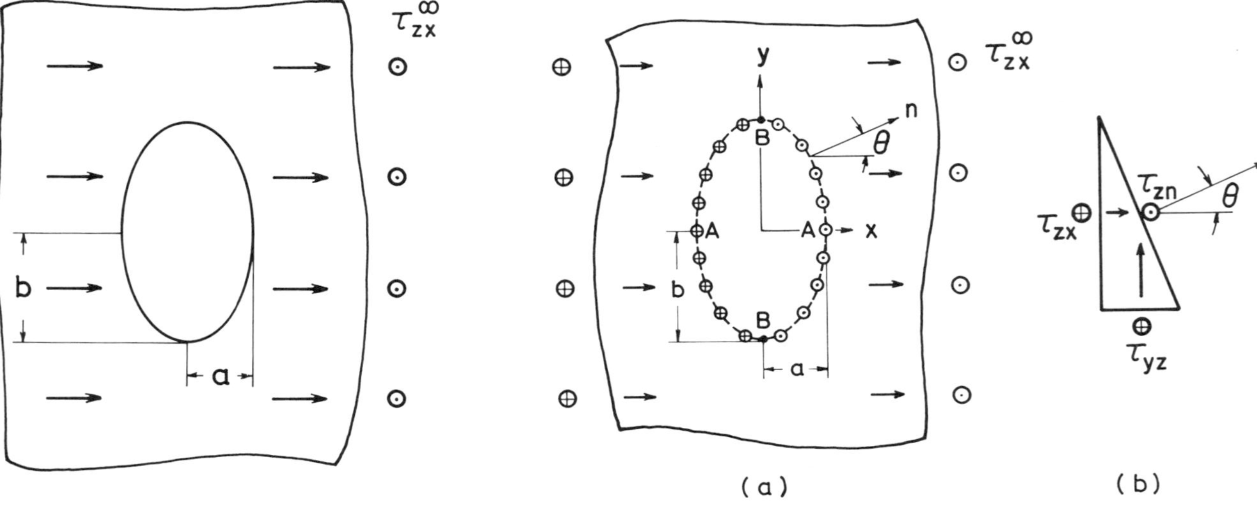

Figure 1.15. Anti-plane shear of an infinite body with an elliptic hole.

Figure 1.16. Imaginary ellipse in an infinite plate.

loading. Substituting equation (1.48) into equation (1.56), τ_{zn} becomes

$$\tau_{zn} = -\oint \frac{\{(x-\xi)\cos\theta + (y-\eta)\sin\theta\}}{2\pi\{(x-\xi)^2+(y-\eta)^2\}}\left(1+\frac{b}{a}\right)\tau_{zx}^{\infty} b\cos\varphi\, d\varphi$$

$$-\frac{1}{2}\left(1+\frac{b}{a}\right)\tau_{zx}^{\infty}\cos\theta + \tau_{zx}^{\infty}\cos\theta = 0 \qquad (1.57)$$

The ellipse in an infinite plate without any hole now becomes traction-free. The maximum stress appears at B ($\varphi_0 = \pm\pi/2$ or $\theta = \pm\pi/2$) in Figure 1.16 and is

$$\tau_{\max} = \tau_{zx}\big|_{\varphi_0=\pi/2} = -\oint \frac{(x-\xi)}{2\pi\{(x-\xi)^2+(y-\eta)^2\}}\left(1+\frac{b}{a}\right)\tau_{zx}^{\infty} b\cos\varphi\, d\varphi + \tau_{zx}^{\infty}$$

$$= \left(1+\frac{b}{a}\right)\tau_{zx}^{\infty} \qquad (1.58)$$

which is a well-known result.

1.5 Fundamental stress field in problems of a semi-infinite plate

Tension [2, 12, 13]. The fundamental stress field in problems of a semi-infinite plate under tension (Figure 1.17) can be obtained by removing the stresses on a straight boundary, which is produced by the fundamental

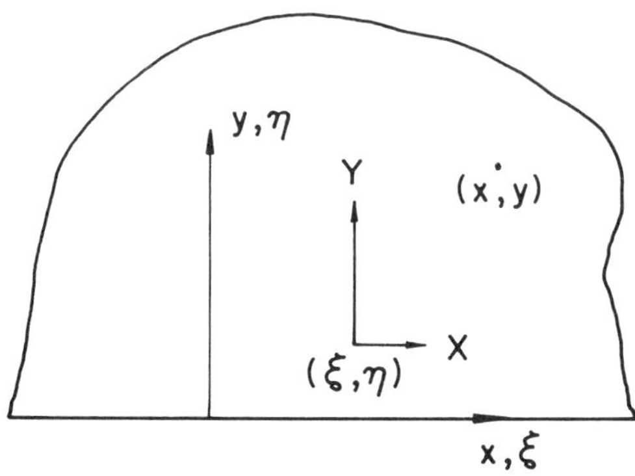

Figure 1.17. Fundamental stress field for the tension of a semi-infinite plate.

stress field of an infinite plate. The results are

$$\begin{aligned}\sigma_x^X = {} & [-Fl\{(3+\nu)l^2+(1-\nu)m^2\}-G_1l\{(3+\nu)l^2+(1-\nu)n^2\} \\ & -G_2\{(1-\nu)(l^5+2(n^2-n)l^3+n^2(n^2-2n)l) \\ & +(1+\nu)(-2(n^2-1)l^3+n^2(-2n^2+8n-6)l)\}]X \end{aligned} \tag{1.59a}$$

$$\begin{aligned}\sigma_y^X = {} & [Fl\{(1-\nu)l^2-(1+3\nu)m^2\}+G_1l\{(1-\nu)l^2-(1+3\nu)n^2\} \\ & -G_2\{(1-\nu)(l^5+2(n^2+n)l^3+n^2(n^2+2n)l) \\ & +(1+\nu)(2(-n^2+2n-1)l^3+n^2(-2n^2-4n+6)l)\}]X \end{aligned} \tag{1.59b}$$

$$\begin{aligned}\tau_{xy}^X = {} & [-Fm\{(3+\nu)l^2+(1-\nu)m^2\}-G_1n\{(3+\nu)l^2+(1-\nu)n^2\} \\ & +G_2\{(1-\nu)(-l^4+n^4)+(1+\nu)(6n(n-1)l^2+n^3(-2n+2))\}]X \end{aligned} \tag{1.59c}$$

and

$$\begin{aligned}\sigma_x^Y = {} & [-Fm\{(1+3\nu)l^2-(1-\nu)m^2\}+G_1n\{(1+3\nu)l^2-(1-\nu)n^2\} \\ & -G_2\{(1-\nu)((n+1)l^4+2n^3l^2+n^3(n^2-n)) \\ & +(1+\nu)(2l^4+n(2n^2+6n-6)l^2+n^3(2n^2-4n+2))\}]Y \end{aligned} \tag{1.60a}$$

$$\begin{aligned}\sigma_y^Y = {} & [-Fm\{(1-\nu)l^2+(3+\nu)m^2\}+G_1n\{(1-\nu)l^2+(3+\nu)n^2\} \\ & -G_2\{(1-\nu)((n-1)l^4+2n^3l^2+n^3(n^2+n)) \\ & +(1+\nu)(n(2n^2-6n+6)l^2+n^3(2n^2+2n-2))\}]Y \end{aligned} \tag{1.60b}$$

$$\begin{aligned}\tau_{xy}^Y = {} & [-Fl\{(1-\nu)l^2+(3+\nu)m^2\}+G_1l\{(1-\nu)l^2+(3+\nu)n^2\} \\ & -G_2\{(1-\nu)(2nl^3+2n^3l)+(1+\nu)(2l^3+n^2(8n-6)l)\}]Y \end{aligned} \tag{1.60c}$$

in which the various parameters in equations (1.59) and (1.60) are given by

$$l=\frac{x-\xi}{y}, \qquad m=\frac{y-\eta}{y}, \qquad n=\frac{y+\eta}{y}$$

$$F=\frac{1}{4\pi y(l^2+m^2)^2}, \qquad G_1=\frac{1}{4\pi y(l^2+n^2)^2}, \qquad G_2=\frac{1}{2\pi y(l^2+n^2)^3} \tag{1.61}$$

Refer to Figures A1.7, A1.8 and A1.19, and Tables A1.10 and A1.12 for the solutions obtained by using equations (1.59) and (1.60).

For a system of periodically spaced point forces as shown in Figure

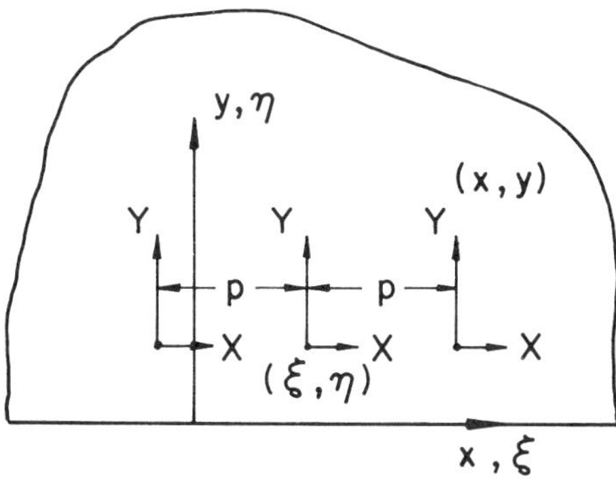

Figure 1.18. Stress field for a periodic array of in-plane forces in a semi-infinite plate.

1.18, the fundamental stress field [10] for $\nu = 0$ is

$$\sigma_x^X = [3K_2 - 2m^2K_4 + 5L_2 - 2(3n^2 + 2n - 2)L_4 + 16n^2(n-1)L_6]X \tag{1.62a}$$

$$\sigma_y^X = [-K_2 + 2m^2K_4 + L_2 - 2(n^2 - 6n + 2)L_4 - 16n^2(n-1)L_6]X \tag{1.62b}$$

$$\tau_{xy}^X = [-3mK_1 + 2m^3K_3 - (3n+2)L_1 + 2n(n^2 + 8n - 6)L_3 - 16n^3(n-1)L_5]X \tag{1.62c}$$

and

$$\sigma_x^Y = [-mK_1 + 2m^3K_3 - (n+6)L_1 - 6n(n^2 - 2)L_3 + 16n^3(n-1)L_5]Y \tag{1.63a}$$

$$\sigma_y^Y = [-mK_1 - 2m^3K_3 - (n-2)L_1 - 2n(n^2 - 4n + 6)L_3 - 16n^3(n-1)L_5]Y \tag{1.63b}$$

$$\tau_{xy}^Y = [K_2 + 2m^2K_4 - L_2 - 2(n^2 - 2n - 2)L_4 + 16n^2(n-1)L_6]Y \tag{1.63c}$$

where $K_1, K_2, \ldots, K_4$ are given by equation (1.47), and

$$L_1 = \frac{\sinh \alpha n}{4pnG}, \qquad L_2 = -\frac{\sin \alpha l}{4pG}$$

$$L_3 = \frac{G \sinh \alpha n - \alpha n H}{8pn^3G}, \qquad L_4 = -\frac{\alpha \sinh \alpha n \sin \alpha l}{8pnG^2}$$

$$L_5 = [\{G \sinh \alpha n - \alpha n H\}\{3G + 2\alpha n \sinh \alpha n\} - \alpha n G\{2 \sinh^2 \alpha n + \alpha n \sinh \alpha n \cos \alpha l\}]/32pn^5G^3$$

$$L_6 = -\frac{\alpha \sin \alpha l\{\alpha n(\sinh^2 \alpha n - H) + G \sinh \alpha n\}}{32pn^3G^3} \tag{1.64a}$$

$$G = \cosh \alpha n - \cos \alpha l$$

$$H = 1 - \cosh \alpha n \cos \alpha l$$

where

$$\alpha = \frac{2\pi y}{p}, \qquad l = \frac{x-\xi}{y}, \qquad m = \frac{y-\eta}{y}, \qquad n = \frac{y+\eta}{y} \tag{1.64b}$$

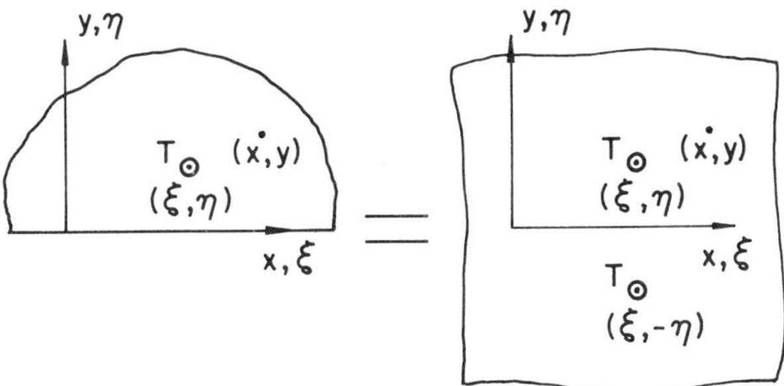

Figure 1.19. Fundamental stress field for the anti-plane shear of a semi-infinite body.

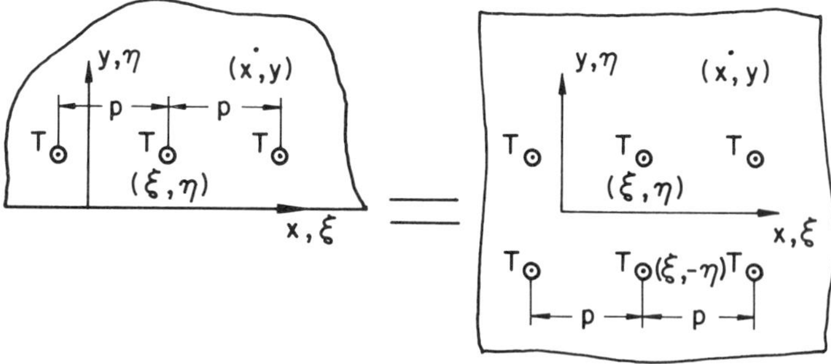

Figure 1.20. Stress field for a periodic array of anti-plane shear forces in a semi-infinite body.

The results in Figure A1.4 and Table A1.13 of the Appendix make use of equations (1.62) and (1.63).

Anti-plane shear. The fundamental stress field for the anti-plane shear problem of a semi-infinite body can be obtained from that of an infinite body subjected to the same load by introducing image point forces to an infinite body and by removing the stress along a straight boundary as illustrated in Figures 1.19 and 1.20.

1.6 Fundamental stress field of the strip problem

Tension [14]. The fundamental stress field for the tension of a strip (Figure 1.21) is composed of two-parts. The first is the stress field caused by point forces acting periodically in an infinite plate as shown in Figure 1.22. This will be denoted by suffix 1. The second is the stress field produced by removing the stresses on the straight boundaries of a strip obtained from the stress field in the first part. The latter will be denoted by suffix 2.

For a point force X or Y acting at a point (ξ, η), the stresses appearing at an arbitrary point (x, y) in a strip may be written as

$$\sigma_x^X = \sigma_x^{X1} + \sigma_x^{X2} \tag{1.65a}$$

$$\sigma_y^X = \sigma_y^{X1} + \sigma_y^{X2} \tag{1.65b}$$

$$\tau_{xy}^X = \tau_{xy}^{X1} + \tau_{xy}^{X2} \tag{1.65c}$$

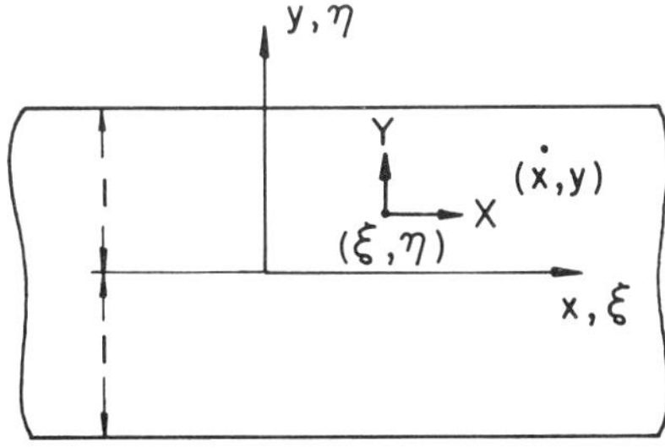

Figure 1.21. Fundamental stress field for the tension of a strip.

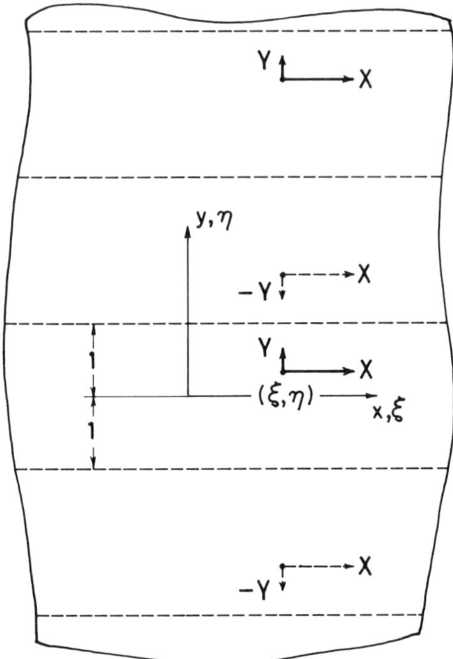

Figure 1.22. Stress field for periodically spaced point forces.

and

$$\sigma_x^Y = \sigma_x^{Y1} + \sigma_x^{Y2} \tag{1.66a}$$

$$\sigma_y^Y = \sigma_y^{Y1} + \sigma_y^{Y2} \tag{1.66b}$$

$$\tau_{xy}^Y = \tau_{xy}^{Y1} + \tau_{xy}^{Y2} \tag{1.66c}$$

The stresses σ_x^{X1}, σ_x^{Y1}, etc. can be obtained from equations (1.45) and (1.46), as

$$\sigma_x^{X1} = \{-(u_1 + u_1') - 2(u_2 + u_2')\}X \tag{1.67a}$$

$$\sigma_y^{X1} = \{-(u_1 + u_1') + 2(u_2 + u_2')\}X \tag{1.67b}$$

$$\tau_{xy}^{X1} = \{-(v_1 + v_1') - 2(v_2 + v_2')\}X \tag{1.67c}$$

and

$$\sigma_x^{Y1} = \{(v_1 - v_1') - 2(v_2 - v_2')\}Y \tag{1.68a}$$

$$\sigma_y^{Y1} = \{-3(v_1 - v_1') + 2(v_2 - v_2')\}Y \tag{1.68b}$$

$$\tau_{xy}^{Y1} = \{-3(u_1 - u_1') + 2(u_2 - u_2')\}Y \tag{1.68c}$$

The quantities u_1, u_2, etc. in equations (1.67) and (1.68) stand for

$$u_1 = \frac{\sinh k\bar{x}}{16(\cosh k\bar{x} - \cos k\bar{y})} \tag{1.69a}$$

$$u_2 = \frac{(\cosh k\bar{x} - \cos k\bar{y}) \sinh k\bar{x} - k\bar{x}(1 - \cosh k\bar{x} \cos k\bar{y})}{32(\cosh k\bar{x} - \cos k\bar{y})^2} \tag{1.69b}$$

and

$$v_1 = \frac{\sin k\bar{y}}{16(\cosh k\bar{x} - \cos k\bar{y})} \tag{1.69c}$$

$$v_2 = \frac{k\bar{x} \sinh k\bar{x} \sin k\bar{y}}{32(\cosh k\bar{x} - \cos k\bar{y})^2} \tag{1.69d}$$

while u_1', v_1', etc. are given by

$$u_1' = u_1|_{\bar{y}=\bar{y}'}, \qquad u_2' = u_2|_{\bar{y}=\bar{y}'} \tag{1.69c}$$

$$v_1' = v_1|_{\bar{y}=\bar{y}'}, \qquad v_2' = v_2|_{\bar{y}=\bar{y}'} \tag{1.69f}$$

In equations (1.69), it is understood that

$$\bar{x} = x - \xi$$

$$\bar{y} = y - \eta, \qquad \bar{y}' = y + \eta - 2$$

and

$$k = \frac{\pi}{2}$$

The stresses on the boundaries $y = \pm 1$ given by equations (1.67) and

(1.68) can be expressed in terms of the Fourier integrals:

$$\sigma_y^{X1}|_{y=\pm 1} = X\int_0^\infty F_{\pm 1} \sin\beta(x-\xi)\,\mathrm{d}\beta \tag{1.70a}$$

$$\tau_{xy}^{X1}|_{y=\pm 1} = 0 \tag{1.70b}$$

and

$$\sigma_y^{Y1}|_{y=\pm 1} = Y\int_0^\infty G_{\pm 1} \cos\beta(x-\xi)\,\mathrm{d}\beta \tag{1.71a}$$

$$\tau_{xy}^{Y1}|_{y=\pm 1} = 0 \tag{1.71b}$$

In equations (1.70) and (1.71), $F_{\pm 1}$ are

$$F_1 = \frac{1}{2\pi\sinh^2 2\beta}[\{\sinh 2\beta - 2\beta\cosh 2\beta\}\cosh(1+\eta)\beta + \{\sinh 2\beta\}(1+\eta)\beta\sinh(1+\eta)\beta] \tag{1.72a}$$

$$F_{-1} = \frac{1}{2\pi\sinh^2 2\beta}[\{\sinh 2\beta - 2\beta\cosh 2\beta\}\cosh(1-\eta)\beta + \{\sinh 2\beta\}(1-\eta)\beta\sinh(1-\eta)\beta] \tag{1.72b}$$

and $G_{\pm 1}$ take the forms

$$G_1 = \frac{1}{2\pi\sinh^2 2\beta}[-\{2\beta\cosh 2\beta + 2\sinh 2\beta\}\sinh(1+\eta)\beta + \{\sinh 2\beta\}(1+\eta)\beta\cosh(1+\eta)\beta] \tag{1.72c}$$

$$G_{-1} = \frac{1}{2\pi\sinh^2 2\beta}[\{2\beta\cosh 2\beta + 2\sinh 2\beta\}\sinh(1-\eta)\beta - \{\sinh 2\beta\}(1-\eta)\beta\cosh(1-\eta)\beta] \tag{1.72d}$$

The stresses σ_x^{X2}, σ_x^{Y2}, etc. are obtained by removing the stresses on the straight boundaries $y = \pm 1$ which is given by equations (1.70) and (1.71) and they may be written as

$$\sigma_x^{X2} = X\int_0^\infty \left[\frac{(F_1 + F_{-1})}{\Sigma_+}S_{x+} + \frac{(F_1 - F_{-1})}{\Sigma_-}S_{x-}\right]\sin\beta(x-\xi)\,\mathrm{d}\beta \tag{1.73a}$$

$$\sigma_y^{X2} = -X\int_0^\infty \left[\frac{(F_1+F_{-1})}{\Sigma_+}S_{y+} + \frac{(F_1-F_{-1})}{\Sigma_-}S_{y-}\right]\sin\beta(x-\xi)\,\mathrm{d}\beta \tag{1.73b}$$

$$\tau_{xy}^{X2} = -X\int_0^\infty \left[\frac{(F_1+F_{-1})}{\Sigma_+}S_{xy+} + \frac{(F_1-F_{-1})}{\Sigma_-}S_{xy-}\right]\cos\beta(x-\xi)\,\mathrm{d}\beta \tag{1.73c}$$

and

$$\sigma_x^{Y2} = Y\int_0^\infty \left[\frac{(G_1+G_{-1})}{\Sigma_+}S_{x+} + \frac{(G_1-G_{-1})}{\Sigma_-}S_{x-}\right]\cos\beta(x-\xi)\,\mathrm{d}\beta \tag{1.74a}$$

$$\sigma_y^{Y2} = -Y\int_0^\infty \left[\frac{(G_1+G_{-1})}{\Sigma_+}S_{y+} + \frac{(G_1-G_{-1})}{\Sigma_-}S_{y-}\right]\cos\beta(x-\xi)\,\mathrm{d}\beta \tag{1.74b}$$

$$\tau_{xy}^{Y2} = Y\int_0^\infty \left[\frac{(G_1+G_{-1})}{\Sigma_+}S_{xy+} + \frac{(G_1-G_{-1})}{\Sigma_-}S_{xy-}\right]\sin\beta(x-\xi)\,\mathrm{d}\beta \tag{1.74c}$$

where

$$\Sigma_+ = \sinh 2\beta + 2\beta, \qquad \Sigma_- = \sinh 2\beta - 2\beta$$

The quantities S_{x+}, S_{y+}, etc. are defined as

$$S_{x+} = \{\beta\cosh\beta - \sinh\beta\}\cosh\beta y - \{\sinh\beta\}\beta y\sinh\beta y \tag{1.75a}$$

$$S_{y+} = \{\beta\cosh\beta + \sinh\beta\}\cosh\beta y - \{\sinh\beta\}\beta y\sinh\beta y \tag{1.75b}$$

$$S_{xy+} = \{\beta\cosh\beta\}\sinh\beta y - \{\sinh\beta\}\beta y\cosh\beta y \tag{1.75c}$$

$$S_{x-} = \{\beta\sinh\beta - \cosh\beta\}\sinh\beta y - \{\cosh\beta\}\beta y\cosh\beta y \tag{1.75d}$$

$$S_{y-} = \{\beta\sinh\beta + \cosh\beta\}\sinh\beta y - \{\cosh\beta\}\beta y\cosh\beta y \tag{1.75e}$$

$$S_{xy-} = \{\beta\sinh\beta\}\cosh\beta y - \{\cosh\beta\}\beta y\sinh\beta y \tag{1.75f}$$

The influence coefficients may be evaluated by means of equations (1.65) to (1.75) and by calculating certain integrals which may be taken as the difference between the integral of a strip and that of a semi-infinite plate. The stress concentration factor for a strip with two edge notches is displayed in Figure A1.15.

Anti-plane shear. In the same way, the anti-plane shear point force problem of a body with two parallel edges free from stresses may also be formulated. Refer to Figure 1.23.

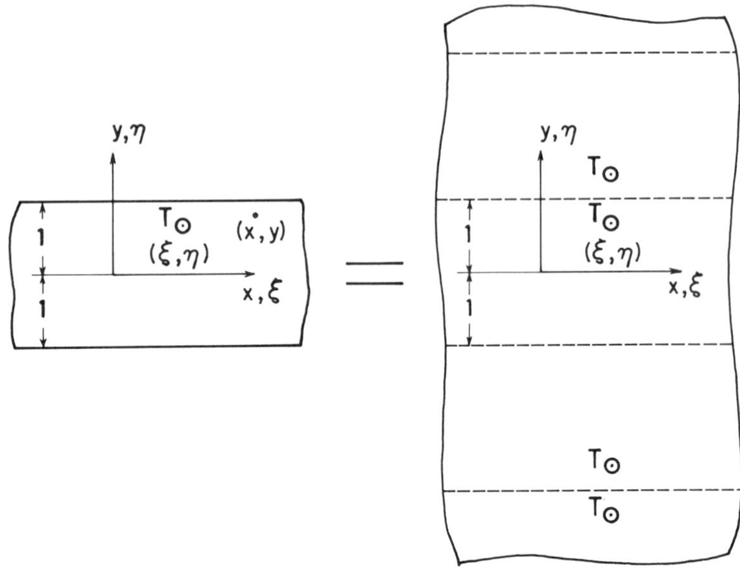

Figure 1.23. Fundamental stress field for the anti-plane shear of a body with two parallel free boundaries.

1.7 Fundamental stress field of the round bar problem

Torsion [15]. Figure 1.24 shows the fundamental stress field in problems of torsion of a round bar. The stress field is composed of two parts. The first consists of a stress field for a ring force in an infinite body. The second stress field produces a stress state that cancels the stresses on the cylindrical surface of the round bar owing to the stress field in the first part. In the following, those quantities pertaining to the first and second parts will be denoted by suffix 1 and 2, respectively.

Following the notations in Figure 1.24, the stresses at (r, z) due to the ring force with magnitude T located at (ρ, ζ) can be written as

$$\tau_{r\theta}^{T} = \tau_{r\theta}^{T1} + \tau_{r\theta}^{T2} \tag{1.76a}$$

$$\tau_{\theta z}^{T} = \tau_{\theta z}^{T1} + \tau_{\theta z}^{T2} \tag{1.76b}$$

All other stress components are equal to zero. The stresses $\tau_{r\theta}^{T1}$ and $\tau_{\theta z}^{T1}$ can be obtained from the stress field for a point force acting in an infinite body. The results can be expressed in terms of the complete elliptic

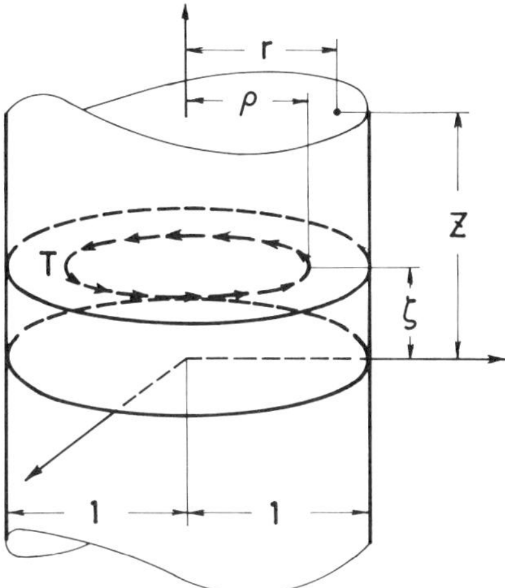

Figure 1.24. Fundamental stress field for the torsion of a round bar.

integrals and they are

$$\tau_{r\theta}^{T1'} = \frac{\rho}{2\pi(2\rho r)^{\frac{3}{2}}}(-\rho I_0 - rI_1 + 2\rho I_2)T \tag{1.77a}$$

$$\tau_{\theta z}^{T1'} = -\frac{\rho(z-\zeta)}{2\pi(2\rho r)^{\frac{3}{2}}} I_1 T \tag{1.77b}$$

in which I_0, I_1, and I_2 are

$$I_0 = -\frac{1}{1-e^2}K_1 \tag{1.78a}$$

$$I_1 = -\frac{e}{1-e^2}K_1 - K_2 \tag{1.78b}$$

$$I_2 = \frac{1-2e^2}{1-e^2}K_1 - 2eK_2 \tag{1.78c}$$

with K_1 and K_2 being given by

$$K_1 = \int_0^{\pi} (e - \cos\varphi)^{\frac{1}{2}}\, d\varphi = \frac{2\sqrt{2}}{k}E(k) \tag{1.78d}$$

$$K_2 = \int_0^{\pi} (e - \cos\varphi)^{-\frac{1}{2}}\, d\varphi = \sqrt{2}kF(k) \tag{1.78e}$$

The complete elliptic integrals

$$E(k)=\int_0^{\pi/2}\sqrt{1-k^2\sin^2\theta}\,\mathrm{d}\theta \tag{1.78f}$$

$$F(k)=\int_0^{\pi/2}\frac{\mathrm{d}\theta}{\sqrt{1-k^2\sin^2\theta}} \tag{1.78g}$$

have the argument

$$k=\sqrt{\frac{2}{e+1}},\qquad e=1+\frac{(r-\rho)^2+(z-\zeta)^2}{2r\rho}$$

The stresses $\tau^S_{r\theta}(\zeta_0, r, z)$ and $\tau^S_{\theta z}(\zeta_0, r, z)$, which are due to a ring force of magnitude S acting tangentially at (h, ζ_0) on the surface of the round bar, are ($h = 1$ in Figure 1.24)

$$\tau^S_{r\theta}(\zeta_0, r, z)=\frac{S}{\pi}\int_0^\infty\left\{\frac{I_2(\beta r)}{I_2(\beta h)}\right\}\cos\beta(z-\zeta_0)\,\mathrm{d}\beta \tag{1.79a}$$

$$\tau^S_{\theta z}(\zeta_0, r, z)=-\frac{S}{\pi}\int_0^\infty\left\{\frac{I_1(\beta r)}{I_2(\beta h)}\right\}\sin\beta(z-\zeta_0)\,\mathrm{d}\beta \tag{1.79b}$$

where $I_1(x)$ and $I_2(x)$ are modified Bessel functions. The integrals in equations (1.79) can be more conveniently expressed in the forms

$$\begin{aligned}\tau^S_{r\theta}(\zeta_0, r, z)=&\frac{S}{\pi}\int_0^\infty\left\{\frac{I_2(\beta r)}{I_2(\beta h)}-\sqrt{\frac{h}{r}}\,\mathrm{e}^{-(h-r)\beta}\right\}\cos\beta(z-\zeta_0)\,\mathrm{d}\beta\\&-\frac{S}{\pi}\sqrt{\frac{h}{r}}\frac{(r-h)}{(r-h)^2+(z-\zeta_0)^2}\end{aligned} \tag{1.80a}$$

$$\begin{aligned}\tau^S_{\theta z}(\zeta_0, r, z)=&-\frac{S}{\pi}\int_0^\infty\left\{\frac{I_1(\beta r)}{I_2(\beta h)}-\sqrt{\frac{h}{r}}\,\mathrm{e}^{-(h-r)\beta}\right\}\sin\beta(z-\zeta_0)\,\mathrm{d}\beta\\&-\frac{S}{\pi}\sqrt{\frac{h}{r}}\frac{(z-\zeta_0)}{(r-h)^2+(z-\zeta_0)^2}\end{aligned} \tag{1.80b}$$

With the aid of equations (1.77) and (1.80), $\tau^{T2}_{r\theta}$ and $\tau^{T2}_{\theta z}$ are obtained:

$$\tau^{T2}_{r\theta}=-\int_{-\infty}^{\infty}\tau^S_{r\theta}(\zeta_0, r, z)|_{S=1}\,\tau^{T1}_{r\theta}|_{\substack{r=h\\z=\zeta_0}}\,\mathrm{d}\zeta_0 \tag{1.81a}$$

$$\tau^{T2}_{\theta z}=-\int_{-\infty}^{\infty}\tau^S_{\theta z}(\zeta_0, r, z)|_{S=1}\,\tau^{T1}_{r\theta}|_{\substack{r=h\\z=\zeta_0}}\,\mathrm{d}\zeta_0 \tag{1.81b}$$

The results for the torsion of a bar with a circumferential notch, obtained by using equations (1.76) to (1.81) are displayed graphically in Figure A1.16.

Tension. The fundamental stress field of this case is the stress field for a ring force with tangential and normal components, as shown in Figure 1.25. This stress field composes two parts. The first is the stress field for a ring force in an infinite body. The second deals with the stress field needed to remove the stresses on the cylindrical surface of the round bar corresponding to those in the first part. Again suffix 1 and 2 will be used to distinguish quantities referred to the first part from those of the second part.

The notations in Figure 1.25 will be used. The stresses at (r, z) for tangential (P) or normal (Q) forces acting on a ring located at (ρ, ζ) are given by

$$\sigma_r^P = \sigma_r^{P1} + \sigma_r^{P2} \tag{1.82a}$$

$$\sigma_\theta^P = \sigma_\theta^{P1} + \sigma_\theta^{P2} \tag{1.82b}$$

$$\sigma_z^P = \sigma_z^{P1} + \sigma_z^{P2} \tag{1.82c}$$

$$\tau_{rz}^P = \tau_{rz}^{P1} + \tau_{rz}^{P2} \tag{1.82d}$$

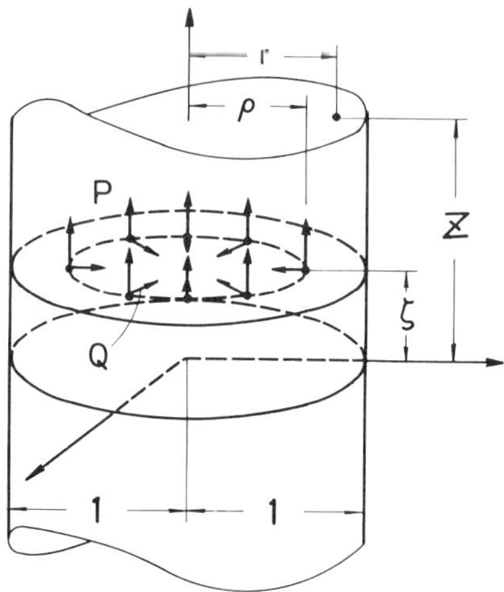

Figure 1.25. Fundamental stress field for the tension of a round bar.

and

$$\sigma_r^Q = \sigma_r^{Q1} + \sigma_r^{Q2} \tag{1.83a}$$

$$\sigma_\theta^Q = \sigma_\theta^{Q1} + \sigma_\theta^{Q2} \tag{1.83b}$$

$$\sigma_z^Q = \sigma_z^{Q1} + \sigma_z^{Q2} \tag{1.83c}$$

$$\tau_{rz}^Q = \tau_{rz}^{Q1} + \tau_{rz}^{Q2} \tag{1.83d}$$

Moreover, the stresses σ_r^{P1}, σ_r^{Q1}, etc. can be obtained from the stress field for a point force acting in an infinite body. They can be expressed in terms of the complete elliptic integrals and are given by

$$\sigma_r^{P1}(\rho, \zeta, r, z) = \frac{\bar{z}\rho}{4\pi(1-\nu)r_m^3}\left[(1-2\nu)I_0 - \frac{3}{r_m^2}(r^2J_0 - 2r\rho J_1 + \rho^2 J_2)\right] \tag{1.84a}$$

$$\sigma_\theta^{P1}(\rho, \zeta, r, z) = \frac{\bar{z}\rho}{4\pi(1-\nu)r_m^3}\left[(1-2\nu)I_0 - \frac{3}{r_m^2}(\rho^2J_0 - \rho^2 J_1)\right] \tag{1.84b}$$

$$\sigma_z^{P1}(\rho, \zeta, r, z) = \frac{\bar{z}\rho}{4\pi(1-\nu)r_m^3}\left[-(1-2\nu)I_0 - \frac{3}{r_m^2}(\bar{z}^2J_0)\right] \tag{1.84c}$$

$$\tau_{rz}^{P1}(\rho, \zeta, r, z) = \frac{\rho}{4\pi(1-\nu)r_m^3}\left[-(1-2\nu)(rI_0 - \rho I_1) - \frac{3\bar{z}^2}{r_m^2}(rJ_0 - \rho J_1)\right] \tag{1.84d}$$

and

$$\sigma_r^{Q1}(\rho, \zeta, r, z) = -\frac{\rho}{4\pi(1-\nu)r_m^3}\Big[(1-2\nu)(-\rho I_0 - rI_1 + 2\rho I_2) + \frac{3}{r_m^2}\{r^2\rho J_0 - r(r^2+2\rho^2)J_1 + \rho(2r^2+\rho^2)J_2 - r\rho^2 J_3\}\Big] \tag{1.85a}$$

$$\sigma_\theta^{Q1}(\rho, \zeta, r, z) = -\frac{\rho}{4\pi(1-\nu)r_m^3}\Big[(1-2\nu)(\rho I_0 + rI_1 - 2\rho I_2) + \frac{3\rho^2}{r_m^2}(\rho J_0 - rJ_1 - \rho J_2 + rJ_3)\Big] \tag{1.85b}$$

$$\sigma_z^{Q1}(\rho, \zeta, r, z) = -\frac{\rho}{4\pi(1-\nu)r_m^3}\Big[(1-2\nu)(-\rho I_0 + rI_1) + \frac{3\bar{z}^2}{r_m^2}(\rho J_0 - rJ_1)\Big] \tag{1.85c}$$

$$\tau_{rz}^{Q1}(\rho, \zeta, r, z) = -\frac{\rho}{4\pi(1-\nu)r_m^3}\left[-(1-2\nu)\bar{z}I_1 + \frac{3\bar{z}}{r_m^2}\{r\rho J_0 - (r^2+\rho^2)J_1 + r\rho J_2\}\right] \tag{1.85d}$$

In the above equations,

$$r_m = \sqrt{2r\rho}, \qquad \bar{z} = z - \zeta$$

and

$$I_n = \int_0^\pi \frac{\cos^n \varphi}{(e-\cos\varphi)^{\frac{3}{2}}}\,\mathrm{d}\varphi \tag{1.86a}$$

$$J_n = \int_0^\pi \frac{\cos^n \varphi}{(e-\cos\varphi)^{\frac{5}{2}}}\,\mathrm{d}\varphi \tag{1.86b}$$

where

$$e = 1 + \frac{(r-\rho)^2 + (z-\zeta)^2}{2r\rho}$$

For $n = 0, 1$ and 2, the results are

$$I_0 = -\frac{1}{1-e^2}K_1 \tag{1.86c}$$

$$I_1 = -\frac{1}{1-e^2}K_1 - K_2 \tag{1.86d}$$

$$I_2 = \frac{1-2e^2}{1-e^2}K_1 - 2eK_2 \tag{1.86e}$$

and

$$J_0 = \frac{4e}{3(1-e^2)^2}K_1 + \frac{1}{3(1-e^2)}K_2 \tag{1.86f}$$

$$J_1 = \frac{(3+e^2)}{3(1-e^2)^2}K_1 + \frac{e}{3(1-e^2)}K_2 \tag{1.86g}$$

$$J_2 = \frac{2e(3-e^2)}{3(1-e^2)^2}K_1 + \frac{(3-2e^2)}{3(1-e^2)}K_2 \tag{1.86h}$$

$$J_3 = \frac{(-3+15e^2-8e^4)}{3(1-e^2)^2}K_1 + \frac{e(9-8e^2)}{3(1-e^2)}K_2 \tag{1.86i}$$

K_1 and K_2 are given by equation (1.78). The stresses σ_r^{P2}, or σ_r^{Q2}, etc. are requested to cancel the stresses $\sigma_r^{P1}(\rho, \zeta, 1, z)$, and $\tau_{rz}^{P1}(\rho, \zeta, 1, z)$ or $\sigma_r^{Q1}(\rho, \zeta, 1, z)$ and $\tau_{rz}^{Q1}(\rho, \zeta, 1, z)$ shown by equations (1.84) or (1.85). Now, writing

$$\sigma_{r0}^{P}(z) = \sigma_r^{P1}(\rho, \zeta, 1, z) \tag{1.87a}$$

$$\tau_{rz0}^{P}(z) = \tau_{rz}^{P1}(\rho, \zeta, 1, z) \tag{1.87b}$$

and

$$\sigma_{r0}^{Q}(z) = \sigma_r^{Q1}(\rho, \zeta, 1, z) \tag{1.88a}$$

$$\tau_{rz0}^{Q}(z) = \tau_{rz}^{Q1}(\rho, \zeta, 1, z) \tag{1.88b}$$

and applying the Fourier integrals, the stresses at the prospective cylindrical surface of a round bar produced by a ring force in an infinite body are

$$\sigma_{r0}^{P}(z) = \int_0^\infty f^P(\beta) \sin \beta(z - \zeta)\, d\beta \tag{1.89a}$$

$$\tau_{rz0}^{P}(z) = \int_0^\infty g^P(\beta) \cos \beta(z - \zeta)\, d\beta \tag{1.89b}$$

in which

$$f^P(\beta) = \frac{2}{\pi} \int_0^\infty \sigma_{r0}^{P}(u) \sin \beta u\, du \tag{1.90a}$$

$$g^P(\beta) = \frac{2}{\pi} \int_0^\infty \tau_{rz0}^{P}(u) \cos \beta u\, du \tag{1.90b}$$

and

$$\sigma_{r0}^{Q}(z) = \int_0^\infty f^Q(\beta) \cos \beta(z - \zeta)\, d\beta \tag{1.91a}$$

$$\tau_{rz0}^{Q}(z) = \int_0^\infty g^Q(\beta) \sin \beta(z - \zeta)\, d\beta \tag{1.91b}$$

with

$$f^Q(\beta) = \frac{2}{\pi} \int_0^\infty \sigma_{r0}^{Q}(u) \cos \beta u\, du \tag{1.92a}$$

$$g^Q(\beta) = \frac{2}{\pi}\int_0^\infty \tau_{rz0}^Q(u) \sin \beta u \, \mathrm{d}u \tag{1.926}$$

Therefore, the stresses σ_r^{P2}, σ_r^{Q2}, etc. are computed as

$$\sigma_r^{P2} = -\int_0^\infty \{F_r f^P(\beta) + G_r g^P(\beta)\} \sin \beta(z - \zeta) \, \mathrm{d}\beta \tag{1.93a}$$

$$\sigma_\theta^{P2} = -\int_0^\infty \{F_\theta f^P(\beta) + G_\theta g^P(\beta)\} \sin \beta(z - \zeta) \, \mathrm{d}\beta \tag{1.93b}$$

$$\sigma_z^{P2} = -\int_0^\infty \{F_z f^P(\beta) + G_z g^P(\beta)\} \sin \beta(z - \zeta) \, \mathrm{d}\beta \tag{1.93c}$$

$$\tau_{rz}^{P2} = -\int_0^\infty \{F_{rz} f^P(\beta) + G_{rz} g^P(\beta)\} \cos \beta(z - \zeta) \, \mathrm{d}\beta \tag{1.93d}$$

and

$$\sigma_r^{Q2} = -\int_0^\infty \{F_r f^Q(\beta) + G_r g^Q(\beta)\} \cos \beta(z - \zeta) \, \mathrm{d}\beta \tag{1.94a}$$

$$\sigma_\theta^{Q2} = -\int_0^\infty \{F_\theta f^Q(\beta) + G_\theta g^Q(\beta)\} \cos \beta(z - \zeta) \, \mathrm{d}\beta \tag{1.94b}$$

$$\sigma_z^{Q2} = -\int_0^\infty \{F_z f^Q(\beta) + G_z g^Q(\beta)\} \cos \beta(z - \zeta) \, \mathrm{d}\beta \tag{1.94c}$$

$$\tau_{rz}^{Q2} = -\int_0^\infty \{F_{rz} f^Q(\beta) + G_{rz} g^Q(\beta)\} \sin \beta(z - \zeta) \, \mathrm{d}\beta \tag{1.94d}$$

The following contractions have been made in equations (1.93) and (1.94):

$$F_r = H(\beta)[\psi\psi' - (\psi - \psi') - 2(1 - \nu) - \beta'^2] \tag{1.95a}$$

$$G_r = \frac{H(\beta)}{\beta}[-(\psi\beta'^2 - \psi'\beta^2) - (1 - 2\nu)(\psi - \psi') - (\beta^2 - \beta'^2)] \tag{1.95b}$$

$$F_\theta = H(\beta)[(\psi - \psi') + 2\nu\psi' + 2(1 - \nu)] \tag{1.95c}$$

$$G_\theta = H(\beta)[(1 - 2\nu)(\psi + \psi') - (1 - 2\nu)\psi\psi' + \beta^2] \tag{1.95d}$$

$$F_z = H(\beta)[-\psi\psi' + 2\psi' + \beta'^2] \tag{1.95e}$$

$$G_z = \frac{H(\beta)}{\beta}[(\psi\beta'^2 - \psi'\beta^2) + 3\psi\psi' - 2(2-\nu)\psi' - \beta'^2] \quad (1.95f)$$

$$F_{rz} = H(\beta)[-\beta'(\psi - \psi')] \quad (1.95g)$$

$$G_{rz} = \frac{\beta' H(\beta)}{\beta}[\psi\psi' + (\psi - \psi') - 2(1-\nu) - \beta^2] \quad (1.95h)$$

The quantities ψ, ψ', etc. stand for

$$\begin{aligned} \psi &= \frac{\beta I_0(\beta)}{I_1(\beta)}, \qquad \psi' = \frac{r\beta I_0(r\beta)}{I_1(r\beta)} \\ \chi &= \frac{I_1(\beta)}{\beta}, \qquad \chi' = \frac{I_1(r\beta)}{r\beta} \end{aligned} \quad (1.96)$$

while β' and $H(\beta)$ are

$$\beta' = r\beta, \qquad H(\beta) = \frac{\chi'}{\chi}\frac{1}{\psi^2 - 2(1-\nu) - \beta^2} \quad (1.97)$$

1.8 Appendix: Stress concentration factor and stress intensity factor solutions for notches and cracks

In this appendix, the stress concentration factor denoted by K_t and stress intensity factor denoted by k_1 for a variety of notch and crack geometries are given. The results are tabulated and presented graphically.

The values of k_1 not relating notches are not included here. Some of them are found in reference [25].

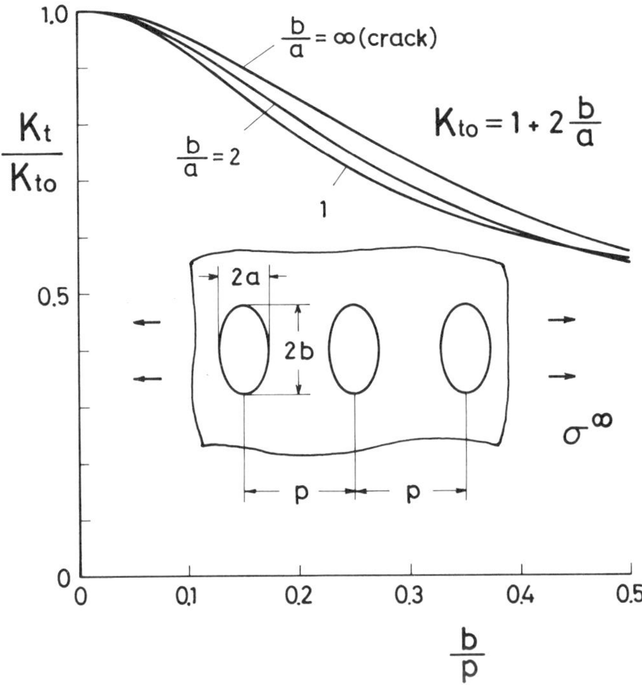

Figure A1.1. Tension of an infinite plate with an infinite row of longitudinal elliptic holes.

TABLE A1.1

Values of K_t in Figure A1.1

$\frac{b}{a}$ \ $\frac{b}{p}$	0	0.1	0.2	0.3	0.4	0.5
1	3.000	2.768	2.326	1.995	1.802	1.676
1	(3.000)	(2.768)	(2.326)	(1.996)	–	–
2	5.000	4.687	4.033	3.439	3.028	2.756

( ): Schultz [16]

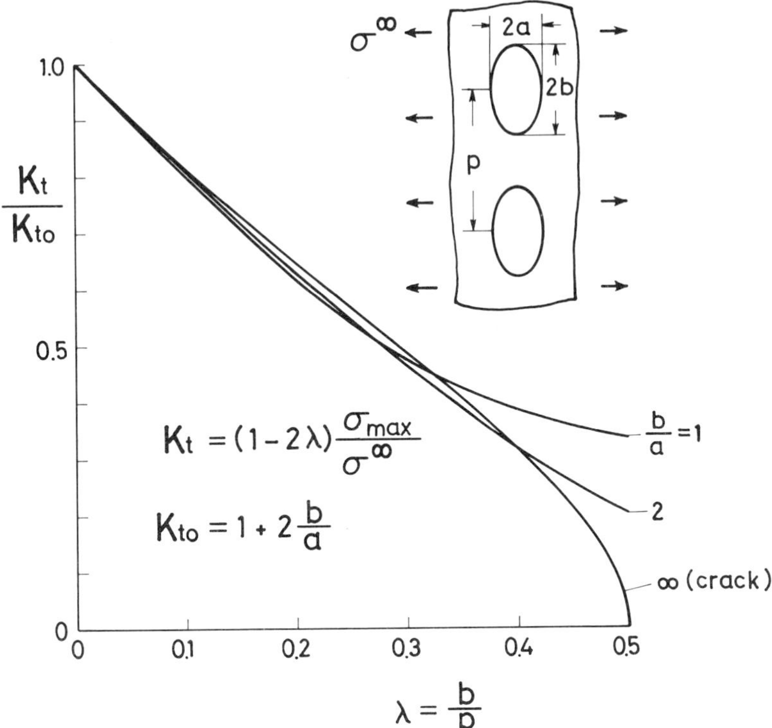

Figure A1.2. Tension of an infinite plate with an infinite row of transverse elliptic holes.

TABLE A1.2

Values of K_t in Figure A1.2

b/a \ b/p	0	0.05	0.1	0.2	0.3	0.4
0.5	2.000	1.791	1.582	1.257	1.102	1.040
1	3.000	2.700	2.405	1.858	1.419	1.151
1	(3.000)	(2.700)	(2.405)	(1.858)	–	–
2	5.000	4.511	4.042	3.151	2.321	1.572
5	11.00	9.936	8.931	7.037	5.216	3.344

( ): Isida [17]

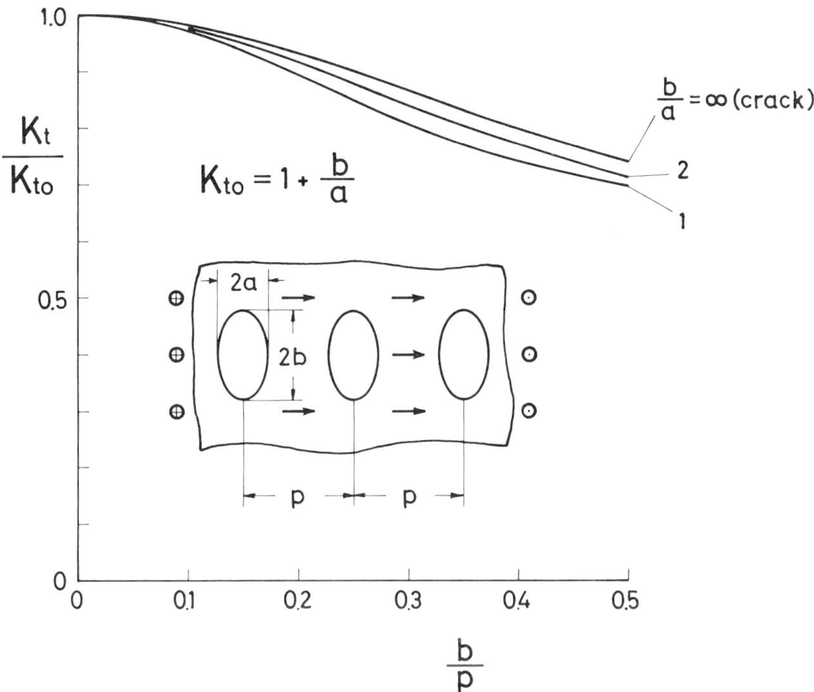

Figure A1.3. Anti-plane shear of an infinite body with an infinite row of longitudinal elliptic holes.

TABLE A1.3

Values of K_t in Figure A1.3

b/a \ b/p	0	0.05	0.1	0.2	0.3	0.4	0.5
1	2.000	1.984	1.938	1.785	1.614	1.481	1.393
2	3.000	2.982	2.929	2.748	2.523	2.310	2.132
5	6.000	5.971	5.886	5.586	5.195	4.795	4.432

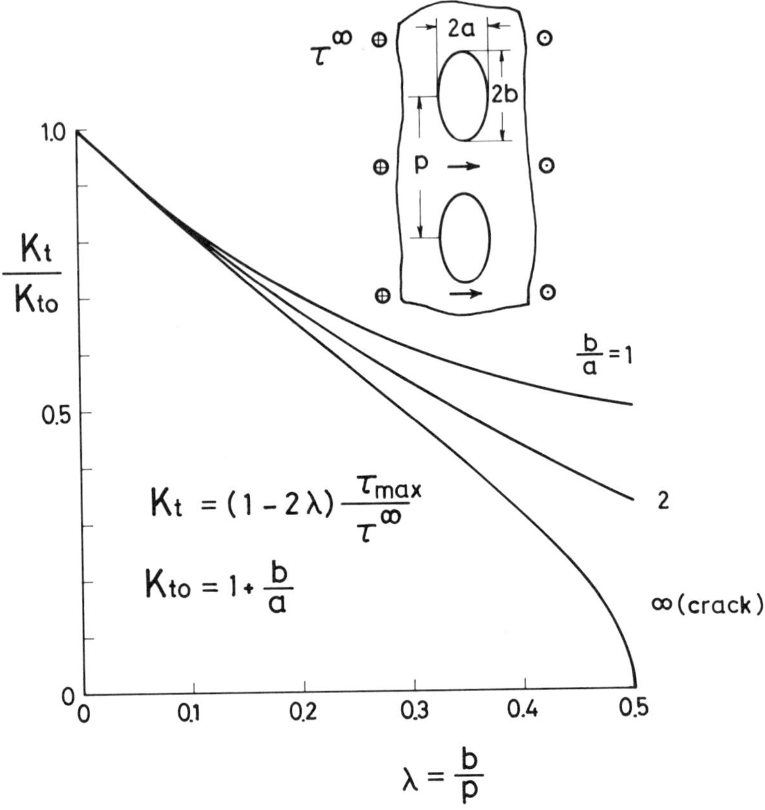

Figure A1.4. Anti-plane shear of an infinite body with an infinite row of transverse elliptic holes.

TABLE A1.4

Values of K_t in Figure A1.4

b/a \ b/p	0	0.05	0.1	0.2	0.3	0.4
0.5	1.500	1.367	1.263	1.126	1.056	1.021
1	2.000	1.815	1.656	1.397	1.209	1.082
2	3.000	2.717	2.462	2.014	1.629	1.293
5	6.000	5.427	4.898	3.933	3.038	2.145

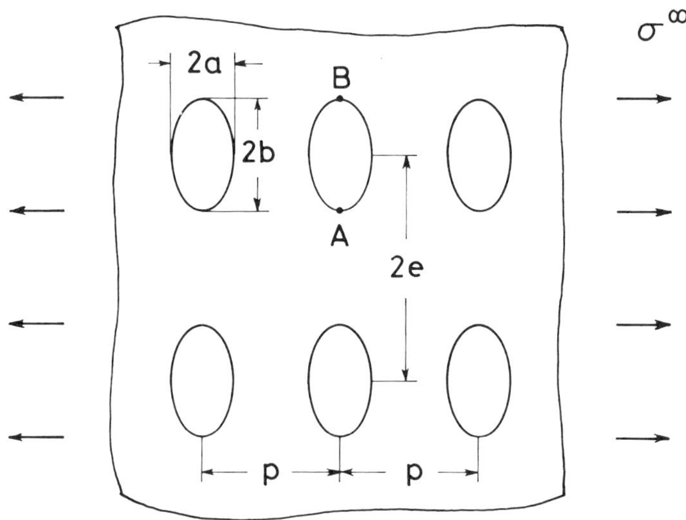

Figure A1.5. Tension of an infinite plate with two infinite rows of elliptic holes.

TABLE A1.5

Values of K_t in Figure A1.5

$\frac{b}{e}$	$\frac{b}{a}$ \ $\frac{b}{p}$	0	0	0.1	0.2	0.3	0.4	0.5
0.5	1	3.020 3.066	(3.020) (3.066)	2.641 2.733	2.170 2.319	1.917 1.995	1.764 1.802	– –
	2	5.132 5.111	– –	4.599 4.639	3.863 4.005	3.343 3.436	2.990 3.028	2.742 2.756

Upper: K_{tA}, Lower: K_{tB}, ( ): Ling [18]

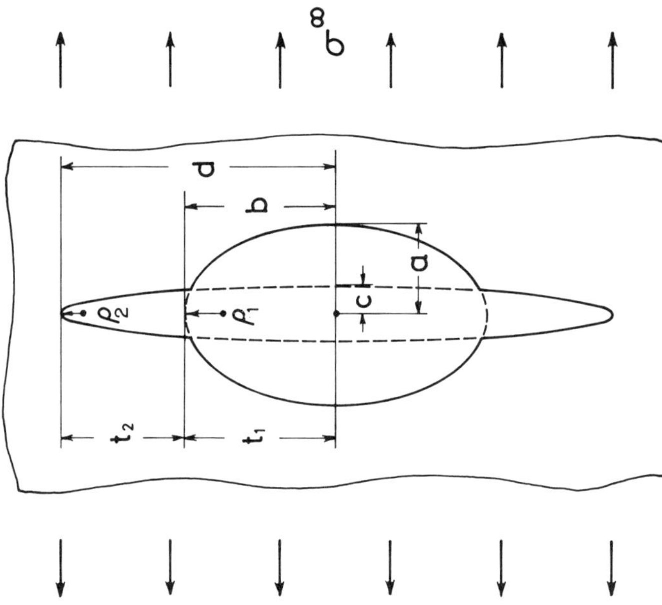

Figure A1.6. Tension of an infinite plate having an elliptic hole with two notches at its apexes.

TABLE A1.6

Values of K_t in Figure A1.6

	$\frac{\rho_2}{t_1}$	0.1				0.05				0.02			
	$\frac{t_2}{t_1}$	0.2	0.4	0.6	1.0	0.2	0.4	0.6	1.0	0.2	0.4	0.6	1.0
$\frac{\rho_1}{t_1}$	$\frac{b}{a}$ \ $\frac{d}{c}$	3.464	3.742	4.000	4.472	4.899	5.292	5.657	6.325	7.747	8.367	8.944	10.00
4.00	0.5	6.757	8.207	9.024	10.17	8.969	11.07	12.33	13.98	12.84	16.81	18.87	21.52
1.00	1	7.801	8.563	9.084	9.972	10.52	11.71	12.48	13.72	15.28	17.95	19.20	21.10
0.25	2	7.949	8.489	9.000	9.942	10.85	11.61	12.32	13.63	15.96	17.79	18.88	20.91

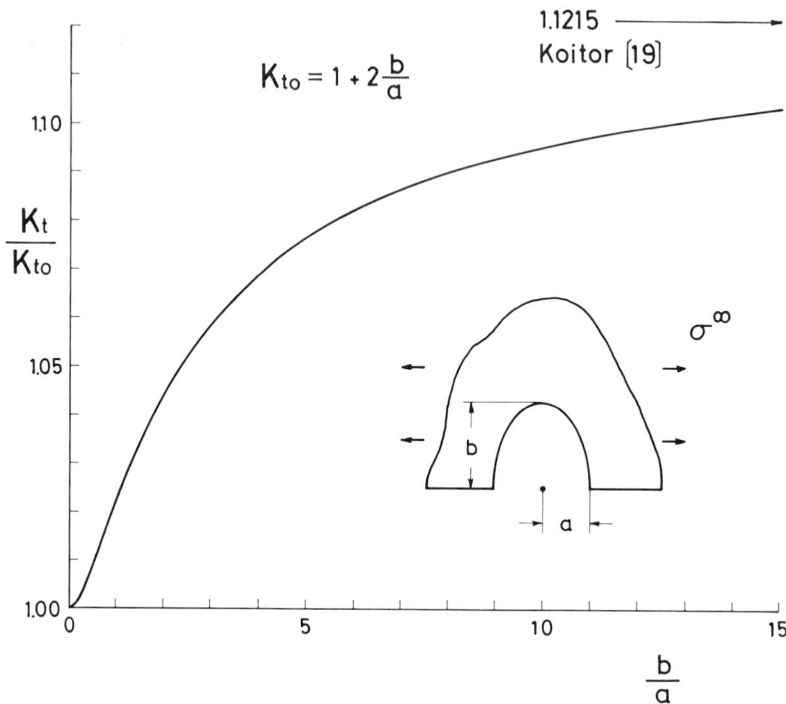

Figure A1.7. Tension of a semi-infinite plate with a semi-elliptic notch.

TABLE A1.7

Values of K_t in Figure A1.7

$\frac{b}{a}$	0.25	0.5	1	2	4	6	8	10	15	20
K_t	1.503	2.016	3.065	5.221	9.623	14.07	18.53	23.00	34.18	45.35

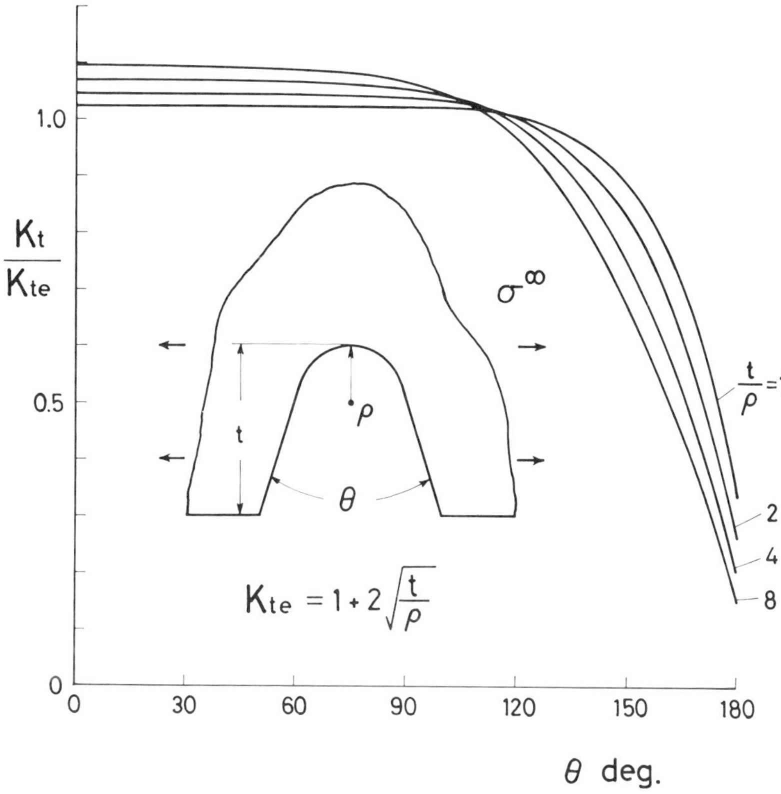

Figure A1.8. Tension of a semi-infinite plate with a V-shaped notch.

TABLE A1.8

Values of K_t in Figure A1.8

t/ρ \ deg.	0	30	60	90	120	150
1	3.065	3.065	3.065	3.062	3.016	2.654
2	3.997	3.997	3.995	3.976	3.839	3.145
4	5.344	5.339	5.331	5.274	4.941	3.725
8	7.284	7.264	7.243	7.097	6.409	4.413

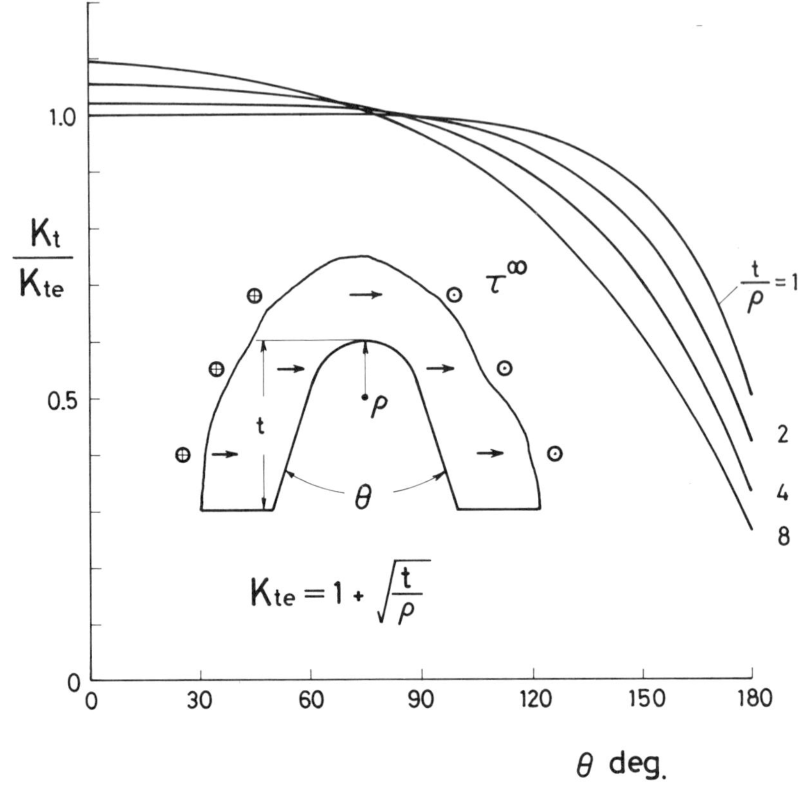

Figure A1.9. Anti-plane shear of a semi-infinite body with a V-shaped notch.

TABLE A1.9

Values of K_t in Figure A1.9

$\frac{t}{\rho}$ \ deg.	0	30	60	90	120	150
1	2.000	2.000	1.998	1.988	1.935	1.729
2	2.467	2.463	2.449	2.403	2.266	1.908
4	3.165	3.141	3.085	2.957	2.676	2.105
8	4.192	4.117	3.965	3.683	3.178	2.321

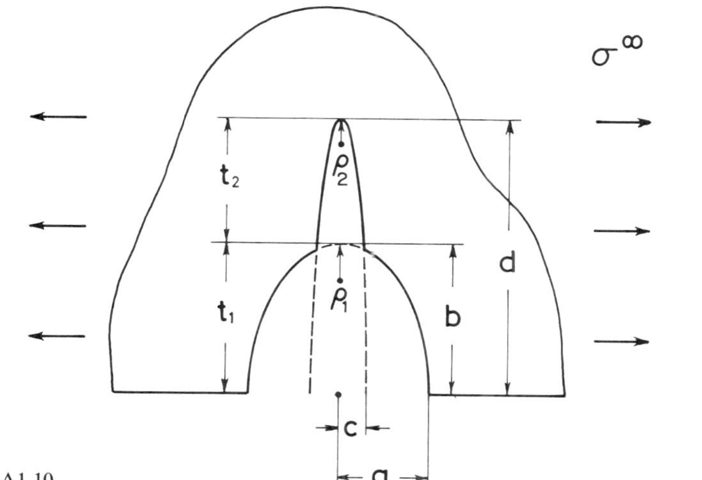

Figure A1.10. Tension of a semi-infinite plate with a double elliptic notch.

TABLE A1.10

Values of K_t in Figure A1.10

	$\frac{\rho_2}{t_1}$	0.1				0.05				0.02			
	$\frac{t_2}{t_1}$	0.2	0.4	0.6	1.0	0.2	0.4	0.6	1.0	0.2	0.4	0.6	1.0
$\frac{\rho_1}{t_1}$	$\frac{b}{a}$ \ $\frac{d}{c}$	3.464	3.742	4.000	4.472	4.899	5.292	5.657	6.325	7.747	8.367	8.944	10.00
4.00	0.5	6.837	8.346	9.204	10.48	9.080	11.25	12.58	14.40	12.99	17.07	19.25	22.19
1.00	1	8.062	8.926	9.545	10.52	10.88	12.23	13.14	14.62	15.80	18.76	20.27	22.55
0.25	2	8.425	9.042	9.618	10.67	11.54	12.44	13.26	14.75	17.03	19.16	20.45	22.81

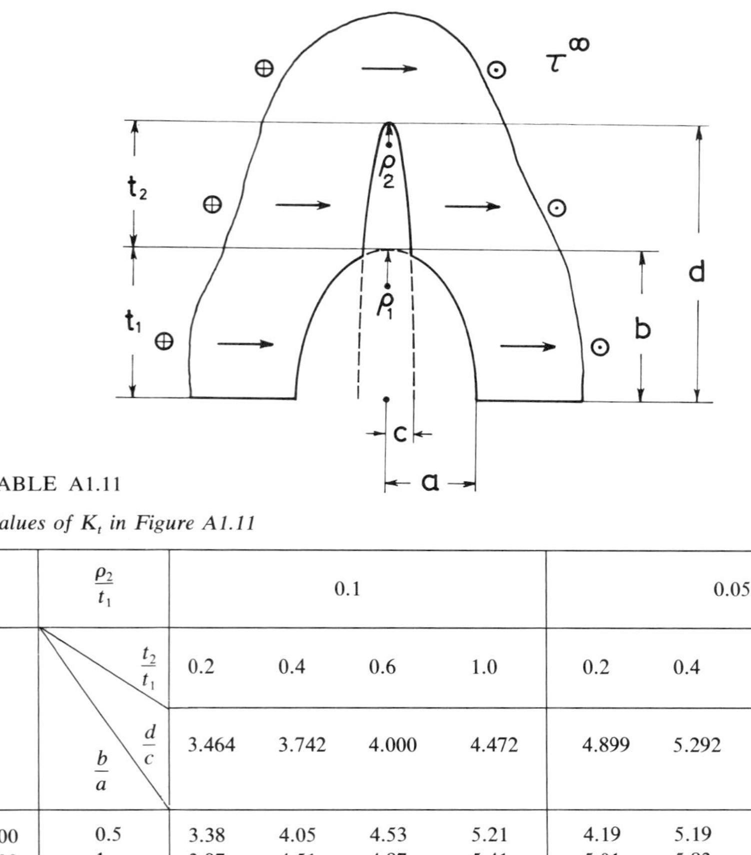

Figure A 1.11. Anti-plane shear of an semi-infinite body with a double elliptic notch.

TABLE A1.11

Values of K_t in Figure A1.11

	$\frac{\rho_2}{t_1}$	0.1				0.05				0.02			
	$\frac{t_2}{t_1}$	0.2	0.4	0.6	1.0	0.2	0.4	0.6	1.0	0.2	0.4	0.6	1.0
$\frac{\rho_1}{t_1}$	$\frac{b}{a}$ \ $\frac{d}{c}$	3.464	3.742	4.000	4.472	4.899	5.292	5.657	6.325	7.747	8.367	8.944	1.000
4.00	0.5	3.38	4.05	4.53	5.21	4.19	5.19	5.88	6.88	5.82	7.50	8.73	10.3
1.00	1	3.97	4.51	4.87	5.41	5.01	5.83	6.38	7.19	7.01	8.42	9.45	10.7
0.25	2	4.41	4.73	5.00	5.47	5.67	6.20	6.61	7.30	8.08	9.07	9.87	10.9

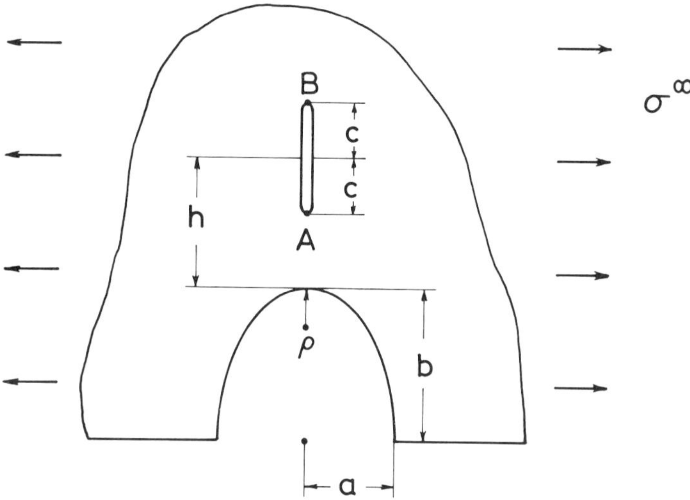

Figure A1.12. Interaction between an elliptic edge notch and a crack in a semi-infinite plate under tension.

TABLE A1.12

Values of F_I in Figure A1.12

b/a \ h/c	$c/\rho = 0.1$			$c/\rho = 0.5$		
	0.1	0.2	0.4	0.1	0.2	0.4
0.5	1.036 1.029	1.159 1.113	1.466 1.304	1.003 1.003	1.013 1.010	1.075 1.040
1	1.266 1.229	1.650 1.530	2.232 1.944	1.070 1.048	1.014 1.011	1.305 1.163
2	2.014 1.924	2.820 2.587	3.859 3.358	1.145 1.123	1.417 1.323	2.055 1.697

Upper: F_{IA}, Lower: F_{IB}, $F_I = \dfrac{k_1}{\sigma^\infty \sqrt{c}}$

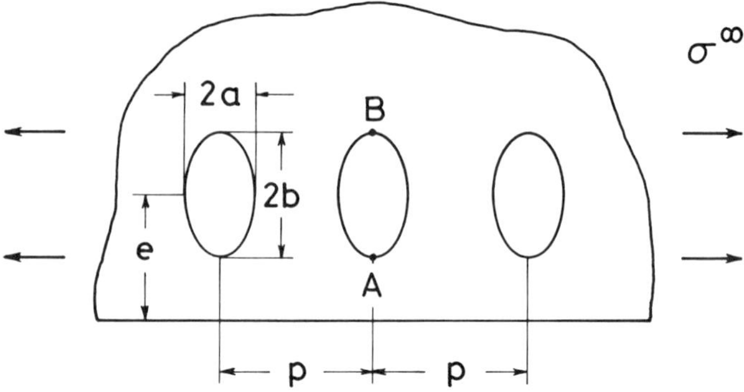

Figure A1.13. Tension of a semi-infinite plate with an infinite row of elliptic holes.

TABLE A1.13

Values of K_t in Figure A1.13

b/e	b/a \ b/p	0	0.1	0.2	0.3	0.4	0.5
0.5	1	3.722 3.212	3.199 2.834	2.609 2.341	2.158 1.995	1.877 1.802	– –
	2	5.753 5.316	5.079 4.783	4.272 4.078	3.605 3.444	3.116 3.028	2.795 2.762

Upper: K_{tA}, Lower: K_{tB}

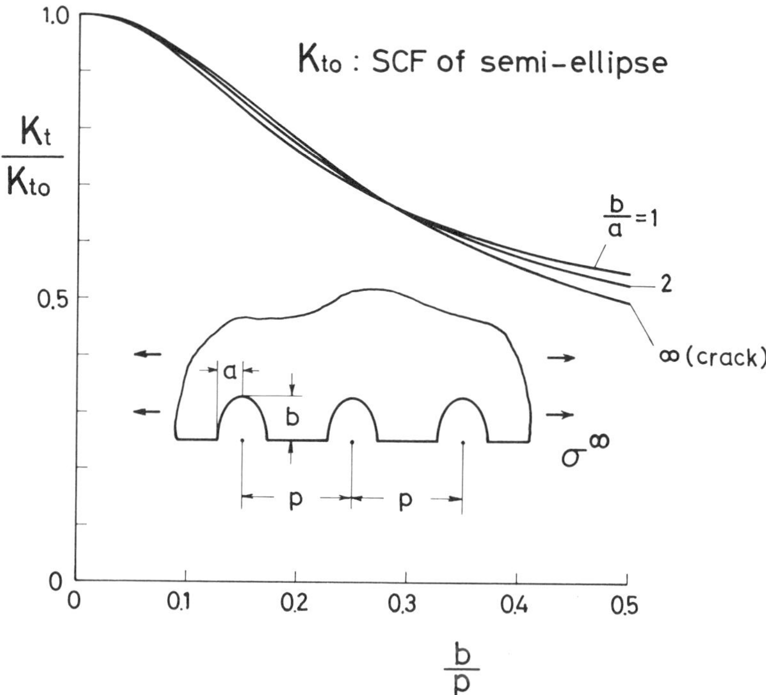

Figure A1.14. Tension of a semi-infinite plate with an infinite row of semi-elliptic notches.

TABLE A1.14

Values of K_t in Figure A1.14

b/a \ b/p	0	0.1	0.2	0.3	0.4	0.5
1	3.065	2.791	2.314	1.992	1.802	1.676
2	5.221	4.805	4.020	3.404	3.015	2.754

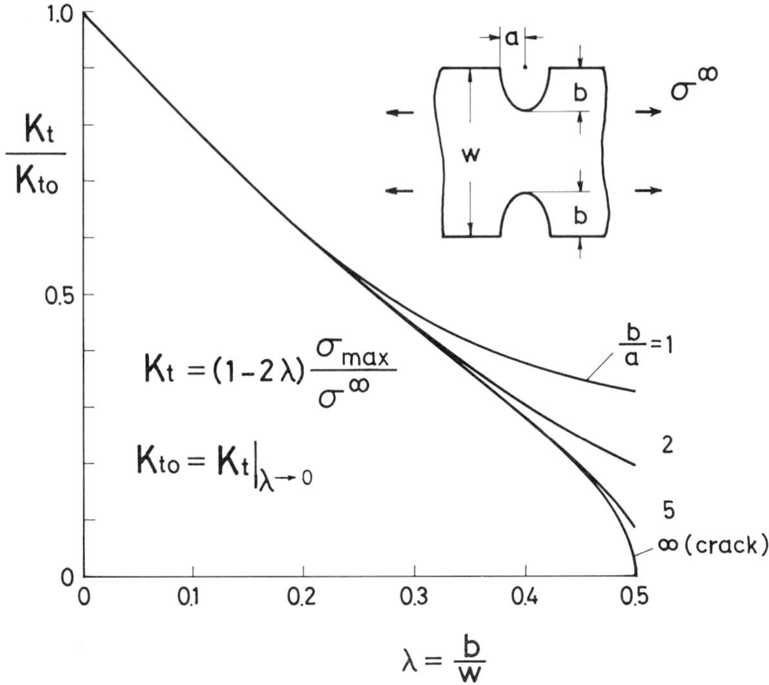

Figure A1.15. Tension of a strip with two semi-elliptic notches located symmetrically.

TABLE A1.15

Values of K_t in Figure A1.15

	K_t					$K_t/K_t\|_{b/w\to 0}$ or $F_I/F_I\|_{b/w\to 0}$				
b/w \ b/a	0.5	1	1	2	5	0.5	1	2	5	
0	2.016	3.065	(3.065)	5.221	11.85	1.000	1.000	1.000	1.000	1.000
0.05	1.803	2.746	(2.745)	4.679	10.62	0.893	0.896	0.896	0.896	0.896
0.1	1.595	2.429	(2.429)	4.141	9.399	0.791	0.792	0.793	0.793	0.793
0.2	1.263	1.864	(1.864)	3.168	7.183	0.626	0.608	0.607	0.606	0.606
0.3	1.102	1.420	–	2.316	5.225	0.547	0.463	0.444	0.441	0.441
0.4	1.040	1.151	–	1.569	3.330	0.516	0.376	0.300	0.281	0.281
0.5	1.000	1.000	–	1.000	1.000	0.496	0.326	0.194	0.084	0

$F_I = \left(1 - \frac{2b}{w}\right)\frac{k_1}{\sigma^\infty\sqrt{b}}$, ( ): Isida [20]

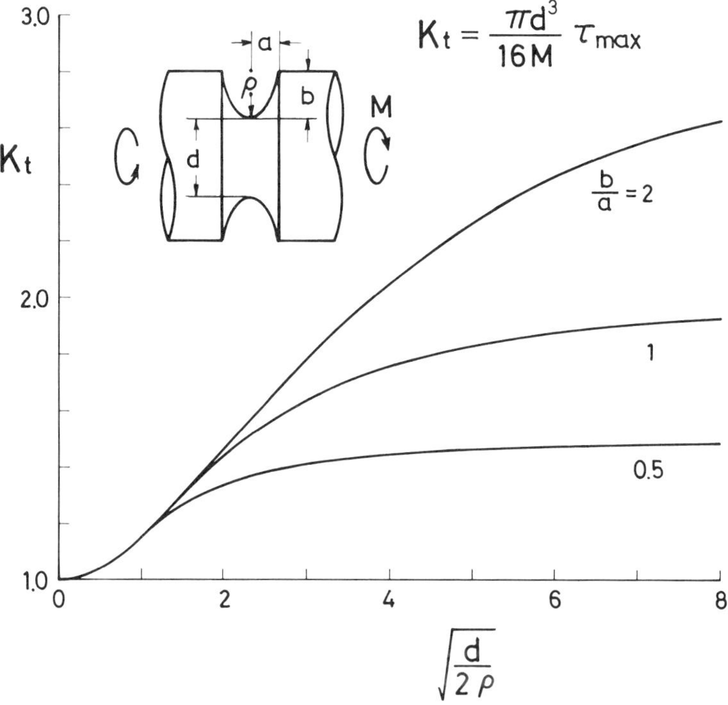

Figure A1.16. Torsion of a round bar with a semi-elliptic circumferential groove.

TABLE A1.16

Values of K_t in Figure A1.16

$\frac{b}{a}$ \ $\sqrt{\frac{d}{2\rho}}$	1	2	4	6	8
0.5	1.151	1.336	1.446	1.474	1.485
1	1.149	1.435	1.760	1.874	1.924
1	(1.14)	(1.43)	–	–	–
2	1.155	1.452	2.052	2.429	2.626

( ): Mathews et al. [21]

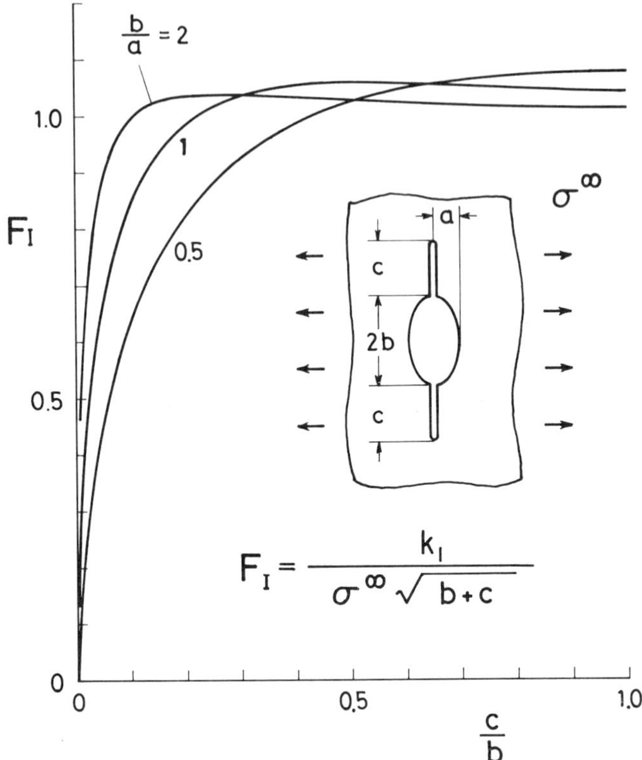

Figure A1.17. Tension of an infinite plate having an elliptic hole with two cracks emanating from its apexes.

TABLE A1.17

Values of F_I in Figure A1.17

b/a \ c/b	0.1	0.2	0.4	0.6	0.8	1.0
0.5	0.640 (0.640)	0.823 (0.824)	0.985 (0.985)	1.047 (1.048)	1.071 (1.071)	1.077 (1.078)
1	0.839 (0.840)	0.984 (0.985)	1.053 (1.054)	1.057 (1.057)	1.049 (1.049)	1.041 (1.041)
2	0.993 (0.993)	1.035 (1.036)	1.031 (1.032)	1.022 –	1.016 –	1.011 –

( ): Newman [23] .

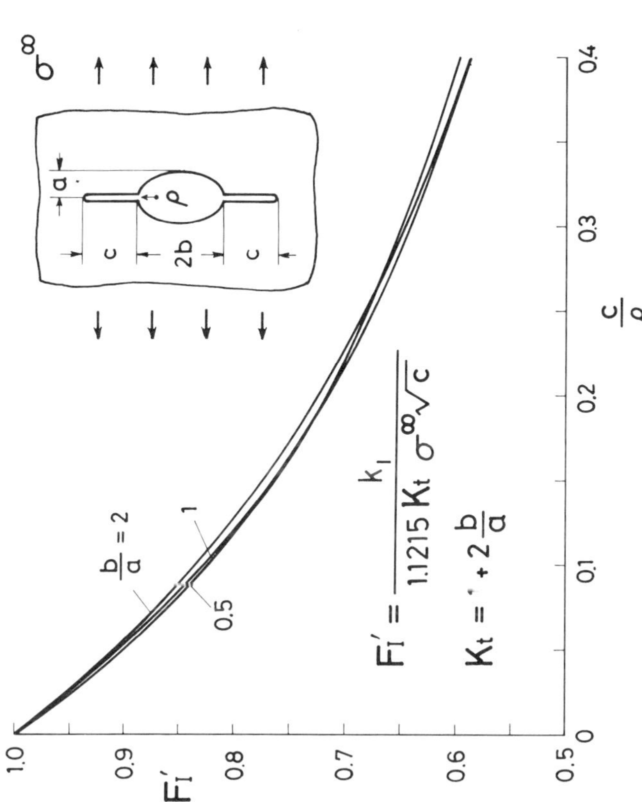

Figure A1.18. Stress intensity factor solution for small cracks for the problem in Figure A1.17.

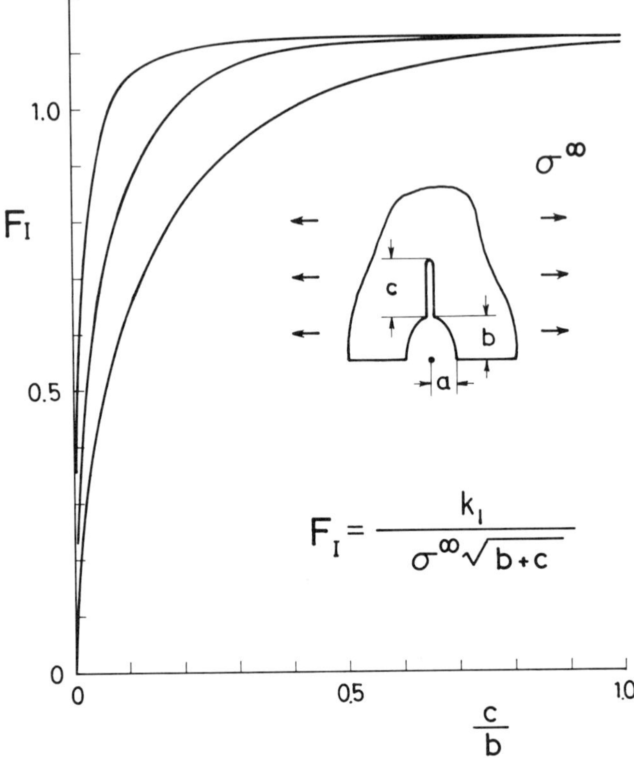

Figure A1.19. Tension of a semi-infinite plate having a semi-elliptic notch with a crack emanating from its apex.

TABLE A1.18

Values of F_I in Figure A1.19

b/a \ c/b	0.1	0.2	0.4	0.6	0.8	1.0
0.5	0.654	0.831	1.000	1.068	1.098	1.111
1	0.862	1.017	1.102	1.118	1.121	1.122
2	1.053	1.109	1.122	1.122	1.121	1.121

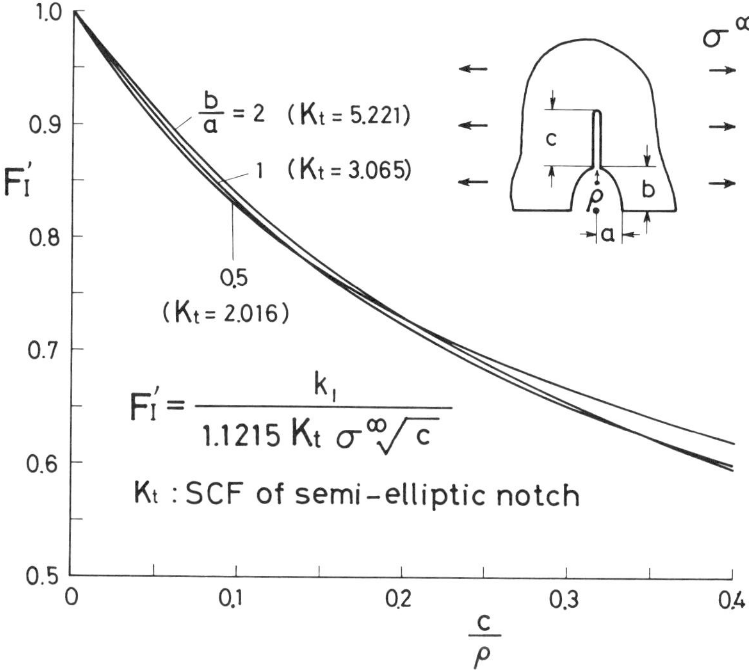

Figure A1.20. Stress intensity factor solution for small cracks for the problem in Figure A1.19.

References

[1] Sih, G. C., ed., *Method of analysis and solutions of crack problems*, Noordhoff International Publishing, Leyden (1973).

[2] Nisitani, H., Two-dimensional problem solved using a digital computer, *Journal of the Japan Society of Mechanical Engineers*, 70, 580, pp. 627–635 (1967).

[3] Nisitani, H. and Murakami, Y., Stress intensity factors of an elliptical crack or a semi-elliptical crack subject to tension, *International Journal of Fracture*, 10, 3, pp. 353–368 (1974).

[4] Timoshenko, S. P. and Goodier, J. N., *Theory of elasticity*, McGraw-Hill Book Company, New York (1951).

[5] Eshelby, J. D., *Elastic inclusions and inhomogeneities, Progress in Solid Mechanics*, Edited by I. N. Sneddon and R. Hill, 2, North-Holland Publishing Company, Amsterdam, pp. 89–140 (1961).

[6] Nisitani, H., Method of approximate calculation for interference of notch effects and its application, *Journal of the Japan Society of Mechanical Engineers*, 71, 589, pp. 209–221 (1968).

[7] Muskhelishvili, N. J., *Some basic problems of the mathematical theory of elasticity*, P. Noordhoff Ltd, Groningen (1953).

[8] Isida, M., On the tension of a semi-infinite plate with an elliptical hole, *Transactions of the Japan Society of Mechanical Engineers*, 22, 123, pp. 803–809 (1956).

[9] Isida, M., Stress-intensity factors for the tension of an eccentrically cracked strip, *Journal of Applied Mechnics*, 33, pp. 674–675 (1966).

[10] Nisitani, H., Suematsu, M. and Saito, K., Tension of a semi-infinite plate with a row of elliptic holes (including cracks) or an infinite plate with two rows of elliptic holes, *Transactions of the Japan Society of Mechanical Engineers*, 39, 324, pp. 2323–2330 (1973).

[11] Nisitani, H., Saito, K. and Hara, N., Stress concentration due to an elliptic hole or crack existing near a notch under tension or longitudinal shear, *Transactions of the Japan Society of Mechanical Engineers*, 39, 324, pp. 2312–2322 (1973).

[12] Melan, E., Der Spannungszustand der durch eine Einzelkraft im Innern beanspruchten Halbscheibe, *Zeitschrift für angewandte Mathematik und Mechanik*, 12, 6, pp. 343–346 (1932).

[13] Nisitani, H. and Murakami, Y., Interaction of elasto-plastic cracks subjected to a uniform tensile stress in an infinite or a semi-infinite plate, *Proceedings of the 1971 International Conference on Mechanical Behavior of Materials*, 2, pp. 346–356 (1972).

[14] Nisitani, H., Tension of a strip with symmetric edge cracks or elliptic notches, *Transactions of the Japan Society of Mechanical Engineers*, 41, 349, pp. 2518–2526 (1975).

[15] Nisitani, H. and Hashimoto, K., Stress concentration factors for the torsion of shafts with a circumferential semi-elliptic groove, Transactions of the Japan Society of Mechanical Engineers, to be published.

[16] Schulz, K. J., Survey of papers on elasticity published in Holland 1940–1946 (written by Biezeno), *Advances in Applied Mechanics*, 1, Academic Press, New York (1948).

[17] Isida, M., On some plane problems of an infinite plate containing an infinite row of circular holes, *Transactions of the Japan Society of Mechanical Engineers*, 25, 159, pp. 1118–1124 (1959).

[18] Ling, C. B., On the stresses in a plate containing two circular holes, *Journal of Applied Physics*, 19, pp. 77–82 (1948).

[19] Koiter, W. T., Rectangular tensile sheet with symmetric edge cracks, *Journal of Applied Mechanics*, 32, pp. 237 (1965).

[20] Isida, M., On the tension of the strip with semicircular notches, *Transactions of the Japan Society of Mechanical Engineers*, 19, 83, pp. 5–10 (1953).

[21] Matthews, G. J. and Hooke, C. J., Solution of axisymmetric torsion problems by point matching, *Journal of Strain Analysis*, 6, 2, pp. 124–133 (1971).

[22] Nisitani, H. and Isida, M., Stress intensity factor for the tension of an infinite plate containing an elliptic hole with two symmetric edge cracks, Preprint of the *Japan Society of Mechanical Engineers*, No. 212, pp. 131–134 (1969).

[23] Newmann, J. C., An improved method of collocation for the stress analysis of cracked plate with various shaped boundaries, *NASA TN D-6376*, pp. 1–45 (1971).

[24] Nisitani, H., Elastic-plastic stress in a semi-infinite plate having an elliptical arc notch with an edge crack under tension or longitudinal shear, *Papers Presented at the Third International Conference on Fracture*, 2, I-513, pp. 1–5 (1973).

[25] Nisitani, H., A method for calculating stress concentrations and its applications, *Strength and Structure of Solid Materials*, edited by H. Miyamoto et al., Noordhoff International Publishing, Leyden, pp. 53–69 (1976).

O. L. Bowie and *C. E. Freese*

2 *Analysis of notches using conformal mapping*

2.1 Introduction

The application of conformal mapping in the stress analysis of two-dimensional notch configurations dates back to 1914 and Kolosoff's solution [1] for an elliptical hole in an infinite sheet under tension. Conformal mapping, as distinguished from curvilinear coordinates, relates of course to transformations in the complex plane and is closely associated with analytic function theory of a complex variable. Kolosoff's complex form of the two-dimensional equations of elasticity [2] has been extended since by the classical works of Muskhelishvili [3]. Utilizing many of the consequences of analytic function theory, Muskhelishvili introduced many effective principles for problem-solving which led to a systematic approach for a wide class of two-dimensional problems. Over the years, many notch configurations have been solved either exactly or approximately using these principles in conjunction with conformal mapping.

Although conformal mapping can be identified with a systematic introduction of curvilinear coordinates, its effectiveness is more subtle and should also be associated with its role in the complex variable formulation of the plane problem. A key property of a conformal mapping function is its analyticity with respect to the complex coordinates of the parameter plane. The two stress functions familiar to the complex variable formulation, analytic with respect to the complex coordinates of the physical plane, can therefore, after transformation, be considered as analytic functions in the parameter plane. Thus, in addition to providing a convenient set of curvilinear coordinates for the imposition of prescribed loading conditions, many of the problem-solving principles such as analytic continuation arguments, Cauchy integral solutions for problems of 'linear relationship', etc., carry over to the parameter plane. With the use of these principles conformal mapping solutions can frequently be derived in a far more efficient manner than with the use of curvilinear coordinates in the real plane.

The classical application of conformal mapping involves first the determination of a functional relationship which transforms a conventional parameter region such as the half plane or interior (exterior) of a circle into a prescribed notch configuration. Although there have been several important notch configurations analyzed by this approach, e.g. elliptical cutouts, hyperbolic and parabolic notches in infinite regions, there are several severe limitations on the practical use of conformal mapping in the classical sense. These limitations include:

(i) Finding 'closed form' mapping functions for specified notch configurations onto the conventional parameter regions.
(ii) Carrying out the stress analysis when the mapping function contains corner-describing singularities (branch points).
(iii) Overcoming the convergence problems intrinsic to the global representation of the solution.
(iv) Cataloguing the numerical solutions in a manner convenient for general use, particularly in the region of chief interest – the notch vicinity.

If the notch geometry is rigidly prescribed, e.g., a constant radius of curvature over a finite interval, only a few 'closed form' mapping functions are known in the classical sense even for infinite or semi-infinite regions. This limitation is additionally complicated if the region is finite and/or multiply-connected. Ironically, the Schwartz–Christoffel transformation provides a much larger class of classical mapping functions for what might appear to be the more difficult class of corner and crack problems. There is a lack of understanding as to the precise nature of the mathematical singularities in the mapping function associated with notch configurations. For example, even the nature of the singularity corresponding to the junction of a circular arc and a tangential straight line has never been resolved. Thus, over the years a variety of numerical and iterative techniques have been invented for finding mapping functions suitably approximating the boundary geometry.

When corner-describing singularities (branch points) occur in the mapping function, difficulties usually are encountered in the stress analysis. Although the mathematical properties of the mapping function are understood in such a case, the corresponding properties of the stress functions can not be easily inferred. In fact the mathematical character of the stress functions obviously depends on both the geometry and the loading (boundary) conditions. Unless numerical methods are employed, e.g., the numerical solution of an integral equation formulation, such mapping functions are of little direct value in problem-solving other than to provide a rational basis for obtaining polynomial approximations of the mapping function.

Implicit to the classical application of conformal mapping is the global representation of the solution. When closed form solutions for the stress functions can be found this situation is ideal. Usually, however, this is not possible and a series or integral representation of the solution must be considered. Owing to rapidly varying stress gradients in the notch vicinity as contrasted to the more remote stress distribution, it is evident that numerical convergence can become a factor in a global representation of the stress functions. When a series representation is taken, it is usually necessary to retain a large number of terms to ensure notch stress accuracy. The classical approach lacks the flexibility to provide effective localized expansions and, at the same time, a less demanding representation of the more uniform portions of the solution.

The matter of cataloguing the numerical solutions is a serious problem in many global solutions. Usually only the maximum stress or the distribution of stress on the notch boundary is presented. On the other hand a knowledge of the sub-surface stresses in the notch vicinity is often of practical interest. Effective local expansions are clearly desirable to document the numerical solution for general use.

The primary emphasis of this chapter is placed on non-classical applications of conformal mapping to notch configurations. The basic plan is the MMC (modified mapping-collocation) method [4, 5] and the MMC method combined with partitioning [6, 7, 8]. These approaches utilize conformal mapping in an unconventional manner, by-passing a global mapping of the entire physical geometry on a rigidly prescribed parameter region. Instead, only subregions involving the notch tip and its vicinity will usually be considered by mapping. With the partitioning concepts, separate expansions can be considered in two or more subregions thus providing effective local expansions in contrast to slowly converging global representations. The need for considering corner-describing singularities no longer arises explicitly. Least square collocation arguments are used for the satisfaction of the loading conditions on the boundary points and for the 'stitching' conditions across internally partitioned solution representations.

When a finite element representation is introduced as part of the partitioning plan, the approach is easily competitive with conventional finite element solutions to notch problems. Finite element methods have difficulties in regions with rapidly varying stress gradients such as notches. Such devices as building in specialized local displacement fields into the elements are often awkward and indecisive. In fact, as we have already remarked, in the case of notches the singular nature of the solution is usually not sufficiently understood to estimate the local character of the displacement fields. These difficulties can be removed by utilizing a continuous representation of the solution locally and a finite element representation elsewhere. Advantage of a 'banded' system of equations corresponding to the finite element conditions can be taken by first computing the 'substructured' system. Parameter studies of the notch geometry can then be carried out without regeneration of the substructured system.

2.2 Basic preliminaries

In this section we shall briefly summarize several topics fundamental to the later applications of conformal mapping to notch problems. For a detailed treatment of these topics, the reader is referred to Muskhelishvili [3].

Complex variable formulation. To summarize the complex variable formulation of Muskhelishvili, we introduce the complex coordinate $z = x + iy$, and henceforth, refer to the z-plane as the physical plane. The formulation depends upon the representation of the Airy stress function, $U(x, y)$, in terms of two analytic functions of the complex variable, z, namely, $\phi(z)$ and $\psi(z)$,

$$U(x, y) = \mathrm{Re}\left[\bar{z}\phi(z) + \int^{z} \psi(z)\,\mathrm{d}z\right]. \tag{2.1}$$

In equation (2.1), Re denotes 'real part' and bars denote complex conjugates.

With this representation and the conventional use of primes to denote differentiation, the stresses and displacements in rectangular coordinates can be written as

$$\begin{aligned}\sigma_y + \sigma_x &= 2[\phi'(z) + \overline{\phi'(z)}] = 4\,\mathrm{Re}[\phi'(z)]\\ \sigma_y - \sigma_x + 2i\tau_{xy} &= 2[\bar{z}\phi''(z) + \psi'(z)]\end{aligned} \tag{2.2}$$

and

$$2\mu(u+iv)=\kappa\phi(z)-z\overline{\phi'(z)}-\overline{\psi(z)} \tag{2.3}$$

where $\mu = E/2(1+\nu)$ and $\kappa = 3-4\nu$ (plane strain) and $\kappa = (3-\nu)/(1+\nu)$ (generalized plane stress). E and ν are Young's modulus and Poisson's ratio, respectively.

The stress resultant acting on an arbitrary arc of the material provides a valuable relationship in the formulation. Consider an arc of the material with element of length ds and normal n drawn to the right of the arc when looking along it in the positive direction. Then, if X_n ds and Y_n ds denote the horizontal and vertical components, respectively, of the force on the element ds (exerted on the side of the positive normal), it can be shown that

$$(X_n+iY_n)\,\mathrm{d}s=-i\,\mathrm{d}\{\phi(z)+z\overline{\phi'(z)}+\overline{\psi(z)}\}. \tag{2.4}$$

Thus

$$\begin{aligned}\phi(z)+z\overline{\phi'(z)}+\overline{\psi(z)}&=i\int^{s}(X_n+iY_n)\,\mathrm{d}s\\&=f_1(s)+if_2(s).\end{aligned} \tag{2.5}$$

When the arc is chosen as the boundary of the physical region, $f_1(s)+if_2(s)$ is known to within a constant on those intervals where the boundary loads are prescribed. We shall henceforth in this chapter refer to equation (2.5) as the 'force' conditions.

Many of the properties of the analytic functions $\phi(z)$ and $\psi(z)$ are systematically deduced by Muskhelishvili from a combination of physical and mathematical considerations. For example, when rigid body motions are not precluded, the analytic functions are obviously not unique. The extent of arbitrariness in their definitions can be systematically considered for the several boundary-value problems of elasticity. Overall equilibrium requirements and the single-valueness of the displacements lead to additional relations. For example, if the arc in equation (2.5) is a closed curve and the boundary of a simply connected region of the material, then, from equilibrium, the integral over the total loop must vanish. Such a condition clearly places specific requirements on multi-valued functions which might be used for stress function representation. When the physical region is infinite or multi-connected further properties of the stress function can be deduced.

Conformal mapping. Let S_z denote an open region in the physical plane and let us introduce a complex parameter plane, the ζ-plane, along with the transformation

$$z = \omega(\zeta). \tag{2.6}$$

It will be assumed that there exists an open region S_ζ in the ζ-plane such that equation (2.6) provides a one-to-one correspondence between the points $z \in S_z$ and $\zeta \in S_\zeta$. Such a transformation is said to be conformal provided the mapping function $\omega(\zeta)$ is analytic and $\omega'(\zeta) \neq 0$ for $\zeta \in S_\zeta$. Singularities in $\omega(\zeta)$ can exist on the boundary of S_ζ for corner-describing properties of the boundary of S_z.

The carry-over of analyticity to the parameter region follows from a well-known property of analytic functions. If $\phi(z)$ is analytic for $z \in S_z$ and $\omega(\zeta)$ is the conformal mapping function in equation (2.6), then $\phi[\omega(\zeta)]$ is an analytic function of ζ for $\zeta \in S_\zeta$. Thus the explicit determination of $\phi(z)$ and $\psi(z)$ can be by-passed and the analyst can seek directly the analytic functions $\phi[\omega(\zeta)]$ and $\psi[\omega(\zeta)]$ as functions of ζ. The necessity for introducing considerable new notation can be avoided by designating $\phi[\omega(\zeta)]$ as $\phi(\zeta)$, etc., which leads to such relationships as $\phi'(z) = \phi'(\zeta)/\omega'(\zeta)$, etc. Thus, the stresses, displacements, etc. can now be written as

$$\sigma_y + \sigma_x = 4\,\mathrm{Re}\{\phi'(\zeta)/\omega'(\zeta)\} \tag{2.7}$$

$$\sigma_y - \sigma_x + 2i\tau_{xy} = 2\{\overline{\omega(\zeta)}[\phi'(\zeta)/\omega'(\zeta)]' + \psi'(\zeta)\}/\omega'(\zeta) \tag{2.8}$$

$$2\mu(u + iv) = \kappa\phi(\zeta) - \omega(\zeta)\overline{\phi'(\zeta)}/\overline{\omega'(\zeta)} - \overline{\psi(\zeta)} \tag{2.9}$$

$$\phi(\zeta) + \omega(\zeta)\overline{\phi'(\zeta)}/\overline{\omega'(\zeta)} + \overline{\psi(\zeta)} = i\int^{s} (X_n + iY_n)\,\mathrm{d}s. \tag{2.10}$$

In the conventional use of conformal mapping, the parameter region S_ζ is often chosen as the interior (exterior) of a circle or the upper (lower) half plane for simply connected regions S_z. In general, the parameter region must enjoy the same connectivity as the physical region. (Exceptions to this remark will be encountered later when certain analytic continuation arguments are made for the stress functions.) The choice of a circular region for S_ζ is particularly convenient for handling boundary conditions with a power series representation of the stress functions. The choice of a half plane is natural to an integral representation of the solution.

Whereas the choice of the parameter region in the conventional use of conformal mapping is geared to analytical arguments for meeting the boundary conditions, when other numerical devices such as boundary collocation are introduced the rigidly prescribed parameter regions above are no longer necessary. This greater flexibility is used in the MMC method to be described later. When, in addition, the partitioning plans of solution are introduced, only subregions of S_z will be described by mapping. It is important intuitively to keep in mind the parametric nature of the ζ-plane. The original physical problem always remains defined in the z-plane. Thus, for example, several different mapping functions could be introduced, each appropriate to a particular subregion of S_z, and used in the solution provided the representations are suitably interrelated. This additional flexibility is utilized in the MMC plus partitioning plan.

Problems of linear relationship, analytic continuation. We shall frequently utilize the continuation arguments of Muskhelishvili in this chapter, thus the basic argument will now be reviewed. First, let us assume the elastic body occupies the lower half plane $S_z^-(y<0)$. Consider $\phi(z)$ as defined in the unoccupied region, $S_z^+(y>0)$, by

$$\phi(z) = -z\bar{\phi}'(z) - \bar{\psi}(z), \quad z \in S_z^+ \tag{2.11}$$

where the bar notation is defined by

$$\bar{f}(z) = \overline{f(\bar{z})}. \tag{2.12}$$

The function $\psi(z)$ can now be expressed as

$$\psi(z) = -\bar{\phi}(z) - z\phi'(z), \quad z \in S_z^-. \tag{2.13}$$

Replacing $\psi(z)$ in the force condition equation (2.5) by its definition equation (2.13), we find

$$\phi(z) - \phi(\bar{z}) + (z-\bar{z})\overline{\phi'(z)} = f_1(s) + if_2(s). \tag{2.14}$$

For most practical loading systems acting on the real axis, $z = x$, one can correctly assume that

$$|\phi'(z)| < A/|z-x|^\alpha, \quad 0 \leqslant \alpha < 1 \tag{2.15}$$

for z near x. Thus the force condition, equation (2.14), on the boundary $z = x$ reduces to

$$\phi^-(x) - \phi^+(x) = f_1(x) + if_2(x) \tag{2.16}$$

where $\phi^-(x)$ and $\phi^+(x)$ denote the values $\phi(x)$ as approached through S_z^- and S_z^+, respectively.

If we assume that the stresses at infinity vanish and the loads prescribed on the real axis are such that $f_1(x)$ and $f_2(x)$ satisfy the Hölder condition, then the solution can be written directly in terms of the Cauchy integral

$$\phi(z) = \frac{1}{2\pi i}\int_L \frac{f_1(x) + if_2(x)}{x - z}\,\mathrm{d}x \tag{2.17}$$

where L denotes the real axis. The requirements of the solution are clearly satisfied by equations (2.13) and (2.17). Problems reducible to solution in this manner are called the problems of 'linear relationship'.

The matter of analytic continuation can be deduced from equation (2.16). On a load-free interval of the x-axis, $f_1(x) + if_2(x)$ is constant which, by adjustment of the constant of integration, can be considered zero. If S_x denotes this load-free interval, then equation (2.16) reduces to

$$\phi^-(x) - \phi^+(x) = 0, \quad x \in S_x. \tag{2.18}$$

From analytic function theory, $\phi(z)$ is therefore analytic for $z \in S_x$ and $\phi(z)$ continues analytically across unloaded intervals of the boundary. Conversely, if we assume a representation of the extended function $\phi(z)$ which is analytic on S_x, this interval can be considered as load-free. This property will be utilized in this chapter to build in load-free conditions over boundary intervals where accuracy is important to the solution.

Similar arguments can be made for other regions with the use of Schwartz's symmetry principle. In addition to circles, it is evident that all regions for which there exists a single-valued mapping function carrying the region onto the upper (lower) half plane or the interior (exterior) of the unit circle in the parameter plane can be handled in a similar manner. For example, consider Kartzivadze's extension [9] for regions mapped onto the exterior (interior) of the unit circle. Equation (2.6) will be considered as a conformal mapping of the exterior of the unit circle S_ζ^- into the region occupied by the elastic material. Points on the unit circle, $\zeta = \sigma$, will be assumed to map into the finite boundary of the physical region. The function $\phi(z)$ is extended by definition into the interior of the unit circle, S_ζ^+, as

$$\phi(\zeta) = -\omega(\zeta)\bar{\phi}'(1/\zeta)/\bar{\omega}'(1/\zeta) - \bar{\psi}(1/\zeta), \quad \zeta \in S_\zeta^+ \tag{2.19}$$

where the bar notation is now defined by

$$\bar{f}(1/\zeta) = \overline{f(1/\bar{\zeta})}. \tag{2.20}$$

The function $\psi(\zeta)$ becomes

$$\psi(\zeta) = -\bar{\phi}(1/\zeta) - \bar{\omega}(1/\zeta)\phi'(\zeta)/\omega'(\zeta), \quad \zeta \in S_\zeta^-. \tag{2.21}$$

The resultant force condition, equation (2.10), when evaluated on the unit circle can be written with obvious notation as

$$\phi^-(\sigma) - \phi^+(\sigma) + [\omega^-(\sigma) - \omega^+(\sigma)]\overline{\phi'(\sigma)/\omega'(\sigma)} = f_1(s) + if_2(s). \tag{2.22}$$

If the mapping function $\omega(\zeta)$ is continuous across the unit circle, then again, equation (2.22) reduces to

$$\phi^-(\sigma) - \phi^+(\sigma) = f_1(s) + if_2(s) \tag{2.23}$$

and the previous continuation arguments are applicable.

Multi-valued mapping function. In the preceding paragraphs, it has been shown that with the application of Muskhelishvili's extension arguments the problems of linear relationship can be solved in a direct manner. The class of configurations for which the mapping function is a polynomial or a rational function fall into this category.

Unfortunately, many notch configurations involve discontinuities in curvature of the boundary geometry which in turn involve singularities in the mapping function incompatible with the properties of linear relationship. A simple example of this situation is provided by the mapping

$$z = \omega(\zeta) = \zeta + \sqrt{\zeta^2 - 1}. \tag{2.24}$$

If we choose the branch such that $z \to 2\zeta$ for large $|\zeta|$, then equation

(2.24) maps the upper half of the ζ-plane into the exterior of a semi-circular notch in a semi-infinite region in the z-plane, Figure 2.1. This multi-valued mapping function has branch points at $\zeta = \pm 1$. In order to apply the extension argument leading to linear relationship it is necessary to consider $\omega(\zeta)$ as defined in the entire ζ-plane. In turn, this requires a choice of branch cut which we choose along the real axis, Figure 2.2.

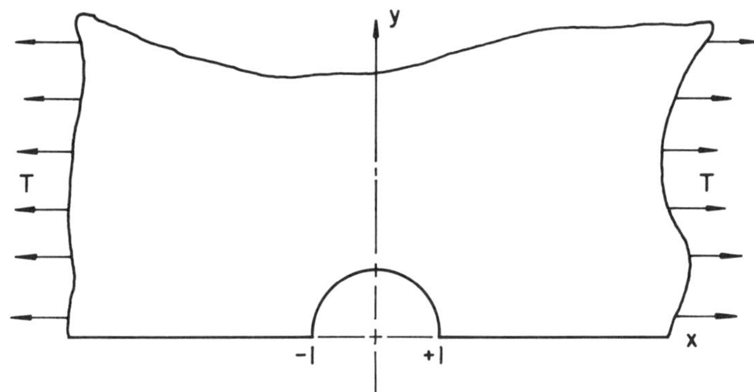

Figure 2.1. Semi-circular notch in a semi-infinite region.

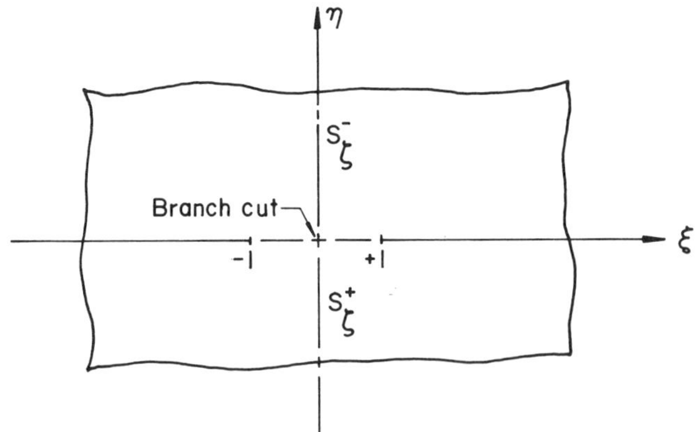

Figure 2.2. Branch cut in the ζ-plane.

Consider the problem of tension $\sigma_x = T$ acting at infinity. If we carry out the extension,

$$\psi(\zeta) = -\bar{\phi}(\zeta) - \bar{\omega}(\zeta)\phi'(\zeta)/\omega'(\zeta), \quad \zeta \in S_\zeta^- \tag{2.25}$$

then the boundary condotion on $\zeta = \xi$ becomes

$$\phi^-(\xi) - \phi^+(\xi) + [\omega^-(\xi) - \omega^+(\xi)]\overline{\phi'^-(\xi)}/\overline{\omega'^-(\xi)} = 0. \tag{2.26}$$

Although equation (26) reduces to

$$\phi^-(\xi) - \phi^+(\xi) = 0, \quad |\xi| \geqslant 1 \tag{2.27}$$

the same is not true across the branch cut where

$$\omega^-(\xi) - \omega^+(\xi) \neq 0, \quad |\xi| \leqslant 1. \tag{2.28}$$

Thus the problem does not reduce to one of linear relationship and a simple direct solution to equations (2.26) and (2.27) is not readily available. (Alternative choices of branch cuts merely transform the difficulty to another form.)

2.3 Approximating polynomial mapping functions

A widely used mapping approach to notch problems has been based on approximating polynomial mapping functions. Except for the toleration of approximations of the physical boundary, this approach can be considered classical in the sense that a global solution defined on a conventional parameter region is sought. A particularly attractive feature of the use of polynomial mappings is the simplicity of the stress analysis. For conventional loading systems acting on such configurations, the problem reduces to one of linear relationship; thus, an exact solution in closed form is obtained for the approximate geometry. The approach is not without its limitations as we shall review and illustrate in this section. On the other hand, these approximate solutions have played an important role for the designer at a time when large scale digital computers were not available.

Truncated expansions of Schwartz–Christoffel transformations. A well-known scheme for obtaining approximating polynomial mapping functions consists of utilizing truncated expansions of appropriate Schwartz–Christoffel transformations. Savin [10] used this approach in 1936 to study filleted cutouts in an infinite sheet. We consider, for example, the case of a square cutout with filleted corners in an infinite sheet (Figure 2.3.)

The well-known Schwartz–Christoffel transformation carrying the unit circle and its exterior in the ζ-plane into a square and its exterior in the z-plane can be written as

$$z = \omega(\zeta) = A \int^{\zeta} t^{-2}(1 + t^4)^{\frac{1}{2}}\, \mathrm{d}t + \text{const.} \tag{2.29}$$

Expanding in a power series, one finds

$$\omega(\zeta) = R[\zeta^{-1} - (1/6)\zeta^3 + (1/56)\zeta^7 - (1/176)\zeta^{11} + \cdots] \tag{2.30}$$

Truncations of equation (2.30) provide approximations of the desired boundary contour. For example, consider

$$\omega(\zeta) = R[\zeta^{-1} - (1/6)\zeta^3]. \tag{2.31}$$

Then R is chosen as $3a/5$ on the basis of the size of the cutout. The 'equivalent' radius of curvature, ρ^*, is computed as the radius of

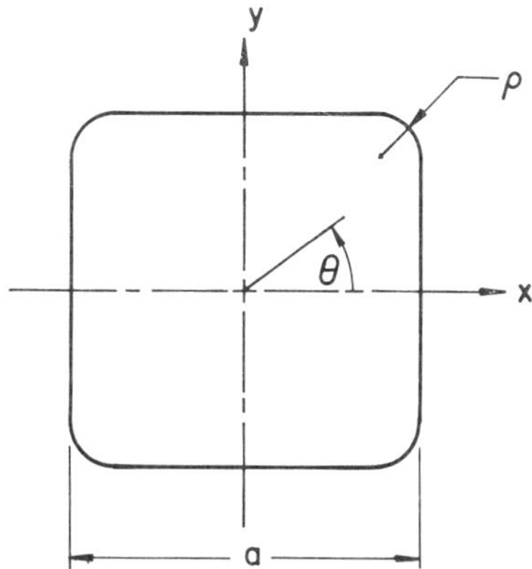

Figure 2.3. Square cutout with filleted corners.

curvature at the point corresponding to $\theta = 45°$. In this case, $\rho^* = R/10 = 3a/50$. Similar considerations hold for truncations of equation (2.30) when additional terms are retained, each truncation requiring an adjustment of R and yielding a new value of ρ^*.

The stress analysis for each of these approximate configurations can be readily carried out using the analytic continuation arguments, equations (2.19)–(2.23). For example, with the loading condition $\sigma_x = T$ at infinity, the solution for the geometry described by equation (2.31) becomes

$$\phi(\zeta) = TR[(1/4)\zeta^{-1} + (3/7)\zeta + (1/24)\zeta^3] \tag{2.32}$$

$$\psi(\zeta) = -TR[(1/2)\zeta^{-1} + (91\zeta - 78\zeta^3)/84(\zeta^4 + 2)]. \tag{2.33}$$

Similarly, for truncations involving additional terms, $\phi(\zeta)$ is a polynomial of the same degree as the mapping polynomial and $\psi(\zeta)$ is a simple rational function in ζ.

In addition to supplying only a discrete set of parameter ratios, ρ^*/a, there is a basic deficiency in this approach. It is impossible to estimate the degree of approximation in the stress distribution relative to the exact geometry described in Figure 2.3. For example, if equation (2.31) is considered as the approximation of Figure 2.3 with a fillet radius of $\rho/a = 0.060$, then there is a variation of ρ/a of as much as five percent in the fillet vicinity. In fact, Savin finds the maximum stress occurring at a point where $\rho/a = 0.057$. The difficulty increases when additional terms are retained in the truncation. Terms with large exponents contribute high frequency ripples in the geometry which, although small in magnitude, can cause considerable local disturbance in the stress distribution.

Modified truncation for crack problems. For the important class of problems when the notch degenerates to a crack, the physical configuration can frequently be represented exactly by a suitable Schwartz–Christoffel transformation. The stress analysis can be conveniently carried out by again considering truncations of the series expansion of the exact transformation. It is interesting to compare this plan with that described above.

A major difference in crack and notch problems is the type of information sought in the solution. In crack problems primary interest centers on the local crack-tip stresses. If a set of polar coordinates (r, θ)

is introduced at the crack tip, the local stresses can be expressed generically as

$$\sigma = Kr^{-\frac{1}{2}}f(\theta) + O(1) \tag{2.34}$$

The distribution of the dominant components of stress can be considered as known and only the stress intensity K must be determined. In contrast, there is no convenient local dominance of the solution for the general notch problem. The stress distribution will vary with the geometric parameters, and it will be necessary to maintain a high degree of accuracy over a considerable interval of the local notch geometry to ensure reliable results.

A modified truncation plan designed for crack problems was introduced by Bowie [11]. Let the expansion of the exact mapping function be given, for example, by the following form:

$$z = \omega(\zeta) = \sum_{n=1}^{\infty} A_n \zeta^n \tag{2.35}$$

where it is presumed that the A_n's can be determined numerically from the exact mapping function. Furthermore, assume $\omega''(\sigma_0) = Q$ (where σ_0 corresponds to the crack tip) can be calculated directly from the exact mapping function. In addition, we know that $\omega'(\sigma_0) = 0$ at a crack tip. Now, the partial sums of the expansions for $\omega'(\zeta)$ and $\omega''(\zeta)$ are given by

$$\begin{aligned} \omega'(\zeta) &= \sum_{n=1}^{M} nA_n \zeta^{n-1} \\ \omega''(\zeta) &= \sum_{n=1}^{M} n(n-1)A_n \zeta^{n-2}. \end{aligned} \tag{2.36}$$

The numerical values of $\omega'(\sigma_0)$ and $\omega''(\sigma_0)$ in (2.36) as a function of the truncation index M will reveal, in general, preferred values, M^*, for which $\omega'(\sigma_0) \approx 0$ and $\omega''(\sigma_0) \approx Q$, simultaneously. Effective truncations can now be constructed as follows: Let

$$\omega_T(\zeta) = \sum_{n=1}^{M^*+2} \varepsilon_n \zeta^n \tag{2.37}$$

where

$$\varepsilon_n = A_n, \quad n = 1, 2, \ldots, M^*; \qquad \varepsilon_{M^*+1} = R; \qquad \varepsilon_{M^*+2} = S. \tag{2.38}$$

The constants R and S are calculated on the basis that $\omega'_T(\sigma_0) = 0$ and $\omega''_T(\sigma_0) = Q$, exactly. Thus, preservation of the geometry at the crack tip is made with little disturbance of the overall configuration.

The modified truncation scheme described above is particularly effective for the computation of the stress intensity factor K. Not only is the cusplike character of the crack-tip preserved but also the correct value of $\omega''_T(\sigma_0)$ which enters directly into the calculation of K. Unfortunately, owing to the non-localized character of general notch problems, the carryover of this truncation scheme to other types of notches does not appear feasible.

Simple polynomial forms. There are several simple polynomial forms of mapping functions which have been used to approximate notch solutions. Classical examples include:

$$z = \omega(\zeta) = R(\zeta + m\zeta^n), \quad R > 0, \quad 0 \leqslant m \leqslant 1/n \tag{2.39}$$

when n is an integer larger than unity. The unit circle and its interior are mapped into an 'epitrochoid' and its interior, respectively, by the mapping (2.39). The mapping (2.40),

$$z = \omega(\zeta) = R(\zeta^{-1} + m\zeta^n), \quad R > 0, \quad 0 \leqslant m \leqslant 1/n \tag{2.40}$$

where n is an integer larger than unity, maps the unit circle and its interior into the exterior of a 'hypotrochoid.'

Morkovin [12] used the mapping function

$$z = \omega(\zeta) = s\zeta + t\zeta^{-1} + r\zeta^{-3} \tag{2.41}$$

(a representation of Greenspan's [13] curvilinear coordinates) in the study of approximate ovaloids in an infinite sheet. (This paper showed the effectiveness of the Muskhelishvili arguments in carrying out the stress analysis compared with an analysis based on the use of the same curvilinear coordinates in the real plane).

These mapping forms, although subject to the limitations already described, have served a useful role in notch stress estimates.

Polynomial approximations by iteration. Several decades ago there was considerable interest in numerical methods for deriving approximate mapping functions, primarily, for applications in hydrodynamics. Bateman [14] provides a brief survey of some of the earlier work in this area. It is somewhat surprising that there has been little interest shown in recent years in these techniques relative to the analysis of notch problems. Several of these methods, although laborious by the standards of hand calculation, are well-suited for large scale digital computers.

A simple iterative technique for obtaining polynomial mappings with a high degree of local accuracy will now be outlined. This approach was used by Hay [15] in the study of ovaloids and slots, Bowie [16] in the stress analysis of gun tube rifling, and Neal [17, 18] for the analysis of U-shaped edge notches. To illustrate the details of the procedure, the problem of a rectangular cutout with filleted corners, Figure 2.4, as solved in [15], is now considered.

We seek a polynomial mapping function which carries the unit circle and its exterior in the ζ-plane into the region described in Figure 2.4. Owing to the symmetries of the boundary contour, C,

$$z = \omega(\zeta) = \alpha\zeta + \sum_{n=1}^{m} \varepsilon_{2n-1}\zeta^{1-2n} \tag{2.42}$$

where m is a positive integer and α and ε_{2n-1} are real constants to be determined. Although the coordinates of C are known explicitly, it is not possible to determine the mapping coefficients directly. The difficulty hinges on the unknown angular relationship between the two planes, otherwise the coefficients could be obtained by a Fourier type of argument.

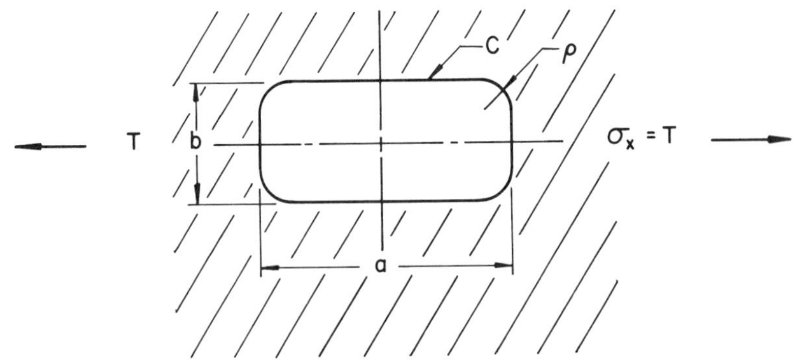

Figure 2.4. Cutout in infinite sheet under tension.

The mapping function $\omega(\zeta)$ will be determined by iteration, that is,

$$\omega(\zeta) \rightarrow \omega^{(n)}(\zeta) \tag{2.43}$$

where $\omega^{(n)}(\zeta)$ is the n-th element of a sequence of mapping approximations derived by a process which is now described.

For reasons which will be apparent, we introduce an additional parameter plane, the ζ_1-plane, and consider the configurations shown in Figure 2.5. It is now assumed that a crude first approximation can be estimated,

$$z = \omega^{(0)}(\zeta_1) = \alpha^{(0)}\zeta_1 + \sum_{n=1}^{m} \varepsilon_{2n-1}^{(0)}\zeta_1^{1-2n}. \tag{2.44}$$

The mapping function $\omega^{(0)}(\zeta_1)$ maps the unit circle γ_1 in the ζ_1-plane into a curve C' adjacent to the desired curve C in the z-plane. Conversely, the curve C can be considered as mapped into a curve γ_1' adjacent to the unit circle γ_1.

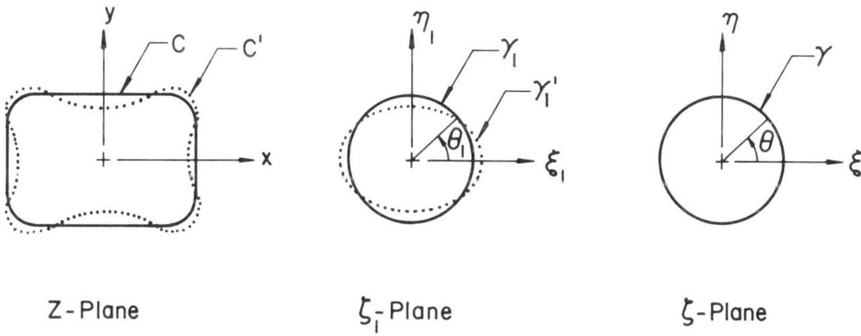

Figure 2.5. Physical and parameter regions.

Although γ_1' could be found directly by solving equation (2.44) for the points z on C, an effective approximation of γ_1' can rapidly be found by utilizing the local properties of a conformal transformation. We denote the polar coordinates of a general point on γ_1' by $(1+\rho, \theta_1)$, thus ρ is a measure of the 'error' in ζ_1-plane and generally $\rho \ll 1$. Similarly, we define $\mathrm{d}(\theta_1)$ for a point (x, y) on C' as distance to C measured along the normal to C. With the convention that d is positive where C' lies inside

C and ρ is positive where γ_1' lies outside γ_1, then

$$\begin{aligned}\rho(\theta_1) &\approx d(\theta_1)/|dz/d\zeta_1| \\ &= d(\theta_1)(\dot{x}^2+\dot{y}^2)^{-\frac{1}{2}}\end{aligned} \tag{2.45}$$

where $\dot{x} = dx/d\theta_1$ and $\dot{y} = dy/d\theta_1$. Although equation (2.45) is an approximation based on the local properties of preservation of angles and magnification of elements of arc length, its accuracy increases as $C' \to C$.

The next step in the iteration is the determination of a mapping function $\omega_1(\zeta)$ which carries the unit circle γ and its exterior in the ζ-plane into the exterior of γ_1' in the ζ_1-plane. Thus, again utilizing symmetry considerations,

$$\zeta_1 = \omega_1(\zeta) = \beta\zeta + \sum_{n=1}^{p} \lambda_{2n-1}\zeta^{1-2n} \tag{2.46}$$

where β and λ_{2n-1} are real constants. With the assumption that γ_1' is almost a circle, i.e. $\rho(\theta_1) \ll 1$, it is reasonable now to assume $\theta \approx \theta_1$. With this assumption, expressions for the coefficients in equation (2.46) can be derived from Fourier considerations. In particular,

$$\begin{aligned}\beta &\approx 1 + (2/\pi)\int_0^{\pi/2} \rho(\theta_1)\, d\theta_1 \\ \lambda_{2n-1} &\approx (2/\pi)\int_0^{\pi/2} \rho(\theta_1)\cos 2n\theta_1\, d\theta_1.\end{aligned} \tag{2.47}$$

The first iteration is completed by utilizing the product transformation

$$\begin{aligned}\omega^{(1)}(\zeta) &= \omega^{(0)}[\omega_1(\zeta)] \\ &\approx \alpha^{(1)}\zeta + \sum_{n=1}^{m(1)} \varepsilon_{2n-1}^{(1)}\zeta^{1-2n}.\end{aligned} \tag{2.48}$$

In the determination of the coefficients in equation (2.48) it is assumed that the coefficients $\lambda_{2n-1} \ll 1$ and, thus, cross products and powers of these terms greater than one can be neglected. Thus,

$$\begin{aligned}\alpha^{(1)} &= \alpha^{(0)}\beta \\ \varepsilon_1^{(1)} &= \alpha^{(0)}\lambda_1 + \varepsilon_1^{(0)}/\beta\end{aligned} \tag{2.49}$$

etc.

The iterative process is obvious with $\omega^{(1)}(\zeta)$ now used in the role of $\omega^{(0)}(\zeta)$ to determine $\omega^{(2)}(\zeta)$, etc. For a fixed number of terms $m+1$, the process is repeated until the coefficients converge to constant values. If the approximation is then inadequate, $m+1$ is increased and the process is repeated.

For notches with fairly large radii of curvature in simply connected regions, the process described above is effective. Accuracy local to the notch can be weighted in the calculation of equation (2.47). For example, the analyst can arbitrarily define $\rho(\theta_1)=0$ on intervals remote to the notch, thus weighting the notch error in the calculation of the improved mapping function.

The stress analysis is carried out along the lines familiar to problems of linear relationship. For the problem illustrated in Figure 2.4

$$\phi(\zeta)=(1/4)\alpha T\zeta+T\sum_{n=1}^{m}a_{2n-1}\zeta^{1-2n} \tag{2.50}$$

$$\psi(\zeta)=-\bar{\phi}(1/\zeta)-\bar{\omega}(1/\zeta)\phi'(\zeta)/\omega'(\zeta). \tag{2.51}$$

The coefficients a_{2n-1} are determined from the set of conditions requiring $\psi(\zeta)\rightarrow(1/2)\alpha\zeta T$ for large $|\zeta|$.

For the particular geometry $a=6$, $b=2$ and $\rho=1/2$, the solution was carried out in [15] for $m=6$, 8, 10. The error measure $\mathrm{d}(\theta)$ and the variation of the tangential stress σ_t along the fillet are shown in Table 2.1 for the three orders of approximation. (Note that θ is the angular measure in the ζ-plane. In Table 2.1a., the corresponding mapping coefficients are listed so that the solution can be identified in terms of the physical coordinates.)

From Table 2.1 the convergence of the solution is evident for this set of parameters. Each truncation yields a relatively smooth error function $\mathrm{d}(\theta)$ which decreases extensively by relatively small contributions of additional terms. For a wide range of the parameters this excellent convergence can be expected to continue. However, if the parameter ρ is taken as very small compared with the other dimensions, convergence difficulties can be expected. Much larger indices m will be required for the same relative accuracy of $\mathrm{d}(\theta)$. With the presence of the higher ordered additional terms, there is a tendency to introduce high frequency ripples in $\mathrm{d}(\theta)$. The corresponding distortions of the local stresses are frequencly difficult to assess.

TABLE 2.1

Fillet stresses, σ_t, $a = 6$, $b = 2$, $\rho = 1/2$

θ (degrees)	$m = 6$		$m = 8$		$m = 10$	
	d(θ)	σ_t/T	d(θ)	σ_t/T	d(θ)	σ_t/T
15	.0012	−0.72	.0009	−0.70	.0007	−0.72
17	.0035	−0.61	.0014	−0.62	.0016	−0.63
19	.0038	−0.45	.0013	−0.45	.0008	−0.46
21	.0031	−0.21	.0007	−0.21	−.0001	−0.21
23	.0021	+0.08	.0003	+0.10	−.0005	+0.01
25	.0012	+0.42	.0005	+0.43	−.0001	+0.44
27	.0005	+0.78	.0010	+0.78	.0007	+0.77
29	.0001	+1.15	.0016	+1.12	.0014	+1.11
31	.0000	+1.53	.0019	+1.46	.0016	+1.44
33	.0004	+1.88	.0020	+1.81	.0012	+1.79
35	.0014	+2.17	.0020	+2.14	.0006	+2.14
37	.0029	+2.38	.0022	+2.41	.0003	+2.44
39	.0047	+2.49	.0029	+2.57	.0008	+2.62
40	.0056	+2.50	.0035	+2.60	.0014	+2.65
41	.0062	+2.48	.0040	+2.59	.0020	+2.63
43	.0064	+2.39	.0047	+2.48	.0031	+2.50
45	.0033	+2.25	.0031	+2.29	.0022	+2.29

TABLE 2.1a

(Mapping Coefficients used in Table 2.1)

α/a	2.23792	2.23792	2.23805
ε_1/a	1.03867	1.03990	1.04002
ε_3/a	−0.23380	−0.23341	−0.23471
ε_5/a	−0.03596	−0.03652	−0.03659
ε_7/a	0.00202	0.00263	0.00279
ε_9/a	−0.00397	−0.00408	−0.00428
ε_{11}/a	−0.00624	−0.00690	−0.00678
ε_{13}/a		−0.00087	−0.00109
ε_{15}/a		0.00188	0.00228
ε_{17}/a			0.00046
ε_{19}/a			0.00001

2.4 The MMC plus partitioning plan

The remainder of this chapter is devoted to outlining and illustrating an effective approach to notch analysis in which conformal mapping is utilized as a tool but not in the classical manner. The approach is an outgrowth of techniques developed by the authors in the course of previous investigations of two-dimensional crack problems. It is not surprising that techniques for crack analysis carry over to notch analysis. On the other hand the emphasis of the solution plan will differ. Whereas the primary interest in crack analysis is the determination of the intensity of the local crack tip distribution, in notch analysis the distribution as well as the intensity must be considered.

Two of the previous crack analysis techniques, the MMC (modified mapping-collocation) and partitioning plans, are joined for the approach to analysis of notches. It will be shown that, by such an approach, the four basic difficulties outlined in 2.1 for the classical application of mapping can be minimized considerably.

The MMC plan. The MMC (modified mapping-collocation) technique depends philosophically on combining the most attractive features of conformal mapping and boundary collocation arguments. The technique was originally presented [4] as a procedure for the analysis of internal cracks in finite geometries for isotropic materials. Its extension to orthotropic materials was carried out shortly thereafter by the authors [5]. Since then, several classes of notch and crack configurations have been successfully analyzed by the MMC method. The approach is described and illustrated in some detail in Reference [19].

In the MMC plan a major modification of classical mapping procedure is made at the outset. No attempt is made to find the mapping function which maps a rigidly prescribed parameter region into the entire physical region. Instead, only a portion of the parameter region is prescribed and simple mapping forms are chosen to describe key portions of the physical boundary at the discretion of the analyst. Algebraically simple forms of mapping functions can therefore be utilized which, in turn, can be inverted to locate the entire geometry of the parameter region.

To illustrate, we consider the simple crack configurations in Figure 2.6. The classical choice of the parameter region would be a concentric ring. For a fixed finite boundary τ, it is obvious that the corresponding mapping function would be extremely difficult to find. It is here that the

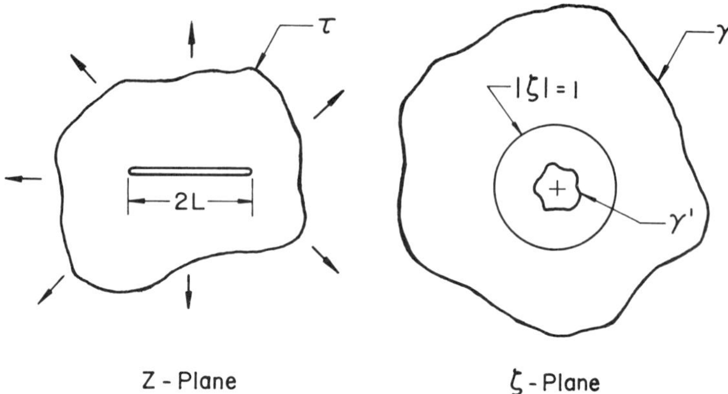

Figure 2.6. Internal crack in a finite region.

analyst makes a simplifying choice in the MMC plan. Only a simple algebraic form of mapping function is chosen to describe a key portion of the physical geometry – in this case, the crack profile. Thus, the analyst chooses the mapping

$$z = \omega(\zeta) = (L/2)(\zeta + \zeta^{-1}). \tag{2.52}$$

The unit circle and its exterior in the ζ-plane are mapped by equation (2.52) into the crack and its exterior in the z-plane. The points in the physical plane can be simply related to the parameter plane by

$$\zeta = (Z/L) + [(Z/L)^2 - 1]^{\frac{1}{2}}. \tag{2.53}$$

In particular, the boundary τ will correspond to some closed curve γ exterior to the unit circle in the ζ-plane (Figure 2.6). The parameter region corresponding to equation (2.52) is therefore the annulus bounded by $|\zeta| = 1$ and γ.

The compromise which is made in the mapping philosophy of the MMC approach does require a non-conventional consideration of the boundary conditions. E.g., in the illustration above γ no longer in general is a circle and the classical arguments for handling the boundary conditions are no longer valid. In the MMC plan a series representation of the stress functions is adopted and a modified plan of boundary collocation is introduced to establish the boundary conditions.

As an illustration we consider again the configuration in Figure 2.6 with traction-free conditions on the crack and a set of applied forces on τ. The advantage of mapping the crack profile onto the unit circle is now evident for the traction-free conditions can be handled by applying the continuation arguments of section 2.2. In particular, if the extension described by equation (2.19)–(2.21) is adopted, traction-free conditions on the crack are ensured if the extended stress function $\phi(\zeta)$ is analytic in the annulus bounded by γ and γ', where γ' is the inverted image of γ with respect to the unit circle. A series representation of $\phi(\zeta)$ converging in this annulus therefore satisfies the crack conditions. Since this is a finite doubly connected region, it is assumed that $\phi(\zeta)$ has a Laurent expansion

$$\phi(\zeta) = \sum_{-\infty}^{\infty} \alpha_n \zeta^n. \tag{2.54}$$

To complete the solution, the α_n's are determined numerically by a modified plan of boundary collocation on γ corresponding to the specified tractions on τ in the physical plane.

The MMC plan of modification of conventional boundary collocation depends on heuristic arguments of assessment of the effects of residual errors in the boundary conditions on the accuracy of the desired information. It is well-known from experience that strict collocation, i.e., matching N boundary conditions at a set of discrete stations with a set of functions with N degrees of freedom can often lead to unreliable solutions. Errors in conditions on the 'off-point' intervals of the boundary are difficult to assess and often 'apparent' convergence with truncation can lead to an incorrect result. An improvement to strict collocation was proposed by Hulbert [20] who suggested over-specification of the conditions and then a satisfaction of these conditions in a least square sense. This approach certainly tends to yield a desirable smoothing of the error for a given number of degrees of freedom of the representation.

In the MMC argument, the effects of residual boundary errors are grouped into two types, namely, remote and local, depending on the information desired. In the problems relating to cracks, it can be assumed that the desired information is the crack-tip stress intensity factor. Errors in the boundary conditions in this case can be considered as local or remote depending on their proximity geometrically from the crack-tip vicinity.

The effects of residual boundary errors of the remote type can be assessed by Saint Venant's argument. In particular consider the resultant force $f_1 + if_2$ in equation (2.5) or (2.10), calculated as a function of the arc length on the boundary. If, in the process of collocation, $f_1 + if_2$ is satisfied at two successive boundary stations, then the residual error in stress boundary conditions in the intermediate boundary interval must correspond to a self-equilibrating distribution of force. A similar calculation can be made for the resultant moment for a strict adherence to the Saint Venant argument. (In practice, the moment conditions are awkward and the effect on most solutions in the collocation plan is negligible.)

The effects of residual boundary errors of the local type must be considered in more detail. Even when the residual boundary errors have been reduced to self-equilibrating distributions by satisfaction of $f_1 + if_2$, local effects of the stress errors must be considered. Thus, in the MMC plan, residual errors of the local type are handled by imposing both the conditions on $f_1 + if_2$ and the local applied stress conditions.

The collocation plan of the MMC method combines least squares with the error measures described above. The plan is versatile and removes many of the difficulties in handling boundary conditions for contours with irregular boundaries. Newman [21, 22] independently studied the use of $f_1 + if_2$ in the least square sense.

Although conceptually the MMC plan applies to both notch and crack problems, there are practical differences to consider. In the original MMC plan a global expansion of the solution is implied. In many crack problems only a few terms are required to find stress intensity factors to within satisfactory accuracy. A plausible explanation of this matter can be made from asymptotic considerations. In crack problems only the dominant intensity of the crack-tip stresses is sought – the corresponding local distribution is assumed known. Thus, it is easy to imagine from a Saint Venant viewpoint many existing solution approximations which yield this information asymptotically. On the other hand for notch problems we generally seek both the intensities and distributions of the stresses in the notch vicinity. The dominance of a single term of the expansion is no longer obvious. This, in turn, is reflected in a global expansion by slowness of convergence. To by-pass the difficulties familiar to global expansion as well as to provide a means for effective local expansion in the notch vicinity, we now introduce the use of partitioning concepts.

Partitioning concepts. The partitioning plan which will now be described was originally presented by the authors in [6]. The plan is based on continuation arguments taken from analytic function theory and applied in the discrete to 'stitch' several power series expansions of the stress function in appropriate subregions of the geometry.

Let us consider the two-dimensional physical region illustrated in Figure 2.7 and the problem of determining the two analytic functions $\phi(z)$ and $\psi(z)$ defined in this region which satisfy prescribed loading conditions on the boundary. Representation of each of the stress functions by a single series expansion (global) relies on the physical region lying within the circle of convergence of each series. It is clear there are problems for which mathematical singularities exist such that no point of expansion can be found for which the circle of convergence overlaps the physical region.

Suppose therefore we consider a partitioning of the region by τ into the two subregions R_1 and R_2. Furthermore, we define

$$\begin{aligned}\phi(z) = \phi_1(z) &= \sum_0^\infty a_n(z - z_1)^n, \quad z \in R_1\\ = \phi_2(z) &= \sum_0^\infty b_n(z - z_2)^n, \quad z \in R_2\end{aligned} \tag{2.55}$$

with a similar representation for $\psi(z)$.

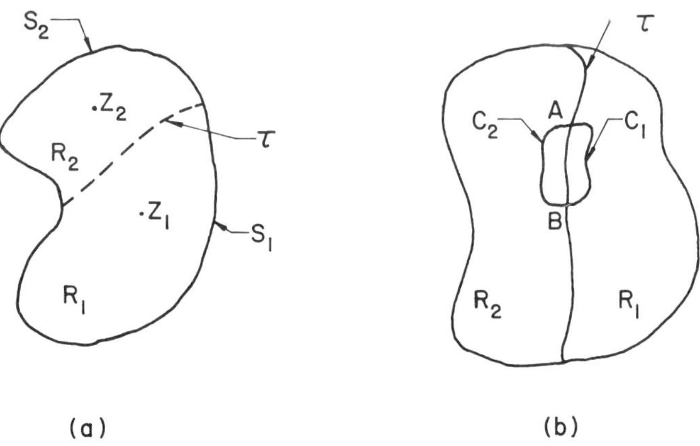

Figure 2.7. Example of partitioning.

The stress functions $\phi(z)$ and $\psi(z)$ are uniquely determined to within a linear function describing a rigid-body motion by the prescribed loading conditions on the boundary $S_1 + S_2$. Clearly, the boundary condition on $S_1(S_2)$ alone are not sufficient to determine ϕ_1 and ψ_1 (ϕ_2 and ψ_2). In addition we must require that ϕ_1 and ϕ_2 are elements of the same function $\phi(z)$. Similarly, ψ_1 and ψ_2 must be elements of the function $\psi(z)$. The conditions under which these elements are analytic continuations of the same analytic function will henceforth be referred to as 'stitching' conditions.

If strictly continuous arguments are used in the solution process, a simple requirement for stitching conditions follows directly from analytic function theory. If $\phi_1(z)$ and $\phi_2(z)$ are analytic in R_1 and R_2, respectively, then a necessary and sufficient condition that $\phi_1(z)$ and $\phi_2(z)$ be the elements of a single analytic function $\phi(z)$ (and similarly for $\psi(z)$) is

$$\phi_1(z) = \phi_2(z); \qquad \psi_1(z) = \psi_2(z), \quad z \in \tau. \tag{2.56}$$

Actually, equation (2.56) can be replaced by continuity across any finite subinterval of τ or, even more generally, continuity across any finite arc common to the two regions of convergence.

In principle, a partitioning plan of solution is evident. The region is partitioned by the analyst in a manner consistent with by-passing anticipated convergence difficulties and local expansions are set up. The set of conditions corresponding to the stitching conditions and the boundary conditions corresponding to each partitioned region will determine the elements of $\phi(z)$ and $\psi(z)$. Clearly, the previous argument extends to any finite number of partitioned elements. Furthermore, any of the elements can be individually handled whenever desired by the mapping techniques in the MMC method.

In practice, a pointwise satisfaction of the stitching conditions is utilized. Since the previous arguments were based on continuity on a continuous interval, it is reasonable to reconsider the implications of equation (2.56) when applied in a discrete sense. It can be seen immediately that the situation is somewhat unsatisfactory. Since the derivatives of the stress functions appear in equations (2.2) and (2.3), continuity of $\phi(z)$ and $\psi(z)$ across discrete points of τ does not assure continuous stresses or even continuous displacements at these points.

On physical grounds we shall require as a minimum the continuity of $u + iv$ and $f_1 + if_2$ at the discrete points on τ. The physical implications of the continuity of the displacements are obvious. The implications of the continuity of the force conditions are illustrated in Figure 2.7b. Consider the interval AB on τ between two successive collocation stations. Let C_1 and C_2 be curves lying in R_1 and R_2, respectively. If the force conditions across τ are continuous at A and B, then the total force resultant over the closed contour $C_1 + C_2$ is zero (as it should be for equilibrium).

From equations (2.3) and (2.5), continuity of $u + iv$ and $f_1 + if_2$ across τ is equivalent to the corresponding continuity of $\phi(z)$ and $\bar{z}\phi'(z) + \psi(z)$. It is suggested therefore, as a minimum requirement for discrete applications, that the stitching conditions in equation (2.56) be replaced by

$$\begin{aligned} &\phi_1(z_k) = \phi_2(z_k), \quad z_k \in \tau \\ &\bar{z}_k\phi_1'(z_k) + \psi_1(z_k) = \bar{z}_k\phi_2'(z_k) + \psi_2(z_k), \quad z_k \in \tau. \end{aligned} \tag{2.57}$$

It is obvious that equations (2.56) and (2.57) are equivalent when z_k represents a continuous description of τ.

In the examples to follow the stitching conditions equation (2.57) will be adopted unless stated otherwise. Actually this condition does not assure stress continuity across τ. Although this matter can be corrected by additional conditions on the derivatives of the stress functions, the disadvantages of the corresponding computational burden must be weighed. If the stitching locus is chosen in such a manner that it is somewhat remote to the region of interest, it is reasonable to assume that local inaccuracies in a stress continuity can be tolerated.

It has already been demonstrated that the MMC plus partitioning plan overcomes the representation difficulties encountered in the original MMC plan alone. An interesting example is provided by the classical problem of a central crack in a rectangular panel under tension, Figure 2.8. A solution to this problem was carried out [23] using the MMC method with a single power series expansion of the solution. Convergence difficulties arise for $h/W < 2.5$, particularly for the deeper cracks. In fact for all h/W values there are convergence difficulties for $2L/W > 0.8$.

In [6], the solution was carried out using the partitioning plan indicated in Figure 2.8. Region I was represented by series for $\phi(z)$ and $\psi(z)$ expanded about z_1. Region II was represented by the MMC plan using the mapping equation (2.52). Traction-free conditions on the crack were imposed by the continuation arguments. By introducing the ob-

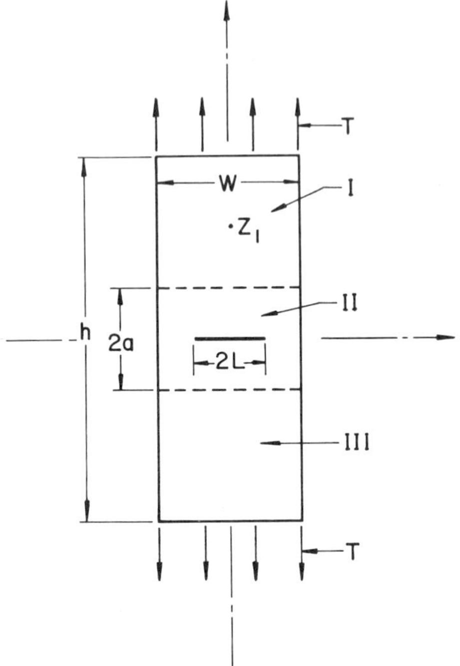

Figure 2.8. Central crack in a rectangular plate.

vious symmetries of the solution into the expansions for Region II, one avoids the necessity of considering Region III explicitly. The results were carried out for $h/W = 3.0$ with the central zone defined by $2a/W = 0.2$. The range of validity of the solution was remarkably increased. Based on the known values for the limiting case, an accuracy of better than one percent was obtained for $2L/W$ values as large as 0.97 using the partitioning plan described above.

Auxiliary use of finite elements. We should consider at this point possible alternative approaches to overcoming the difficulties of global representation. Such approaches as finite-difference of finite element techniques obviously do not depend directly on global representation. However, such methods have accuracy problems in regions with rapidly varying stress gradients. In crack problems some of these difficulties have been overcome by building in specialized local displacement fields in the local elements based on the known character of the local crack-tip displacements. For the general notch problem, however, this device is not obvious as the singular nature of the solution is not precisely understood. The alternative of flooding the notch region with small elements would appear to be uneconomic as well as indecisive.

On the other hand the use of finite elements in conjunction with the MMC plus partitioning plan is a feasible approach. Freese [7] has demonstrated that regions of localized high-stress gradients, e.g. crack tips, elliptical notches, etc. can be effectively handled by using a continuous representation locally and joining (stitching) to a finite element representation elsewhere. Another illustration of this approach is the solution [8] of a square cut-out with circular fillets in a tensile field, Figure 2.9.

The partitioning plan indicated in Figure 2.9 utilizes obvious symmetries so that only a quarter of the panel need be explicitly considered. Continuous representations of the solution were taken in regions I, II, III. Region IV was represented by a set of cubic isoparametric finite

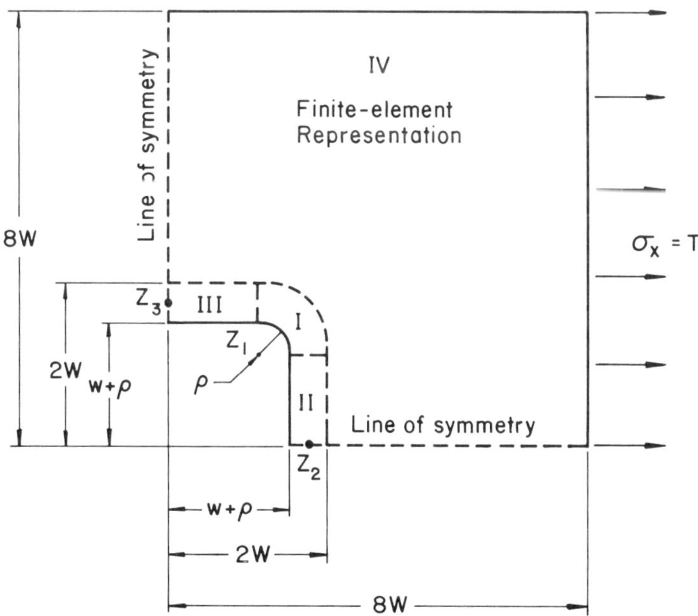

Figure 2.9. Partitioning of the cut-out problem.

elements. Regions II and III were represented by Taylor expansions about z_2 and z_3, respectively, for the stress functions $\phi(z)$ and $\psi(z)$. Suitable properties of the coefficients in these expansions were chosen to ensure symmetries across the real axis in region II and across the imaginery axis in region III.

Conformal mapping was introduced to facilitate the respresentation in region I. The mapping

$$z = \omega(\zeta) = i\rho\, e^{-i\zeta} + z_1 \tag{2.58}$$

maps the rectangular region S_1 in Figure 2.10 into the region I in Figure 2.9. In particular, the fillet maps into the interval of the real axis, $0 \leq \xi \leq \pi/2$. This particular choice of mapping is very useful for notches with constant radii of curvature over an interval of the tip. Load-free conditions on the fillet are ensured by the continuation of $\phi(\zeta)$ across the real axis into S_2 by defining

$$\psi(\zeta) = -\bar{\phi}(\zeta) - \bar{\omega}(\zeta)\phi'(\zeta)/\omega'(\zeta). \tag{2.59}$$

A power series expansion about ζ_0 was then chosen for $\phi(\zeta)$.

Although we have already discussed the stitching conditions between continuous representations of the solution, the joining of continuous representations to the finite-element description in region IV requires further consideration. The situation is complicated by the difficulty of establishing a precise equivalence of $f_1 + if_2$ between the continuous and finite-element representations. Although such an equivalence can be approximated by interpolation, this approach is awkward at best.

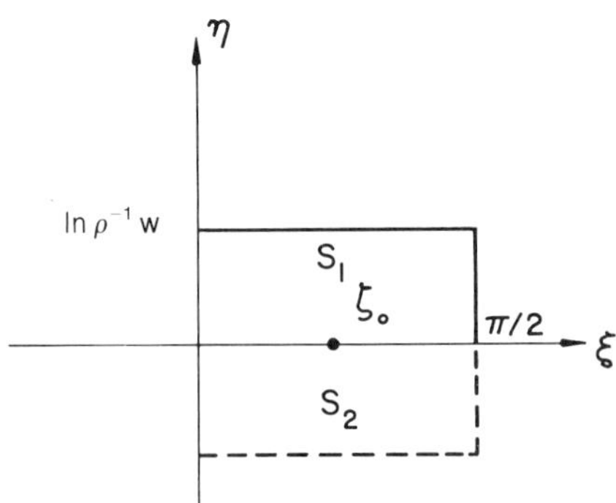

Figure 2.10. Parameter region S_1 in the ζ-plane corresponding to region I.

Freese [7] suggested a procedure for joining the two types of representations which is simple, yet effective. At the boundary of the finite-element and continuous representations, one can extend 'fictitiously' the finite elements one more layer of elements into the 'continuous' regions. The stitching condition is then taken as the continuity of $2\mu(u + iv)$ at *all* nodal points of the fictitious elements. Conceptually, this argument is similar to the use of fictitious points in specifying normal derivatives on the boundary in finite-difference methods.

In the original partitioning plan, the stitching and boundary conditions were solved in a least-square collocation sense. Although the complete system of equations in the present case could also have been set up in a least-square sense, a considerable loss of computational time would result from not utilizing the 'bandedness' of the system corresponding to the finite element equations. In [7], Freese adopted the procedure of first computing the 'sub-structured' system from the finite-element conditions. The stitching condition, i.e. continuity of displacements in the continuous and finite-element representations at all nodal points of the fictitious elements, was then prescribed on the substructured system. To this system was added the remaining boundary and stitching conditions between the continuous representations, and, this was the system which was solved in a least-square sense. Thus, not only was the bandedness of the finite-element matrix utilized; a further economy was realized by the fact that the substructured system required no regeneration in parameter studies of the fillet geometry.

2.5 Semi-elliptical notch in a semi-infinite sheet

As a prelude to cataloguing several notch solutions in 2.6, we now describe in some detail the two approaches, MMC and MMC plus partitioning, for the classical problem of a semi-elliptical notch in a semi-infinite sheet under tension, Figure 2.11. These solutions afford a striking comparison of the effectiveness of localized expansions permitted by partitioning in contrast with a global representation. Considerable insight can also be gained as to the nature of the singularities involved and their effects on solution convergence.

The MMC solution. The MMC solution was carried out utilizing a reflection argument originally proposed by Bowie [24] for handling edge notches in a semi-infinite region. Consider the material with the edge notch as occupying the lower half plane S_z^- with the boundary of the notch denoted as C^-. Now extend $\phi(z)$ into S_z^+ by the definition,

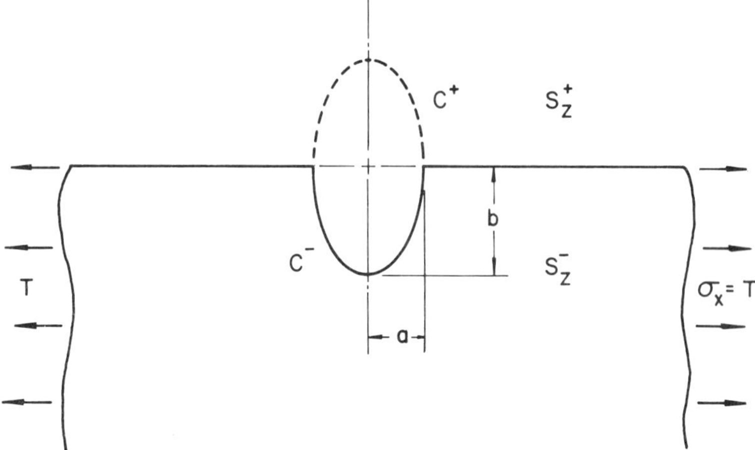

Figure 2.11. Semi-elliptical notch in a semi-infinite sheet.

equation (2.11). $\phi(z)$ is now defined in the region exterior to $C = C^+ + C^-$ where C^+ is the reflection of C^- across the real axis. If $\phi(z)$ is now considered as analytic exterior to C (except at infinity), then, from equation (2.14),

$$\phi(z) - \phi(\bar{z}) + (z - \bar{z})\overline{\phi'(z)} = f_1(s) + if_2(s) \tag{2.14'}$$

That portion of the boundary lying along the real axis can be considered as traction-free. Apart from the conditions at infinity, the remaining boundary conditions correspond to traction-free (or $f_1 + if_2 = 0$) conditions on C^-. Thus, we require

$$\phi(z) - \phi(\bar{z}) + (z - \bar{z})\overline{\phi'(z)} = 0, \quad z \in C^-. \tag{2.60}$$

It is at this point we introduce the auxiliary ζ-plane and the mapping

$$z = \omega(\zeta) = R\zeta + S\zeta^{-1} \tag{2.61}$$

where $R = (a + b)/2$ and $S = (a - b)/2$. The unit circle $\zeta = \sigma = e^{i\theta}$ and its exterior are mapped into C and its exterior, respectively, by equation (2.61). The stress function $\phi(z)$ can now be considered as $\phi(\zeta)$ in the usual manner and condition (2.60) becomes

$$\phi(\sigma)-\phi(1/\sigma)+(R-S)(\sigma-1/\sigma)\overline{\phi'(\sigma)}/\overline{\omega'(\sigma)}=0,\quad \pi\leqslant\theta\leqslant 2\pi. \qquad (2.62)$$

The loading conditions at infinity and the symmetries of the solution are ensured by choosing

$$\phi(\zeta)=(T/4)\omega(\zeta)+\phi_1(\zeta) \qquad (2.63)$$

where

$$\phi_1(\zeta)=T\left\{\sum_{n=1}^{\infty}a_{2n}\zeta^{1-2n}+i\sum_{n=1}^{\infty}a_{2n+1}\zeta^{-2n}\right\} \qquad (2.64)$$

and the coefficients a_K are real. Defining $G(\sigma)$ as the left hand side of equation (2.62) and substituting equation (2.63), the boundary condition on the notch becomes

$$\begin{aligned} G(\sigma)=\phi_1(\sigma)-\phi_1(1/\sigma)+(R-S)(\sigma-1/\sigma)\overline{\phi_1'(\sigma)}/\overline{\omega'(\sigma)}\\ +(T/2)(R-S)(\sigma-1/\sigma)=0,\quad \pi\leqslant\theta\leqslant 2\pi. \end{aligned} \qquad (2.65)$$

In [24], the condition (2.65) was satisfied by minimizing the square of $G(\sigma)$ over the notch interval with respect to the coefficients a_K in equation (2.64). The system of equations

$$\frac{\partial}{\partial a_K}\left\{\int_{\pi}^{2\pi}[G(\sigma)]^2\,d\theta\right\}=0,\quad K=1,2,3,\ldots \qquad (2.66)$$

provides a linear system of simultaneous equations for the determination of the unknowns, a_n. In the MMC method, the vanishing of $G(\sigma)$ was imposed in a least square boundary collocation sense. The two plans yield nearly identical results even for the deeper notches.

(It should be noted here that the choice of mapping function plays a considerable role in boundary collocation arguments. Since

$$dz=\omega'(\zeta)\,d\zeta, \qquad (2.67)$$

it is clear that in intervals where $\omega'(\zeta)\approx 0$ a natural 'packing' or weighting of points of collocation results. For deep notches or cracks this situation exists in intervals where the corresponding accuracy is most important.)

The maximum stress, which occurs of course at the notch tip, is listed in Table 2.2 below for several $\lambda = a/b$ ratios. For purposes of comparison, σ^*_{max}, the peak stress for an elliptical hole in an infinite sheet with a corresponding uniaxial tension at infinity is presented, where

$$\sigma^*_{max} = T(1 + 2/\lambda). \tag{2.68}$$

By calculating the ratio Q, where

$$Q = \sigma_{max}/\sigma^*_{max} \tag{2.69}$$

and utilizing Koiter's result for the limiting case of an edge crack, $\lambda \to 0$, it is possible to provide an accurate interpolation over the total range, $0 \leqslant \lambda \leqslant 1$.

TABLE 2.2

Maximum stresses for semi-elliptic notches

$\lambda = a/b$	σ_{max}/T	σ^*_{max}/T	Q
1.0000	3.065	3.000	1.022
0.8182	3.540	3.444	1.028
0.6667	4.136	4.000	1.034
0.5358	4.910	4.714	1.042
0.4286	5.948	5.667	1.050
0.3333	7.412	7.000	1.059
0.2500	9.625	9.000	1.069
0.1765	13.320	12.333	1.080
0.1111	By interpolation	→	1.092
0.0526	By interpolation	→	1.106
0.0000	Koiter's result	→	1.1215

Convergence difficulties were encountered for $\lambda < 1/4$. In order to obtain the tabulated value for $\lambda = 0.1765$, double precision on a large scale digital computer and approximately one hundred series terms were required. The cause of the difficulty is apparent in this formulation. In the function $G(\sigma)$ the function $\omega'(\sigma)$ occurs as the denominator of one of the terms. The zeroes of $\omega'(\zeta)$ occur at

$$\zeta = \pm i\sqrt{\frac{1-\lambda}{1+\lambda}}. \tag{2.70}$$

For the deeper notches as $\lambda \to 0$, the zeros of $\omega'(\zeta)$ approach the unit circle. A nearby pole in $\phi(\zeta)$ can be inferred in such a situation. For the limiting case of a crack, $\lambda = 0$, the function $\omega'(\zeta)$ has zeroes at $\zeta = \pm i$. Since $\phi_1(-i)$ must be finite, it is clear that $\phi(\zeta)$ must have a simple pole at $\zeta = i$. For the limiting case of a crack, the series (2.64) would diverge. The nearby presence of a comparable singularity in $\phi(\zeta)$ for the deep notches is reflected by the slow convergence of the global expansion.

The MMC plus partitioning solution. The same problem will now be considered with the use of the MMC plus partitioning plan of solution. The configuration and partitioning plan is indicated in Figure 2.12.

The region S_z^+ is considered as partitioned by the semi-circular arc $\widehat{AB}$. Zone I is the intersection of S_z^+ and $|z| \geq 2b$. Zone II is the simply connected region *ABEDCA*.

In zone I, reflection is used to ensure loadfree conditions on the real axis, $|x| \geq 2b$, thus,

$$\psi(z) = -\bar{\phi}(z) - z\phi'(z), \quad z \in S_z^+. \tag{2.71}$$

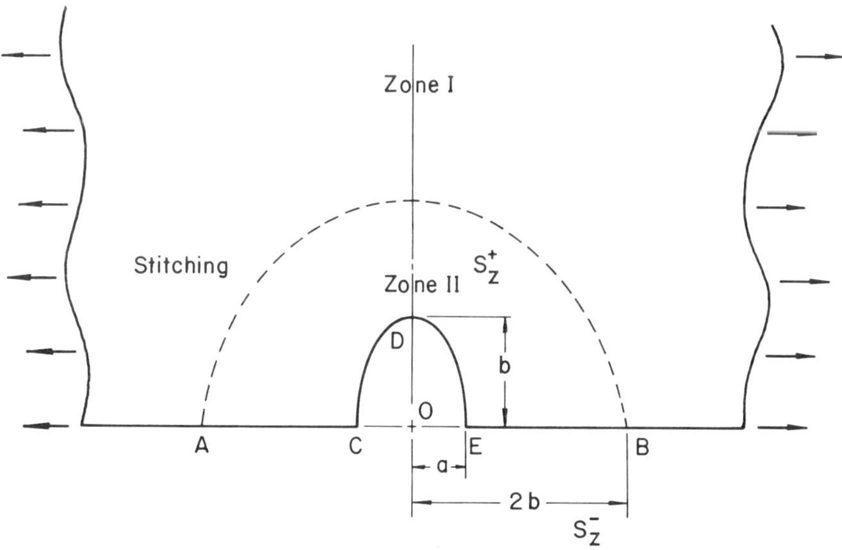

Figure 2.12. Partitioned region for semi-elliptical notch.

The extended stress function $\phi(z)$ is now represented as

$$\phi(z) = Tz/4 + T\sum_{n=0}^{\infty} \alpha_n z^{-n}, \quad |z| \geq 2b \tag{2.72}$$

where the α_n's are purely real (odd powers) or purely imaginery (even powers). The symmetries and loading conditions in zone I have been accounted for by equations (2.71) and (2.72).

For the representation in zone II, the mapping function

$$z = i\{R\,\mathrm{e}^{-i\zeta} - S\,\mathrm{e}^{i\zeta}\} \tag{2.73}$$

is introduced where $R = (a+b)/2$ and $S = (a-b)/2$. (This mapping is essentially the product transformation of equations (2.58) and (2.61).) A typical shape of the parameter region, S_ζ^+, is shown in Figure 2.13. The elliptical notch maps on the real interval, $-\pi/2 \leq \xi \leq \pi/2$. Load-free conditions on the notch are ensured by reflection across the ξ-axis. Thus, the representation in zone II corresponds to

$$\psi(\zeta) = -\bar{\phi}(\zeta) - \bar{\omega}(\zeta)\phi'(\zeta)/\omega'(\zeta), \quad \zeta \in S_\zeta^+ \tag{2.74}$$

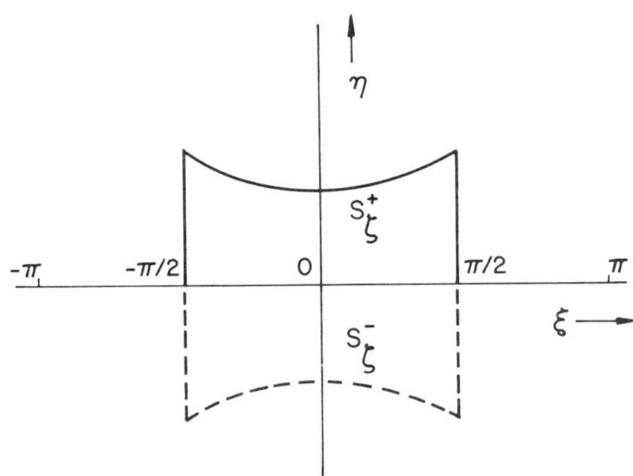

Figure 2.13. Parameter region for zone II.

and

$$\phi(\zeta) = bT \sum_{n=1}^{\infty} \beta_n \zeta^n, \quad \zeta \in S_\zeta^+ + S_\zeta^- \tag{2.75}$$

where the β_n's are purely real (odd powers) or purely imaginery (even powers). With this representation the symmetries with respect to the imaginery axis are ensured as well as the loading conditions on the notch. To complete the solution, we must impose the force conditions $f_1 + if_2 = 0$ on EB using the representation in zone II and the stitching conditions on the circular arc AB using the expansions in zone I and zone II.

The expansion (2.75) is particularly effective for a study of the local notch stresses. The point $\zeta = 0$ about which the expansion is taken corresponds to the notch tip, D, and, in this case coincides with the maximum stress. In fact the maximum stress is determined from the determination of the coefficient β_1 in equation (2.75).

The analysis leads to two rapidly converging series for equations (2.72) and (2.75). In the numerical calculations equal powers were retained in the truncated forms of the two expansions. The maximum stress σ_θ/T at D as computed from the expansion in zone II is listed in Table 2.3 as a function of the truncation index, M, representing the number of coefficients used in equation (2.75). Listed also are the values $\sigma_{max} = Q\sigma^*_{max}$ taken from Table 2.2. It is evident that even for the deep notches an extremely effective solution is obtained. For $\lambda = 0.01$ it is

TABLE 2.3

Maximum stresses for semi-elliptic notches with number of coefficients, M, in (2.75)

$\lambda = a/b$	$M = 10$	$M = 12$	$M = 15$	$Q\sigma^*_{max}/T$
1.00	3.069	3.063	3.065	3.065
0.50	5.229	5.217	5.220	5.220
0.25	9.643	9.619	9.621	9.621
0.20	11.870	11.840	11.840	11.840
0.10	23.060	23.000	23.000	23.000
0.05	45.520	45.390	45.400	45.360
0.02	113.000	112.700	112.700	112.600
0.01	225.400	224.800	224.800	224.900

doubtful whether the global solution could be obtained at all by the MMC method, yet, the local solution above with only twelve terms gives an accurate solution for the maximum stress. An additional indication of convergence of the local expansion is that the stress at point E is very nearly zero as it should be for a free corner. The series (2.75) with twelve terms can be expected therefore to provide an accurate description of the stresses in zone II except perhaps in the vicinity of the stitching locus AB. (It will be recalled that the stitching plan does not ensure continuity of stresses across the stitching locus of the two expansions.)

The effectiveness of the expansion in zone II is an indication that the singularities which led to the convergence difficulties in the global expansion have been by-passed by the argument above. Apparently no singularity in the stress function occurs at points C and E and the difficulty lies in the type of singularities described in equation (2.70). It is clear that the mapping (2.73) keeps these points well-removed from point of expansion. For the limiting case of an edge crack, a singularity in $\phi(z)$ would be expected at $z = -ib$ according to the previous discussion. This point (a branch point in the mapping (2.73)) would map into $\zeta = \pm\pi$ in Figure 2.13. The expansion on $S_\zeta^+ + S_\zeta^-$ is clearly well-removed from these points of singularity. The procedure can therefore be expected to be valid for the calculation of the stress intensity factor for an edge crack of length b. In fact, the stress intensity factor K is given by

$$K = 2\phi'(0)/\sqrt{i\omega''(0)}. \qquad (2.76)$$

The values 1.1246 and 1.1216 for $K/T\sqrt{b}$ were obtained for $M = 10$ and $M = 12$, respectively, which can be compared with the well-known value of 1.1215 for this problem.

The coefficients β_n for several of the deeper notches are recorded in Table 2.4. It is evident that approximately seven terms of the series are required for the calculation of the stresses of interest to within one percent accuracy. This provides a convenient manner of cataloguing the solution for the region of interest in contrast with recording the many terms required in a global expansion.

2.6 Several notch solutions

It was demonstrated in section 2.5 that the MMC plus partitioning plan can lead to remarkably effective solutions to two-dimensional notch

TABLE 2.4

Coefficients β_n, semi-elliptical notch, $M = 12$

β_n	$\lambda = 0.20$	$\lambda = 0.10$	$\lambda = 0.05$	$\lambda = 0.02$	$\lambda = 0.01$
β_1	.59192	.57492	.56740	.56330	.56201
β_2/i	.07603	.07152	.06885	.06707	.06645
β_3	−.07678	−.07048	−.06708	−.06495	−.06422
β_4/i	−.01680	−.01696	−.01702	−.01704	−.01705
β_5	.00783	.00791	.00797	.00802	.00803
β_6/i	−.00015	−.00001	.00006	.00011	.00013
β_7	.00027	.00030	.00032	.00032	.00032
β_8/i	−.00003	.00001	.00003	.00004	.00005
β_9	.00013	.00010	.00009	.00008	.00008
β_{10}/i	−.00004	.00004	.00004	.00004	.00004
β_{11}	−.00002	−.00001	−.00001	−.00001	−.00001
β_{12}/i	.00000	.00000	.00000	.00000	.00000

problems. A global representation with its associated convergence difficulties is by-passed and rapidly converging localized expansions of the solution can be obtained. These expansions local to the notch provide a convenient plan for the cataloguing of the essential information derived numerically for each set of parameters. With this information not only the peak notch stresses but also the distribution and the local subsurface stresses can readily be evaluated without recourse to the global solution.

In this section, the cataloguing plan is illustrated by providing the local expansions for several well-known notch configurations. Edge notches in a semi-infinite sheet and notched cutouts in an infinite sheet under tension will be considered. In most cases the peak stresses have been recorded elsewhere, e.g. Peterson [25], but effective localized expansions are not available. The problem choice is dictated primarily by the space limitations in a chapter of this type. Alternative loading systems or choice of specimen geometry, although causing little disturbance in the overall solution plan, would obviously lead to an unwieldy set of solution parameters.

Primary emphasis is placed on configurations in which the notch tip is an interval of a circular arc. The mapping of the type in equation (2.58) provides a convenient parametric representation of the region local to the notch tips. The coefficients of the corresponding local expansions are catalogued numerically in the Appendix for each problem.

Circular-arc edge notch. We consider the solution of a circular-arc edge notch in a semi-infinite sheet under tension, $\sigma_x = T$, as indicated in Figure 2.14. When $\alpha = \pi/2$ the configuration in Figure 2.14 reduces to the classical problem of a semi-circular edge notch in a semi-infinite sheet.

The partitioning plan for the range $0 < \alpha \leqslant \pi/2$ is indicated in Figure 2.14. The stitching locus CB was taken as a semi-circle with radius

$$|z| = S = 2\rho(1 + \cos\alpha), \quad 0 < \alpha \leqslant \pi/2 \tag{2.77}$$

which divides the physical region into the two indicated zones. For the range $\pi/2 < \alpha < \pi$ (notches with a 'thumbnail' character) the partitioning plan was slightly altered. Again, the stitching locus was taken as a semi-circle, but, in this case the radius was taken as

$$|z| = S = \rho \sin\alpha, \qquad \pi/2 < \alpha < \pi. \tag{2.78}$$

In zone II, the solution for $|z| \geqslant S$ was represented in a form identical with equations (2.71) and (2.72). With this representation only the stiching conditions on CB require satisfaction.

For the representation in zone I, the mapping function

$$z = \omega(\zeta) = i\rho(e^{-i\zeta} + \cos\alpha) \tag{2.79}$$

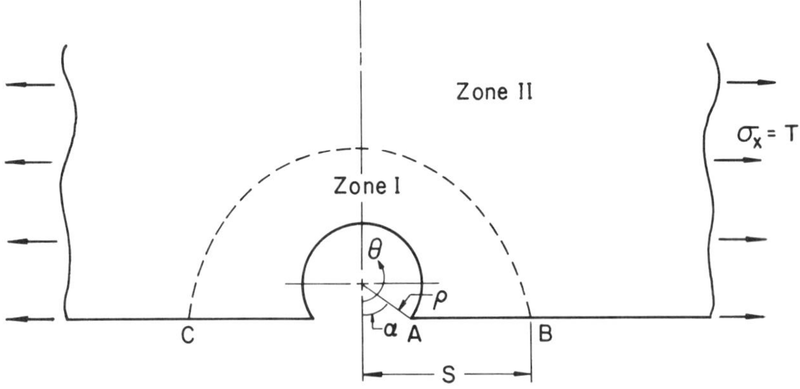

Figure 2.14. Circular-arc edge notch.

was introduced. The notch corresponds to the real interval $-\pi + \alpha \leqslant \xi \leqslant \pi - \alpha$ in the ζ-plane. Load-free conditions on the notch are ensured by reflection across the ξ-axis, i.e.,

$$\psi(\zeta) = -\bar{\phi}(\zeta) - \bar{\omega}(\zeta)\phi'(\zeta)/\omega'(\zeta). \tag{2.80}$$

Representation of the extended definition of $\phi(\zeta)$ is taken as a series expansion about the point $\zeta = 0$, thus

$$\phi(\zeta) = \rho T \sum_{n=1}^{\infty} \beta_n \zeta^n \tag{2.81}$$

where, from symmetry considerations, the β_n's are purely real (odd powers) or purely imaginery (even powers).

To complete the solution the force conditions $f_1 + if_2 = 0$ on AB using the expansion in zone I must be satisfied along with the stitching conditions on the circular arc CB using the expansions in zone I and zone II. For the parameter range $\pi/2 < \alpha < \pi$, due to the alternate choice of partitioning, the segment $AB = 0$ and only the stitching requirements are necessary for the solution.

Calculation of the stresses on the notch boundary is simplified by the fact that the normal stress is zero, thus, from equation (2.7)

$$\sigma_\theta = 4\,\mathrm{Re}\{\phi'(\zeta)/\omega'(\zeta)\}. \tag{2.82}$$

The maximum stress, $\sigma_{\max}$, occurs at $\xi = 0$ and it can be calculated from the simple relation

$$\sigma_{\max}/T = 4\beta_1. \tag{2.83}$$

The subsurface stresses local to the notch tip can be calculated from a knowledge of the coefficients β_n, using equations (2.7), (2.8), (2.79), (2.80) and (2.81).

Numerical results are presented in the Appendix. In Figure 2.24 the peak notch stress, $\sigma_{\max}/T$, is presented as a function of the parameter α. Typical notch stress distributions are included to indicate the change in character of the distribution as the notch approaches the thumbnail type. For $\alpha > 140°$ it will be noted that the point A ceases to act as a 'free' corner. In Table 2.5, the coefficients β_n are tabulated for a ten term expansion of equation (2.81). Although the coefficients appear to be diverging for $\alpha = 150°$ and $175°$, it should be considered that in the range of validity of this local expansion the corresponding values of ζ are very small and, indeed, the expansion converges in the region of interest.

U-notch in semi-infinite sheet. We consider now the problem of a U-notch in a semi-infinite sheet under tension, Figure 2.15. A partitioning plan is shown which, utilizing symmetry, requires representation of the solution in three zones. In zone I, equations (2.71) and (2.72) were again used and validity of this representation was assumed for $|z| \geqslant 2b$. In zone III, reflection across the real axis was assumed and a series expansion for $\phi(z)$ was taken using positive powers of $z - z_0$. For zone II, the mapping function

$$z = \omega(\zeta) = i\rho(\mathrm{e}^{-i\zeta} + a\rho^{-1}) \tag{2.84}$$

was introduced. The notch interval maps into the segment of the real axis $-\pi/2 \leqslant \xi \leqslant \pi/2$ in the ζ-plane and the point $\zeta = 0$ maps into the notch tip, A. Reflection across the real axis of the parameter plane ensures load-free boundary conditions on the notch, thus

$$\psi(\zeta) = -\bar{\phi}(\zeta) - \bar{\omega}(\zeta)\phi'(\zeta)/\omega'(\zeta). \tag{2.85}$$

The stress function $\phi(\zeta)$ is again represented in the extended parameter

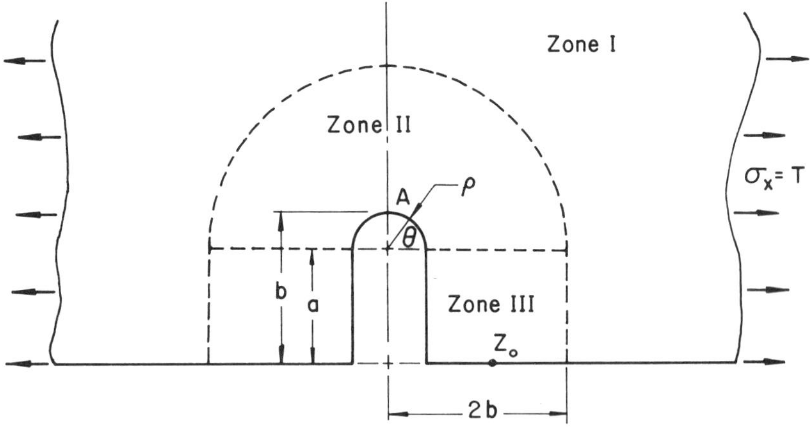

Figure 2.15. U-notch in a semi-infinite sheet.

region by the series

$$\phi(\zeta) = \rho T \sum_{n=1}^{\infty} \beta_n \zeta^n \tag{2.86}$$

where the β_n's are purely real (odd powers) or purely imaginery (even powers).

The results are catalogued in the Appendix. Typical notch stress distributions are shown in Figure 2.25. Again, since the maximum stress occurs at the notch tip, A, equation (2.86) is particularly effective for this calculation. In fact,

$$\sigma_{\max}/T = 4\beta_1 \tag{2.87}$$

where β_1 can be found in Table 2.6. In Table 2.6, the first ten coefficients of the expansion (2.86) as a function of b/ρ are catalogued. The validity of the local expansion for the subsurface stresses is confined to zone II with some allowance made for local inaccuracies near the stitching locus.

The solution above is particularly interesting with regard to a fairly large class of 'equivalent' notch configuration, Figure 2.16. The figure

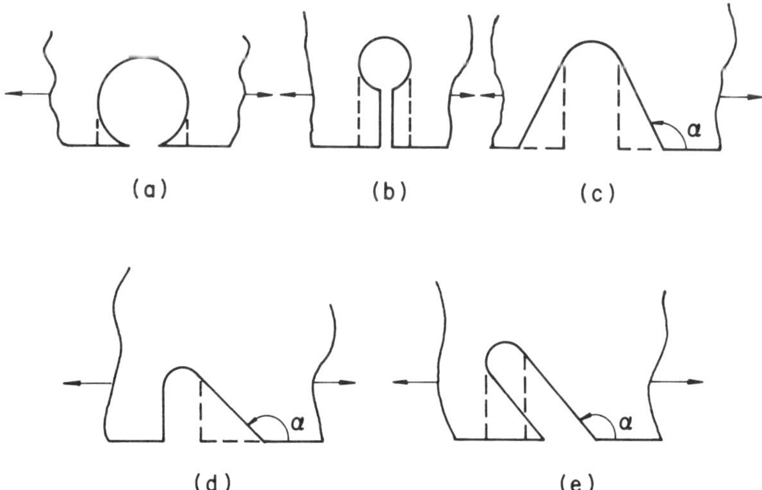

Figure 2.16. 'Equivalent' notch configurations.

illustrates the notch type and the equivalent U-notch. Although the equivalence of $\sigma_{\max}$ for these configurations has been observed previously, e.g. [25], a study of these configurations was carried out by the MMC plus partitioning plan. On the common notch intervals in configurations a, b, and c ($\alpha < 3\pi/4$), not only were the peak stresses essentially equal, but, in addition, the distribution of stresses as well. For configurations d and e with $\alpha < 3\pi/4$ the stress distributions were somewhat distorted on the common notch intervals but the maximum stresses differed by approximately only one percent with the U-notch results serving as an upper bound. This equivalence can only be inferred for the case of tension and edge notches in large panels. Obviously it would be easy to contruct loading situations or geometrical configurations where the equivalence no longer would hold.

Edge crack emanating from a semi-elliptical edge notch. An interesting application of the partitioning plan occurs in the solution for an edge crack emanating from a semi-elliptical edge notch in a semi-infinite sheet under tension, Figure 2.17. The partitioning plan indicated breaks the physical region into three zones.

Zone I, the region in the upper half plane exterior to

$$|z| = t > b + L, \tag{2.88}$$

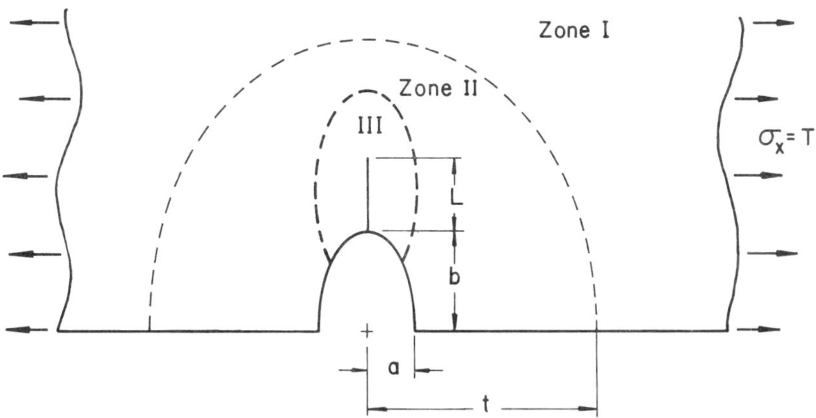

Figure 2.17. Edge crack emanating from a semi-elliptical edge notch.

was considered as previously with equations (2.71) and (2.72) representing the stress functions $\phi(z)$ and $\psi(z)$.

For the representation in zone II, the mapping function

$$z = \omega_2(\zeta_2) = i\{R\,e^{-i\zeta_2} - S\,e^{i\zeta_2}\} \tag{2.89}$$

was introduced where again $R = (a+b)/2$ and $S = (a-b)/2$. The parameter region corresponding to zone II is the shaded region shown in Figure 2.18a. The semi-circular boundary, with $|\zeta_2| = s$, maps into the boundary common to zone II and zone III in Figure 2.17. The radius s was chosen sufficiently large to ensure that the crack tip was located well within zone III. The reason for avoiding the region local to $\zeta_2 = 0$ is evident. The point $\zeta_2 = 0$ corresponds to the tip of the semi-ellipse which in this case would necessarily be a branch point of the stress function for physical compatibility of the solution. Reflection with respect to the real axis of the ζ_2-plane ensures load-free conditions on the elliptical intervals of zone II, thus

$$\psi(\zeta_2) = -\bar{\phi}(\zeta_2) - \bar{\omega}_2(\zeta_2)\phi'(\zeta_2)/\omega_2'(\zeta_2). \tag{2.90}$$

The stress function $\phi(\zeta_2)$ in the doubly-connected extended parameter region can be represented as

$$\phi(\zeta_2) = T \sum_{n=-\infty}^{\infty} \alpha_n \zeta_2^n. \tag{2.91}$$

In zone III, the mapping function

$$z = \omega_3(\zeta_3) = i(L/2)[e^{-i\zeta_3} + e^{i\zeta_3}] + ib \tag{2.92}$$

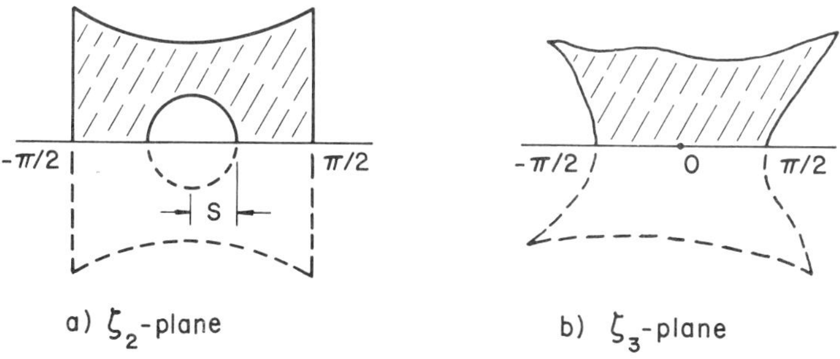

Figure 2.18. Parameter regions for zone II and zone III.

was introduced. The shaded region in Figure 2.18b maps into zone III with the point $\zeta_3 = 0$ mapping into the crack tip. Reflection with respect to the real axis of the ζ_3-plane ensures load-free conditions on the crack, thus

$$\psi(\zeta_3) = -\bar{\phi}(\zeta_3) - \bar{\omega}_3(\zeta_3)\phi'(\zeta_3)/\omega_3'(\zeta_3). \tag{2.93}$$

The stress function $\phi(\zeta_3)$ in the extended parameter region can be represented as

$$\phi(\zeta_3) = TL \sum_{n=1}^{\infty} \beta_n \zeta_3^n. \tag{2.94}$$

The solution was carried out in the usual manner with the appropriate force and stitching conditions imposed on representations above. Numerical results were obtained for $b/a = 1.0$ and $b/a = 2.0$ for various crack lengths and the corresponding values of β_n are catalogued in Tables 2.7 and 2.8, respectively, in the Appendix. The expansion (2.94) can be considered valid for $|z - i(b+L)| < L$ in zone III.

The crack tip stress intensity factors, K, are of particular interest for this problem class. The particular expansion we have taken in (2.94) leads to the very simple calculation for K,

$$K = 2\phi'(0)/\sqrt{i\omega_3''(0)} = 2T\beta_1\sqrt{L}. \tag{2.95}$$

The non-dimensional values for $K/T\sqrt{L}$ are presented in Figure 2.26 for the two cases. The limiting values as $L/b \to 0$ were obtained from

$$K/T\sqrt{L} \to 1.12\sigma_{\max} \tag{2.96}$$

where $\sigma_{\max}$ can be found in Table 2.2.

Rectangular edge notch with corner fillets. In the solution for a rectangular edge notch with corner fillets in a semi-infinite sheet under tension, Figure 2.19, the effects of the loss of notch symmetry will now be considered. Whereas the previous problems enjoyed a symmetry about the notch tip with the maximum stress occurring at the tip itself, in the present case no such symmetry exists. In fact the maximum stress in this problem shifts with the choice of parameters and often falls fairly close to the stitching locus between the partitioned zones II and IV.

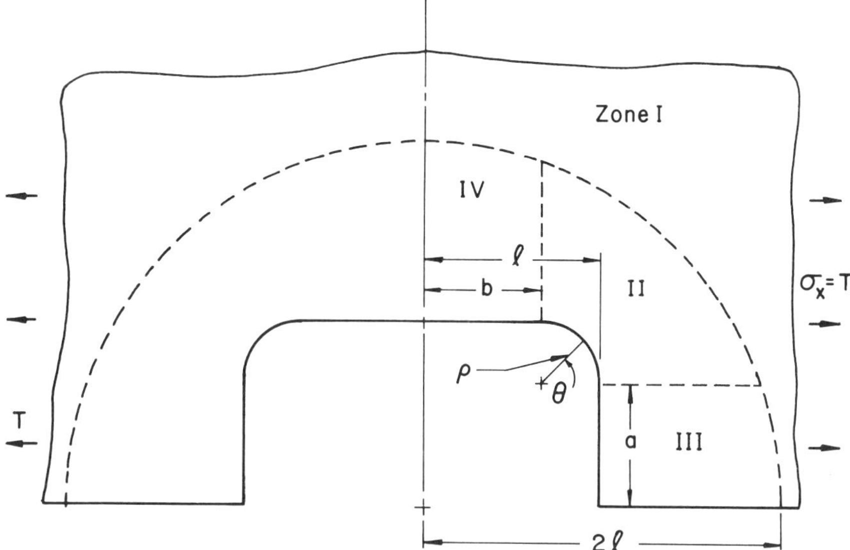

Figure 2.19. Rectangular edge notch with corner fillets.

Convergence in zone II can therefore be expected to be slower and more terms of the expansion will be necessary to retain for a reliable evaluation of the maximum stress.

Utilizing the obvious symmetries of the problem and the partitioning plan indicated in Figure 2.19, only the four zones shown need be considered explicitly. In zones I and III reflection with respect to the real axis was again utilized along with appropriate series expansion for the extended function $\phi(z)$. In zone IV, series expansions for $\phi(z)$ and $\psi(z)$ were chosen to reflect the problem symmetries with respect to the imaginery axis. In zone II the mapping function

$$z = \omega(\zeta) = i\rho\, \mathrm{e}^{-i\zeta} + b + ia \tag{2.97}$$

was introduced. The notch interval maps into the real interval $0 \leqslant \xi \leqslant \pi/2$ in the ζ-plane. Load-free conditions on the notch interval were again ensured by reflection with respect to the real axis of the ζ-plane, thus

$$\psi(\zeta) = -\bar{\phi}(\zeta) - \bar{\omega}(\zeta)\phi'(\zeta)/\omega'(\zeta). \tag{2.98}$$

The expansion in the parameter region corresponding to zone II was taken about the center of the extended region, $\zeta = \pi/4$, thus,

$$\phi(\zeta) = \rho T \sum_{n=0}^{\infty} \beta_n(\zeta - \pi/4)^n. \tag{2.99}$$

With the loss of symmetry in zone II, the coefficients β_n are now complex numbers in general.

The numerical results for $b/a = 1.0$ are summarized in the Appendix. In Figure 2.27 typical notch stress distributions are shown and the maximum notch stress, σ_{max}, is tabulated for several values of l/ρ. Owing to space limitations only a representative listing of the coefficients β_n in equation (2.99) is included in Table 2.9. Predictably, the matter of convergence was found to be a more sensitive consideration than in the previous problems particularly for the smaller l/ρ values. In fact, as shown in Figure 2.27, the maximum stress for $l/\rho = 2$ occurs at approximately $\theta = 75°$ which is relatively close to the stitching locus between zones II and IV. For the larger l/ρ values the difficulty disappears as the peak stress shifts towards the notch center. By comparing the results with solutions in which more degrees of freedom were used, it was found that the listed solution was well within an accuracy of one percent.

Cutout formed by two equal arcs of a circle. We consider now the problem of a cutout formed by two equal arcs of a circle in an infinite sheet under tension, Figure 2.20. The solution of this problem has been considered previously by Ling [26] with the use of bi-polar coordinates.

The partitioning plan for the range $0 < \alpha \leq \pi/2$ is indicated in Figure 2.20. The stitching locus dividing the region into zones I and II is the circle

$$|z| = S = 2\rho(1 + \cos\alpha). \tag{2.100}$$

In zone I, a series expansion for $\phi(z)$ and $\psi(z)$ was chosen which was consistent with the problem symmetries and the loading conditions at infinity. In zone II, the mapping function equation (2.79) was again introduced. Load-free conditions on the upper portion of the notch were ensured by the continuation argument, thus $\psi(\zeta)$ was again defined by equation (2.80). Representation of the extended definition of $\phi(\zeta)$ was again taken as a series expansion about the point $\zeta = 0$, thus

$$\phi(\zeta) = \rho T \sum_{n=0}^{\infty} \beta_n \zeta^n. \tag{2.101}$$

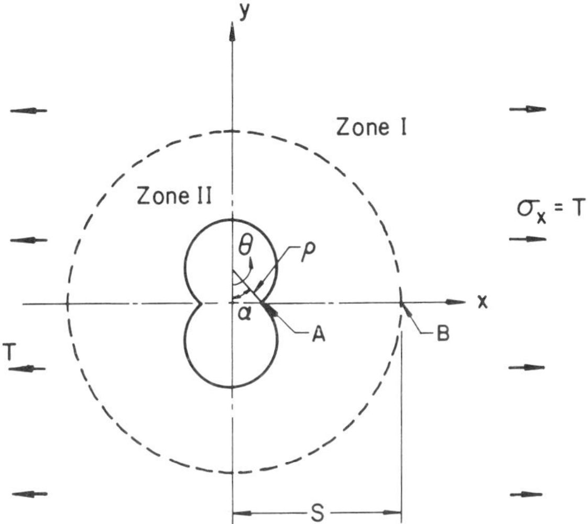

Figure 2.20. Cutout formed by two equal arcs of a circle.

where, from symmetry considerations, the β_n's are purely real (odd powers) or purely imaginery (even powers). In addition to the stitching conditions the expansion in zone II must satisfy the conditions $\tau_{xy} = 0$ and the vertical displacement $v = 0$ on the interval AB. With the symmetries already built into the expansion forms the lower half of zone II need not be considered explicitly.

Typical notch stress distributions and a table of maximum notch stresses are given in the Appendix. Numerical values of the coefficients β_n in equation (2.101) are catalogued in Table 2.10. Ling found the values 3.869 and 3.493 for σ_{max}/T for $\alpha = 0$ and 60°, respectively. The comparable values found in the present calculation are 3.868 and 3.468, respectively.

Slot in an infinite sheet under tension, $\sigma_y = T$. A problem of somewhat surprising difficulty is the case of a slot in an infinite sheet under the loading $\sigma_y = T$ as indicated in Figure 2.21. The location of the maximum stress presents a practical difficulty in the analysis of this problem. For the limiting case $l/\rho = 1.0$, the peak obviously occurs at $\theta = 0$ (or π). As l/ρ increases, the locations of the peak stress shift somewhat but generally remain close to the junction of the straight line and circular arc. It will be recalled that mention was made previously of the unresolved nature of the corresponding local solution.

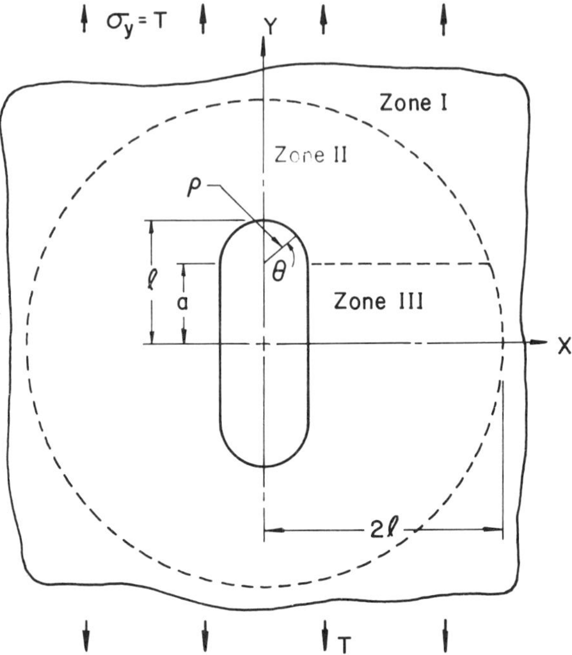

Figure 2.21. Slot in an infinite sheet under tension, $\sigma_y = T$.

A solution plan obvious from the partitioning indicated in Figure 2.21 was carried out. In zone II, equations (2.84) and (2.85) were again used in the representation and the stress function $\phi(\zeta)$ was again represented in the extended parameter region by a series

$$\phi(\zeta) = \rho T \sum_{n=0}^{\infty} \beta_n \zeta^n. \tag{2.102}$$

For l/ρ values near unity the maximum stress occurs at points near the stitching locus and cannot be considered reliable although the solution is accurate over most of the notch interval.

It was found for $l/\rho \geq 2$ the maximum stress has shifted to $\theta \approx 10°$ (or 170°) and the peaks can be calculated with some reliability. Notch stress distributions for the range $2 \leq l/\rho \leq 6$ are shown in Figure 2.29 in the Appendix. The corresponding coefficients β_n are listed in Table 2.11. Despite the relatively poor convergence in the vicinity of $\theta \approx 0$, it is estimated that the listed maximum stresses are correct to at least two percent accuracy.

It is evident from Figure 2.29 that the solution converges rapidly with respect to l/ρ. On the other hand with the present solution plan only an estimate can be legitimately claimed in the vicinity of the stitching locus. Based on extrapolation of results obtained by comparing several truncations of the representation, the authors suggest that to within one percent accuracy, $\sigma/T \to 2.0$ at $\theta = 0$ and $\sigma_{max}/T \to 2.2$ for large l/ρ.

Slot in an infinite sheet under tension, $\sigma_x = T$. With only trivial changes in the solution plan, the more conventional solution for the slot problem with tension $\sigma_x = T$ was carried out, Figure 2.22. Only the representation in zone I is altered to reflect the change in loading. Again in zone II, equation (2.84) and (2.85) were used and the stress function $\phi(\zeta)$ was represented by

$$\phi(\zeta) = \rho T \sum_{n=0}^{\infty} \beta_n \zeta^n. \tag{2.103}$$

The numerical results are presented in Figure 2.30 and Table 2.12 in the Appendix. The convergence difficulty encountered previously is now

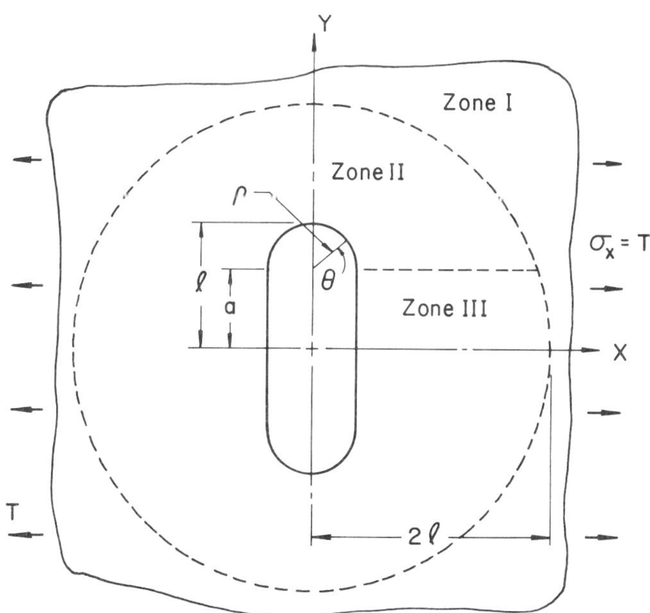

Figure 2.22. Slot in an infinite sheet under tension, $\sigma_x = T$.

of little consequence as region of greatest stress occurs at $\theta = \pi/2$. The tabulated values for the expansion (2.103) can be expected to yield stresses to within one percent in zone II away from the stitching locus. On the notch, this interval of validity can be assumed as $15° \leq \theta \leq 165°$.

Rectangular cutout with corner fillets. As the last illustrative example in this chapter, we consider the well-known problem class of rectangular cutouts with corner fillets in an infinite sheet under tension, Figure 2.23. The partitioning plan is indicated where advantage is taken of the problem symmetries. The solution representation is obvious and again we shall be primarily concerned with zone II. In zone II, equations (2.97) and (2.98) are again utilized and, again $\phi(\zeta)$ is expanded in the parameter region as the series

$$\phi(\zeta) = \rho T \sum_{n=0}^{\infty} \beta_n(\zeta - \pi/4)^n \tag{2.104}$$

With no symmetry properties with respect to the notch center for this loading situation, the coefficients β_n are complex numbers in general.

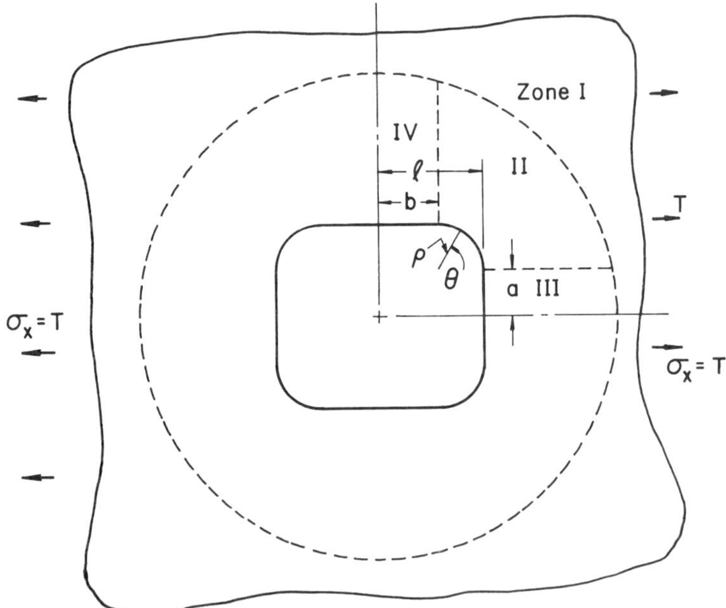

Figure 2.23. Rectangular cutout with corner fillets.

Numerical results for $\lambda = 1.0$ and 3.0 where

$$\lambda = (b + \rho)/(a + \rho) \tag{2.105}$$

are presented in the Appendix. The maximum stress as a function of l/ρ is given in Figure 2.31. Values of the coefficients β_n in equation (2.104) are listed for representative values of l/ρ in Tables 2.13 and 2.14.

It is interesting to compare the results for the maximum stress for $\lambda = 3$ and $l/\rho = 6$ with Hay's results in Table 2.1 obtained by polynomial mapping approximation. The value $\sigma_{max}/T = 2.73$ in Figure 2.31 is consistent with the maximum in Table 2.1 allowing for an extrapolation as $d(\theta) \to 0$.

2.7 Observations

In addition to a brief review of the classical application of conformal mapping to the solution of two-dimensional notch problems, the authors have proposed and illustrated an alternative mapping plan for the effective solution for most notch configurations. Several deficiencies in the classical mapping approach were outlined in section 2.1 and illustrated in later sections. In addition to the difficulty of finding suitable mapping functions carrying classical parameter regions into most notch configurations, one encounters convergence difficulties associated with the corresponding global representation of the solution. On the other hand, when mapping is used in the non-classical sense of the modified mapping-collocation plan the former difficulty is by-passed. When the MMC method is combined with partitioning, the difficulties of global representation are avoided and effective expansions local to the notch can be found and catalogued in a concise manner.

Curiously, the solution of general notch problems presents more difficulties than the limiting case of cracks. This is due primarily to the difference in the understanding of the nature of the mathematical singularities intrinsic to the solution. Crack tips are boundary points of the physical configuration and the associated mathematical singularities can be introduced into the solution representation. On the other hand the mathematical singularities associated with notch configurations are not well-understood and frequently occur in the unoccupied region of the plane. An illustration of this matter was provided in section 2.5 in the solution for a semi-elliptical notch in a semi-infinite sheet. In that case it was possible to forecast singularities which, although outside the physical region, contribute to the slow convergence of a global expansion.

With the MMC plus partitioning plan of solution these convergence difficulties were removed by a choice of mapping which permitted expansions relatively remote from the singularities. Notch and crack analyses differ too in the type of information sought. In crack analysis, the distribution of the dominant components of stress at the crack tip can be considered as known and only the intensity of this distribution is of interest. In contrast, there is no convenient local dominance in the general notch problem. Both the intensity and the distribution of notch stresses is of interest which of course is a more demanding requirement on the local solution.

With the choice of examples illustrating the MMC plus partitioning plan, extensive parameter studies were avoided, yet the problem class provided a range of difficulties common to most notch problems. The use of local analytic continuation arguments plays an important role in the solution plan particularly in intervals including the notch tip. A strict satisfaction of the boundary conditions in the region of greatest stress gradients leads to generally smooth distributions of notch stresses which aids considerably in the study of convergence with respect to degrees of freedom and collocation stations. The anticipation of singularities both of the remote and the local types and the corresponding partitioning plan was illustrated in several of the problems. The parametric nature of the use of mapping is particularly well-emphasised in the solution of the edge crack emanating from a semi-elliptical notch in a semi-infinite sheet. Two mapping functions were used in the solution plan for that problem. Another feature frequently illustrated in the solution plan was the reduction of computation by utilizing the solution symmetries in the structure of the series expansions. Finally, it was demonstrated that effective expansions local to the notch can often be found by a suitable mapping of the local notch geometry into the real axis of the parameter region. Only a few terms of these expansions are necessary for the cataloguing of sufficient information for the calculation of the local notch and subsurface stresses.

The carryover of the solution plan to other notch configurations presents no major difficulty. Displacement or mixed boundary conditions can easily be adopted in the MMC plan. Finite geometries can also be handled by series representation. When the region remote to the notch is awkward to represent by single series expansions, e.g. narrow strips, the region can be partitioned, or, preferably a finite element representation can be introduced for the representation. In the class of problems considered in this chapter the use of finite elements was unnecessary for the range of parameters considered.

2.8 Appendix

(1) *Circular-arc edge notches in a semi-infinite sheet under tension, T.*

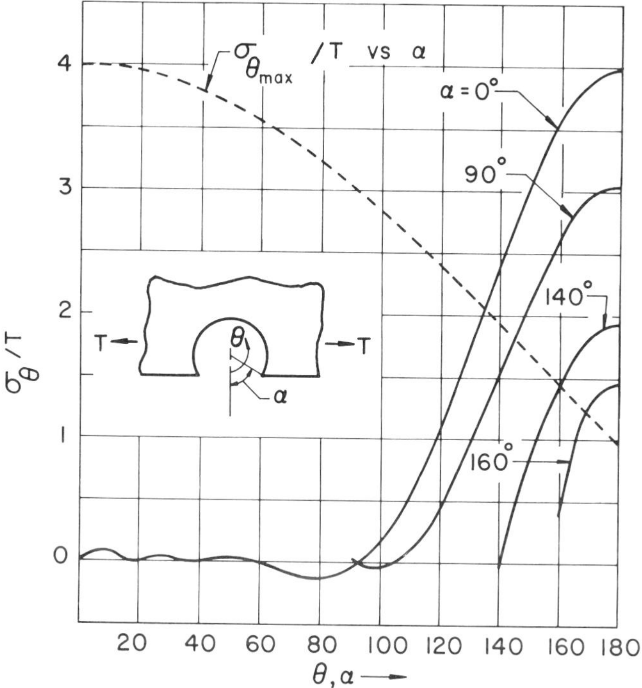

Figure 2.24. Maximum stress and typical distributions.

TABLE 2.5

Coefficients β_n in equation (2.81)

β_n \ α	0°	30°	60°	90°	120°	150°	175°
β_1	.99912	.97160	.89189	.76639	.60603	.42668	.27815
β_2/i	.07614	.07827	.08511	.09575	.10827	.11973	.12489
β_3	−.10584	−.10645	−.10971	−.11881	−.14408	−.23555	−1.2395
β_4/i	−.01523	−.01520	−.01566	−.01667	−.01850	−.00734	.17338
β_5	.00775	.00784	.00821	.00877	.00662	−.13286	−29.596
β_6/i	.00015	.00005	−.00002	.00054	.01084	−.19214	131.17
β_7	−.00006	−.00004	.00003	−.00047	−.01306	.39184	−4603.6
β_8/i	.00001	.00001	−.00006	−.00116	−.03966	1.7039	−33973.
β_9	−.00001	.00000	.00003	.00056	.01442	−2.6185	−43996.
β_{10}/i	.00000	.00000	.00001	.00061	.06253	−4.6369	−908230

(2) *U-notch in a semi-infinite sheet under tension.*

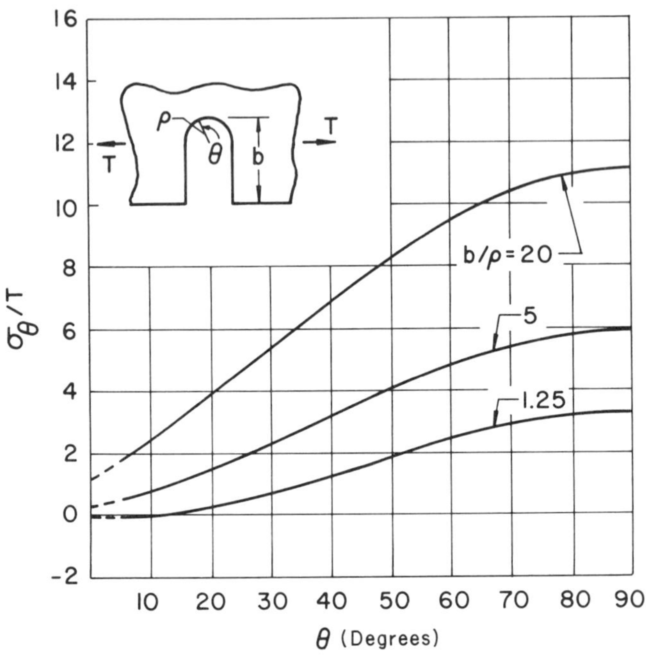

Figure 2.25. Typical notch stress distribution.

TABLE 2.6

Coefficients β_n in equation (2.86)

β_n \ b/ρ	1.25	2.00	3.00	4.00	5.00	10.0	20.0
β_1	.83205	.99891	1.18036	1.33461	1.47125	2.01091	2.78623
β_2/i	.09013	.07595	.06042	.04729	.03579	−.00825	−.06790
β_3	−.11315	−.10576	−.10351	−.10419	−.10602	−.11967	−.14248
β_4/i	−.01584	−.01514	−.01442	−.01383	−.01332	−.01151	−.00886
β_5	.00829	.00774	.00693	.00632	.00587	.00466	.00506
β_6/i	−.00021	.00011	.00033	.00040	.00044	.00045	.00073
β_7	.00011	−.00007	−.00018	−.00022	−.00022	−.00022	−.00001
β_8/i	−.00013	+.00001	.00001	−.00000	−.00001	−.00002	.00006
β_9	.00007	−.00000	−.00001	−.00000	.00000	.00000	.00002
β_{10}/i	.00003	.00000	−.00000	−.00000	−.00000	−.00000	.00000

(3) *Crack emanating from a semi-elliptical edge notch.*

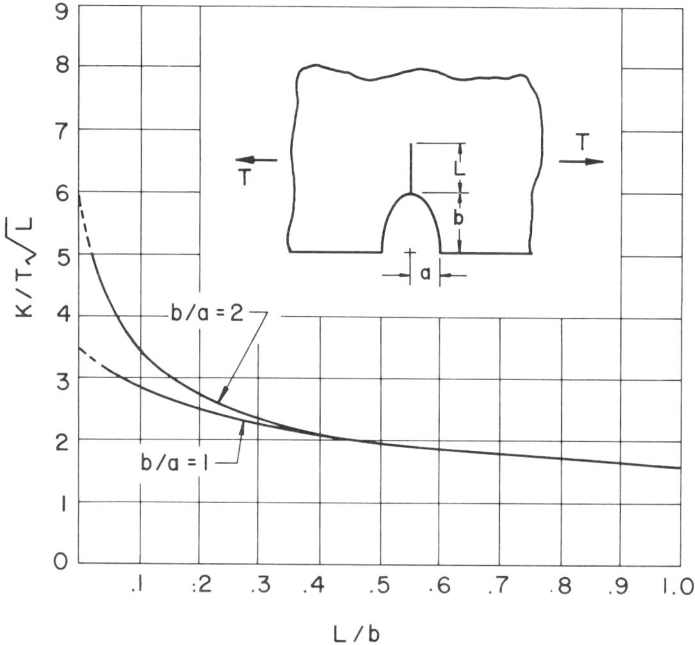

Figure 2.26. Stress intensity factors for a crack emanating from a semi-elliptical edge notch.

TABLE 2.7

Coefficients β_n in equation (2.94) *for $b/a = 1.00$*

β_n \ L/b	0.02	0.05	0.10	0.20	0.40	0.60	0.80	1.00
β_1	1.66244	1.56072	1.43150	1.24751	1.03163	.91472	.84219	.79350
β_2/i	.19181	.17038	.13983	.10158	.07004	.06111	.06018	.06218
β_3	−.18697	−.16077	−.13213	−.09794	−.07029	−.06229	−.06052	−.06065
β_4/i	−.03876	−.03214	−.02337	−.01450	−.01075	−.01158	−.01244	−.01273
β_5	.01413	.01188	.00768	.00386	.00334	.00438	.00505	.00540
β_6/i	−.00022	−.00121	−.00190	−.00172	−.00056	−.00002	.00018	.00034
β_7	.00032	.00100	.00123	.00088	.00019	.00006	.00004	−.00001
β_8/i	.00006	.00011	.00004	−.00014	−.00018	−.00006	−.00002	−.00002
β_9	−.00001	.00004	.00011	.00018	.00015	.00006	.00003	.00002
β_{10}/i	.00000	.00002	.00003	.00003	.00001	.00000	.00000	−.00001

TABLE 2.8

Coefficients β_n in equation (2.94) for $b/a = 2.00$

β_n \ L/b	0.02	0.05	0.10	0.20	0.40	0.60	0.80	1.00
β_1	2.53021	2.14717	1.75266	1.36204	1.04988	.91555	.84031	.79211
β_2/i	.26620	.16302	.09980	.06346	.06044	.06202	.06316	.06407
β_3	−.25753	−.16952	−.11169	−.07577	−.06431	−.06217	−.06165	−.06183
β_4/i	−.04247	−.02134	−.01178	−.00903	−.01006	−.01091	−.01149	−.01213
β_5	.01187	.00413	.00208	.00248	.00395	.00446	.00483	.00529
β_6/i	−.00312	−.00278	−.00171	−.00055	.00040	.00046	.00046	.00055
β_7	.00133	.00094	.00054	.00007	−.00011	−.00012	−.00011	−.00012
β_8/i	−.00003	−.00017	−.00022	−.00017	−.00002	−.00001	.00002	.00002
β_9	.00008	.00012	.00013	.00014	.00001	.00000	−.00000	.00001
β_{10}/i	.00001	.00001	.00000	.00002	−.00000	−.00000	−.00000	.00002

(4) *Rectangular edge notch with corner fillets, $b/a = 1.0$*

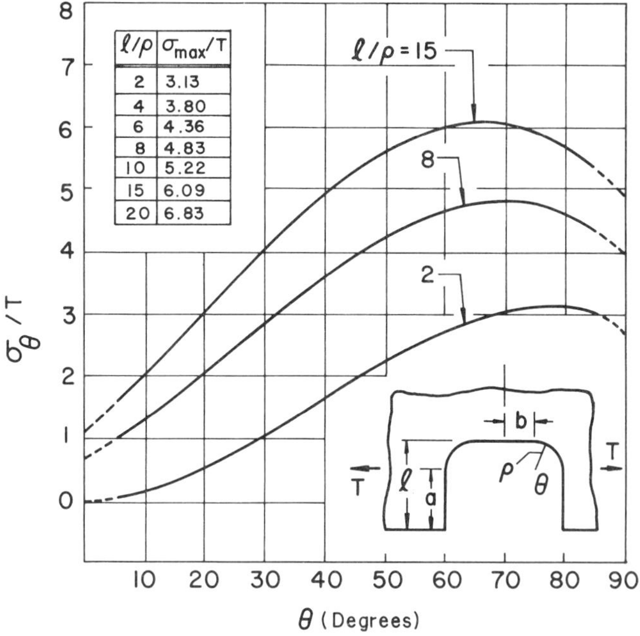

Figure 2.27. Typical notch stress distributions for a rectangular edge noteh with corner fillets.

TABLE 2.9

Coefficients β_n in equation (2.99) for $b/a = 1.0$

	$l/\rho = 2$		$l/\rho = 10$		$l/\rho = 20$	
n	Re β_n	Im β_n	Re β_n	Im β_n	Re β_n	Im β_n
0	1.1221	−.0603	5.3331	−.6697	10.4600	−2.5825
1	.7462	.0597	1.3469	−.2023	1.7252	−.4017
2	−.1574	.0427	−.0585	.0385	−.0374	.0398
3	−.0913	−.0542	−.1542	−.0342	−.1838	−.0200
4	.0475	−.0099	.0347	−.0044	.0226	.0006
5	−.0048	.0008	−.0089	.0071	−.0038	.0054
6	.0067	−.0000	.0021	−.0001	−.0010	.0014
7	−.0053	.0016	−.0033	.0019	−.0011	.0002
8	.0032	−.0012	.0005	−.0004	−.0001	.0001
9	−.0023	.0016	−.0006	.0004	−.0001	−.0000
10	.0011	−.0010	.0000	−.0001	−.0000	.0000
11	−.0009	.0008	−.0001	.0001	.0000	−.0000
12	.0003	−.0005	−.0000	−.0000	.0000	−.0000
13	−.0001	.0003	−.0000	.0000	.0000	−.0000
14	−.0001	−.0001	−.0000	−.0000	.0000	−.0000

(5) *Cutout formed by two equal arcs of a circle.*

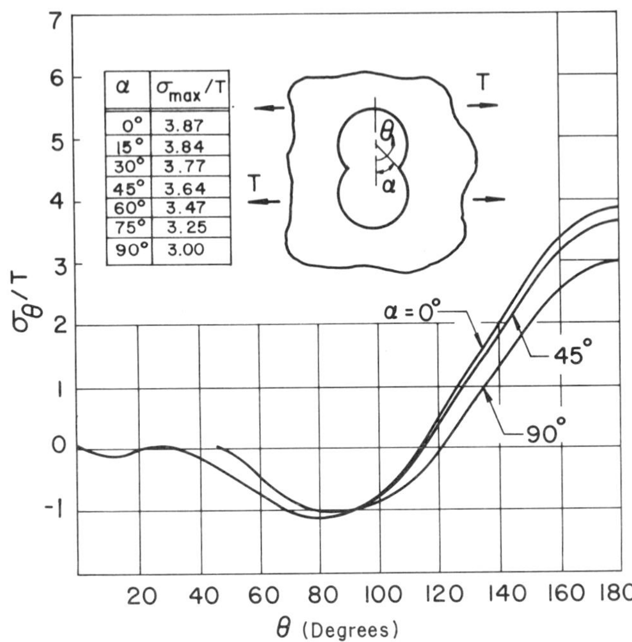

Figure 2.28. Typical notch stress distributions for cutout formed by two equal arcs of a circle.

TABLE 2.10

Coefficients β_n in equation (2.101)

β_n \ α	0°	15°	30°	45°	60°	75°	90°
β_0/i	−.45111	−.44476	−.42607	−.39535	−.35453	−.30537	−.25000
β_1	.96663	.96040	.94160	.91018	.86706	.81339	.75000
β_2/i	.13166	.13172	.13154	.13045	.12875	.12704	.12500
β_3	−.11292	−.11321	−.11426	−.11544	−.11724	−.12048	−.12501
β_4/i	−.00936	−.00947	−.01006	−.01022	−.01030	−.01087	−.01045
β_5	.00652	.00652	.00676	.00683	.00686	.00714	.00628
β_6/i	.00002	.00006	.00030	.00036	.00040	.00094	.00039
β_7	−.00013	−.00012	−.00014	−.00015	−.00018	−.00041	−.00016
β_8/i	.00004	.00004	−.00003	−.00005	−.00008	−.00040	−.00004
β_9	.00001	.00001	−.00000	.00001	.00002	.00005	−.00001
β_{10}/i	−.00002	−.00001	−.00000	.00001	.00001	.00012	.00001

(6) *Slot in an infinite sheet under tension, $\sigma_y = T$.*

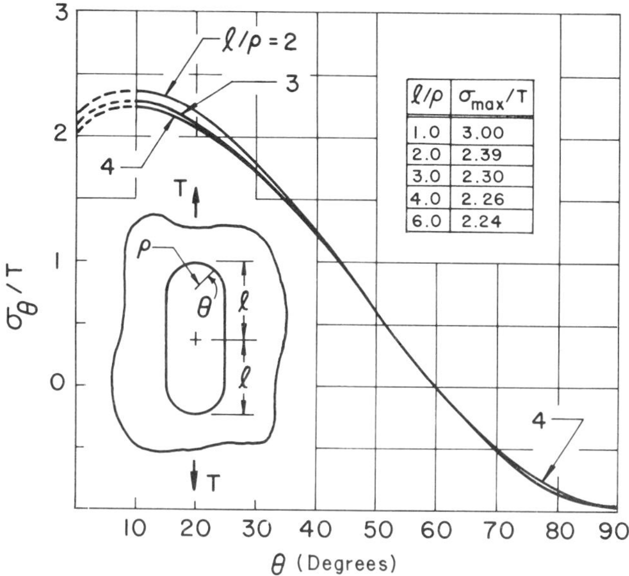

Figure 2.29. Notch stresses for a slot in an infinite sheet under tension, $\sigma_y = T$.

TABLE 2.11

Coefficients β_n for equation (2.102)

β_n \ l/ρ	2.0	3.0	4.0	6.0
β_0/i	1.07137	1.36551	1.64890	2.19879
β_1	−.24200	−.23885	−.23713	−.23537
β_2/i	−.35544	−.34761	−.34319	−.33843
β_3	.04374	.04375	.04359	.04329
β_4/i	.03503	.03550	.03556	.03537
β_5	−.00188	−.00194	−.00200	−.00204
β_6/i	−.00050	−.00048	−.00053	−.00062
β_7	.00010	.00005	.00004	.00005
β_8/i	.00016	.00011	.00009	.00007
β_9	−.00001	−.00001	−.00001	−.00000
β_{10}/i	.00001	.00001	.00001	.00001

(7) *Slot in an infinite sheet under tension*, $\sigma_x = T$.

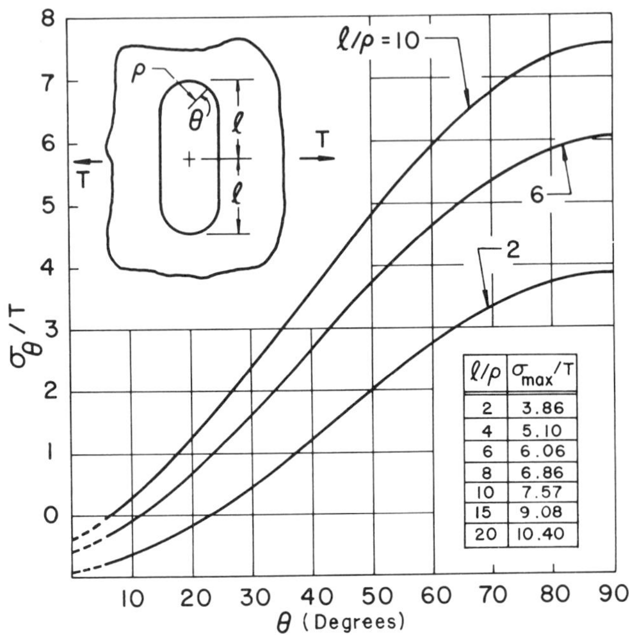

Figure 2.30. Typical notch stress distributions for slot in an infinite sheet under tension, $\sigma_x = T$.

TABLE 2.12

Coefficients β_n for equation (2.103)

β_n \ l/ρ	2	4	6	8	10	15
β_0/i	−.45950	−.86875	−1.28208	−1.69900	−2.12031	−3.19053
β_1	.96622	1.27550	1.51397	1.71465	1.89164	2.26933
β_2/i	.12972	.12893	.12355	.11704	.11030	.09355
β_3	−.11393	−.11316	−.11749	−.12275	−.12814	−.14084
β_4/i	−.01083	−.01286	−.01434	−.01524	−.01579	−.01647
β_5	.00667	.00583	.00531	.00497	.00476	.00454
β_6/i	.00015	−.00014	−.00015	−.00008	.00002	.00023
β_7	−.00018	−.00021	−.00021	−.00021	−.00021	−.00019
β_8/i	−.00008	−.00007	−.00005	−.00003	−.00001	.00002
β_9	.00000	.00000	.00000	−.00000	−.00000	−.00000
β_{10}/i	−.00001	−.00001	−.00000	−.00000	.00000	.00000

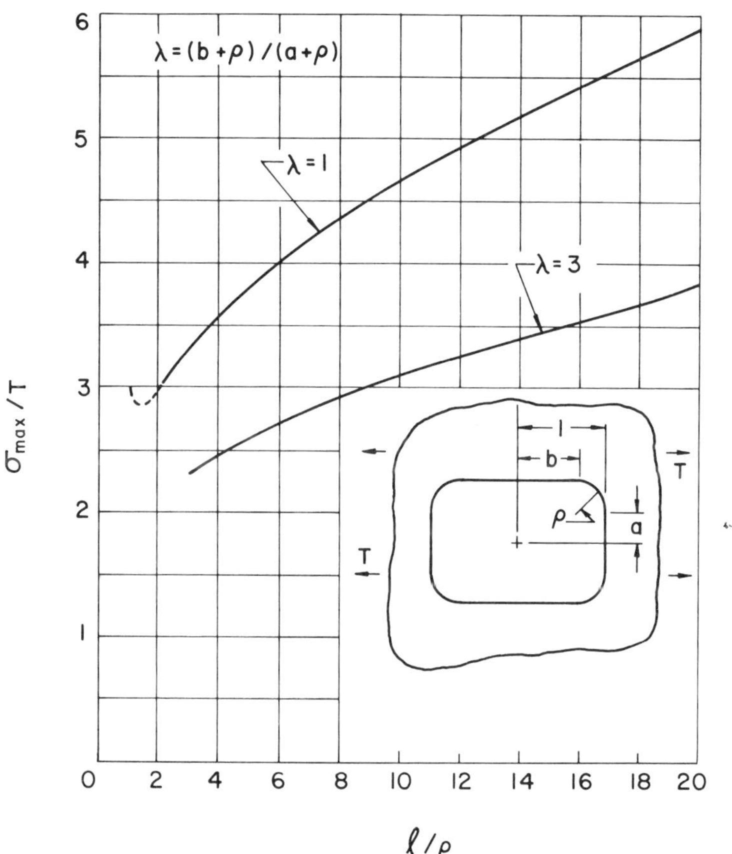

Figure 2.31. Maximum notch stresses for a rectangular cutout with corner fillets in an infinite sheet under tension.

TABLE 2.13

Coefficients β_n for equation (2.104), $\lambda = 1.00$

l/ρ	2.0		6.0		10.0	
n	Re β_n	Im β_n	Re β_n	Im β_n	Re β_n	Im β_n
0	1.09150	−.36622	3.18262	−1.09682	5.21925	−1.82475
1	.70066	.16386	1.04149	.10922	1.24144	.05816
2	−.18471	.10088	−.11426	.09290	−.09336	.08473
3	−.10467	−.04021	−.13606	−.05222	−.15329	−.05579
4	.04073	−.00987	.03772	−.01679	.03120	−.01609
5	−.00649	−.00338	−.00715	−.00149	−.00516	−.00299
6	.00623	−.00298	.00425	−.00249	.00168	−.00160
7	−.00531	−.00178	−.00403	.00005	−.00210	−.00254
8	.00414	−.00248	.00184	−.00125	.00128	−.00109
9	−.00305	−.00043	−.00145	.00023	−.00026	−.00114
10	.00293	−.00220	.00056	−.00062	.00062	−.00047
11	−.00158	.00043	−.00038	.00009	.00004	−.00029
12	.00163	−.00166	.00015	−.00022	.00019	−.00012
13	−.00066	.00045	−.00006	.00002	.00003	−.00004
14	.00079	−.00103	.00003	−.00005	.00004	−.00002
15	−.00005	.00036	−.00000	.00000	.00000	−.00000

TABLE 2.14

Coefficients in equation (2.104), $\lambda = 3.00$

l/ρ	4.00		6.00		8.00	
n	Re β_n	Im β_n	Re β_n	Im β_n	Re β_n	Im β_n
0	1.55622	−.23007	2.32818	−.35209	3.09003	−.47160
1	.54581	.16013	.64709	.15725	.72463	.15207
2	−.18945	.09679	−.16022	.09787	−.14449	.09760
3	−.10514	−.03282	−.10277	−.03550	−.10451	−.03852
4	.03834	−.00975	.03501	−.01144	.03175	−.01227
5	−.00327	.00133	−.00003	−.00105	.00052	−.00195
6	.00373	−.00371	.00138	−.00184	.00038	−.00132
7	−.00348	.00024	−.00092	−.00081	−.00059	−.00089
8	.00140	−.00311	.00050	−.00055	.00027	−.00028
9	−.00096	−.00004	−.00009	−.00014	−.00003	−.00013
10	.00043	−.00166	.00009	−.00009	.00004	−.00003
11	−.00001	−.00017	−.00000	−.00001	−.00000	−.00001
12	.00016	−.00059	.00000	−.00001	.00000	−.00000
13	.00010	−.00009	.00000	−.00000	.00000	−.00000
14	.00005	−.00013	.00000	−.00000	.00000	−.00000
15	.00004	−.00002	.00000	−.00000	.00000	−.00000

References

[1] Kolosoff, G., Uber einige eigenschaften des ebenen problems der elastizitatstheorie, *Zeitachs. Math. Physik*, 62, pp. 383–409 (1914).

[2] Kolosoff, G., Doctoral dissertation, Dorpat (1909).

[3] Muskhelishvili, N. I., *Some basic problems of mathematical theory of elasticity*, Noordhoff, Groningen, Holland (1953).

[4] Bowie, O. L. and Neal, D. M., A modified mapping-collocation technique for accurate calculation of stress intensity factors, *International Journal of Fracture Mechanics*, 6, pp. 199–206 (1970).

[5] Bowie, O. L. and Freese, C. E., Central crack in plane orthotropic rectangular sheet, *International Journal of Fracture Mechanics*, 8, pp. 49–57 (1972).

[6] Bowie, O. L., et al., Solution of plane problems of elasticity utilizing partitioning concepts, *Journal of Applied Mechanics*, 95, pp. 767–772 (1973).

[7] Freese, C. E., Collocation and finite elements – a combined method, *AMMRC-TR73-28, Army Materials & Mechanics Research Center, Watertown, Mass.* (1973).

[8] Freese, C. E. and Bowie, O. L., Stress analysis of configurations involving small fillets, *Journal of Strain Analysis*, 10, pp. 53–58 (1975).

[9] Kartzivadze, I. N., The fundamental problems of the theory of elasticity for the elastic circle, *Comptes rendus de l'academie des sciences de l'U.R.S.S.*, 20, pp. 95–104 (1943).

[10] Savin, G. N., *Stress Concentration Around Holes*, Pergamon, New York (1961).

[11] Bowie, O. L., Rectangular tensile sheet with symmetric edge cracks, *Journal of Applied Mechanics*, 31, pp. 208–212 (1964).

[12] Morkovin, V., Effect of a small hole on the stresses in a uniformly loaded plate, *Quarterly of Applied Mathematics*, 2, pp. 350–352 (1945).

[13] Greenspan, M., Effect of small holes on stresses in plates, *Quarterly of Applied Mathematics*, 2, pp. 60–71 (1944).

[14] Bateman, H., *Partial Differential Equations of Mathematical Physics*, Dover Publications, New York (1944).

[15] Hay, G. E., *Stress analysis of an infinite elastic solid with a rectangular filleted hole*, WAL Report No. 893-40, Watertown, Mass. (1945).

[16] Bowie, O. L., *Effect of sharpness of groove fillet on elastic stress concentration in standard rifling of 3 inch gun tubes*, WAL Report No. 730-504, Watertown, Mass. (1949).

[17] Bowie, O. L. and Neal, D. M., The effective crack length of an edge slot in a semi-infinite sheet under tension, *International Journal of Fracture Mechanics*, 3, pp. 111–119 (1967).

[18] Baratta, F. I. and Neal, D. M., Stress concentration factors in U-shaped and semi-elliptical edge notches, *Journal of Strain Analysis*, 5, pp. 121–127 (1970).

[19] Bowie, O. L., *Methods of Analysis and Solutions of Crack Problems* (Chapter I), (Edited by G. C. Sih), Noordhoff International Publishing, Leyden (1973).

[20] Hulbert, L. E., *The Numerical Solution of Two-dimensional Problems of the Theory of Elasticity*, Bulletin 198, Ohio State University, Columbus, Ohio (1963).

[21] Newman, J. C., Jr., *Stress analysis of simply and multiply connected regions*, Master's thesis, Virginia Polytechnical Institute, Blacksburg, Va. (1969).

[22] Newman, J. C., Jr., *An improved method of collocation for the stress analysis of cracked plates with various shaped boundaries*, NASA TN. D-6376, Langley Research Center, Hampton, Va. (1971).

[23] Bowie, O. L., and Neal, D. M., A note on the central crack in a uniformly stressed strip, *Engineering Fracture Mechanics*, 2, pp. 181–182 (1970).
[24] Bowie, O. L., Edge notches in a semi-infinite region, *Journal of Mathematics and Physics*, 45, pp. 356–366 (1966).
[25] Peterson, R. E., *Stress Concentration Factors*, John Wiley and Sons, New York (1974).
[26] Ling, C. B., The stresses in a plate containing an overlapped circular hole, *Applied Physics*, 4, pp. 405–411 (1948).

Chih-Bing Ling

3 Stress analysis of edge notches

3.1 Introduction

In this Chapter, we shall endeavor to undertake the task of introducing a method to analyze the stresses in a tensile plate or strip which contains one or more edge notches. We begin with the fundamental problem of a single edge notch in a plate. In order to render such a problem amenable to exact mathematical treatment, it is assumed that a single notch of general shape is contained at the edge of a semi-infinite plate such that no other adjacent boundary is present to interact on the notch. It is seen that the edge of such a plate consists of three intersecting parts, two semi-infinite straight lines joined by a finite curve of notch in the middle.

The plate is further assumed to be of homogeneous, isotropic and elastic material in a state of generalized plane stress under a uniform unit tension T parallel to the edge. As shown in Figure 3.1, the mid-plane of the plate occupies the upper half of a complex z-plane or $y \geqslant 0$ and the notch encloses the origin. For convenience, dimensionless variables x, y and z $(= x + iy)$ are used in the analysis such that the dimension of the plate is measured by referring to a certain typical length b of the plate.

In subsequence, we shall first introduce a method of solution for the fundamental problem and then give certain particular solutions in the case when the notch is symmetrical. Next, the method is extended to

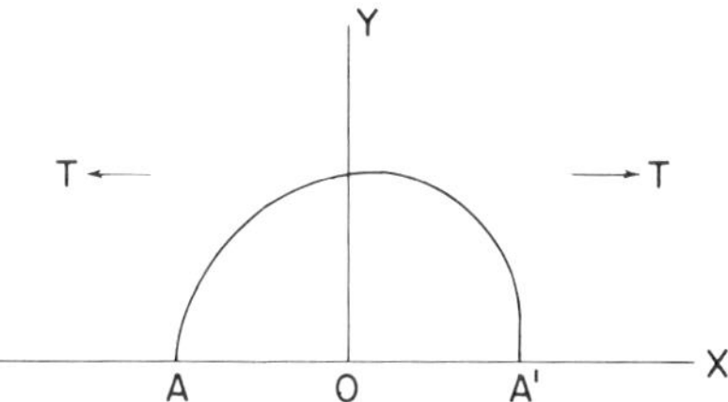

Figure 3.1. A single edge notch in a semi-infinite plate $y \geqslant 0$; dimension being measured in unit b.

problems in which neighboring boundaries are present so that their interactions are taken into consideration.

3.2 General solution of a single edge notch

Some time ago, Maunsell [1] formulated a solution for a tensile plate containing a semicircular notch. The Airy's stress function in his solution is given in the form:

$$\chi = Tb^2\left\{\chi_{00} + \sum_{m=1}^{\infty} A_m\chi_{2m} + \sum_{m=0}^{\infty} B_m\chi'_{2m}\right\}, \tag{3.1}$$

where the factor Tb^2 is introduced so as to render the real parametric coefficients A_m and B_m dimensionless. In Maunsell's solution, the functions χ_{00}, χ_{2m} and χ'_{2m} are expressed in terms of real variables. When expressed in terms of a complex variable z, they are

$$\begin{aligned}
\chi_{00} &= \tfrac{1}{4}\mathrm{Re}[z\bar{z} - z^2],\\
\chi_{2m} &= \mathrm{Re}\left[\frac{i\bar{z}}{z^{2m}} - \frac{(2m+1)i}{(2m-1)z^{2m-1}}\right],\\
\chi'_{2m} &= \mathrm{Re}\left[\frac{1}{z^{2m}} - \frac{\bar{z}}{z^{2m+1}}\right],
\end{aligned} \tag{3.2}$$

where the bar denotes complex conjugate. Note that here a constant factor has been removed from the original expression of χ_{2m} for convenience. The stress function χ satisfies all the boundary conditions of the notched plate under consideration, except that the boundary conditions of zero traction along the curve of notch remain to be adjusted. It appears that this solution can be extended to solve the problem of a single notch of general shape. The method of solution may consist of the following steps:

(i) Convert the Airy's stress function in Maunsell's solution into two analytic functions $\Phi(z)$ and $\Psi(z)$ in Muskhelishvili's complex variable solution.

(ii) Transform the curve of notch in the z-plane into a portion of the circumference of a unit circle in a complex ζ-plane.

(iii) Split, truncate and then solve the set of equations resulted from adjusting the boundary conditions of no traction along the curve of notch.

(iv) Devise a suitable method of computation to cope with the inaccuracy introduced into the solution due to the effect of truncation.

According to Muskhelishvili [2], the stresses referred to the coordinates x and y are given by

$$\sigma_{xx}+\sigma_{yy}=2T\{\Phi(z)+\bar{\Phi}(\bar{z})\}, \tag{3.3a}$$

$$\sigma_{yy}-\sigma_{xx}+2i\tau_{xy}=2T\{\bar{z}\Phi'(z)+\Psi(z)\}, \tag{3.3b}$$

where the prime denotes derivative and T is introduced for convenience. Now, from the preceding Airy's stress function, the corresponding expressions of stress are

$$\begin{aligned}\sigma_{xx}+\sigma_{yy}&=\frac{1}{b^2}\left(\frac{\partial^2}{\partial x^2}+\frac{\partial^2}{\partial y^2}\right)\chi\\&=T-2T\sum_{m=1}^{\infty}\left(\frac{2mi}{z^{2m+1}}-\frac{2mi}{\bar{z}^{2m+1}}\right)A_m\\&\quad+2T\sum_{m=0}^{\infty}\left(\frac{2m+1}{z^{2m+2}}+\frac{2m+1}{\bar{z}^{2m+2}}\right)B_m,\end{aligned} \tag{3.4a}$$

$$\begin{aligned}\sigma_{yy}-\sigma_{xx}+2i\sigma_{xy}&=\frac{1}{b^2}\left(\frac{\partial}{\partial x}-i\frac{\partial}{\partial y}\right)^2\chi\\&=-T+2T\sum_{m=1}^{\infty}\frac{2m(2m+1)i\bar{z}A_m}{z^{2m+2}}\\&\quad-2T\sum_{m=0}^{\infty}\frac{(2m+1)(2m+2)\bar{z}B_m}{z^{2m+2}}\\&\quad-2T\sum_{m=1}^{\infty}2m(2m+1)\left(\frac{iA_m}{z^{2m+1}}-\frac{B_m}{z^{2m+2}}\right).\end{aligned} \tag{3.4b}$$

Comparison gives at once

$$\Phi(z)=\tfrac{1}{4}-\sum_{m=1}^{\infty}\frac{2miA_m}{z^{2m+1}}+\sum_{m=0}^{\infty}\frac{(2m+1)B_m}{z^{2m+2}}, \tag{3.5a}$$

$$\Psi(z)=-\tfrac{1}{2}-\sum_{m=1}^{\infty}2m(2m+1)\left(\frac{iA_m}{z^{2m+1}}-\frac{B_m}{z^{2m+2}}\right). \tag{3.5b}$$

Next, consider a closed curve formed partly by the curve of notch in the upper half plane and partly by a curve in the lower half plane which is chosen to suit our convenience. Suppose that the exterior of the closed curve in the z-plane is transformed conformally into the *exterior* of a unit circle in a complex ζ-plane by the mapping function

$$z = \omega(\zeta). \tag{3.6}$$

Define a pair of polar coordinates (ρ, ϕ) in the ζ-plane by

$$\zeta = i\rho\, \mathrm{e}^{-i\phi}. \tag{3.7}$$

Suppose further that the curve of notch is transformed into a portion of the circumference of the unit circle $\rho = 1$ in the ζ-plane from $\phi = \alpha$ to $\phi = \beta$.

The stresses referred to the pair of polar coordinates (ρ, ϕ) are

$$\sigma_{\rho\rho} + \sigma_{\phi\phi} = 2T\{\Phi(z) + \bar{\Phi}(\bar{z})\}, \tag{3.8a}$$

$$\sigma_{\phi\phi} - \sigma_{\rho\rho} + 2i\sigma_{\rho\phi} = -\frac{2Tz'}{\bar{z}'}\{\bar{z}\Phi'(z) + \Psi(z)\}\, \mathrm{e}^{-2i\phi}, \tag{3.8b}$$

where

$$z' = \frac{\mathrm{d}z}{\mathrm{d}\zeta} = \omega'(\zeta), \qquad \bar{z}' = \frac{\mathrm{d}\bar{z}}{\mathrm{d}\bar{\zeta}} = \bar{\omega}'(\bar{\zeta}). \tag{3.9}$$

Note the differences in sign in the second expression of stress when compared with that given by Muskhelishvili. The differences are due to the present definition of the polar coordinates. Subtraction of the expressions yields

$$\sigma_{\rho\rho} + i\sigma_{\rho\phi} = T\left[\Phi(z) + \bar{\Phi}(\bar{z}) + \frac{z'}{\bar{z}'}\{\bar{z}\Phi'(z) + \Psi(z)\}\, \mathrm{e}^{-2i\phi}\right]. \tag{3.10}$$

Expand this expression at $\rho = 1$ into a Fourier series of ϕ over the interval (α, β). For convenience, a real factor $z'\bar{z}'$ or the Jacobian of transformation may be included in the expansion. In complex form, we can write

$$[J^2(\sigma_{\rho\rho} + i\sigma_{\rho\phi})]_{\rho=1} = T_n \sum_{n=-\infty}^{\infty} C_n \exp\left(\frac{2n\pi i\phi}{\beta - \alpha}\right) \tag{3.11}$$

where J^2 stands for $z'\bar{z}'$. By inversion, the complex coefficient C_n is

$$C_n = \frac{1}{(\beta-\alpha)T}\int_\alpha^\beta [J^2(\sigma_{\rho\rho}+i\sigma_{\rho\phi})]_{\rho=1}\exp\left(-\frac{2n\pi i\phi}{\beta-\alpha}\right)\mathrm{d}\phi. \tag{3.12}$$

Hence, the boundary conditions of no traction along the curve of notch are satisfied provided that for $n = 0, \pm 1, \pm 2, \ldots,$

$$C_n = 0, \tag{3.13}$$

which gives a set of equations for determining A_m and B_m. Therefore, the functions Φ and Ψ are fully determined if A_m and B_m are solved. Subsequently, the stresses can be calculated from equations (3.3) or (3.8).

3.3 Solution of set of equations

Referring to the functions Φ and Ψ in equations (3.5), the preceding set of equations may be written in the following form:

$$\sum_{m=1}^{\infty} {}^n\mu_m A_m + \sum_{m=0}^{\infty} {}^n\nu_m B_m = c_n, \quad (n = 0, \pm 1, \pm 2, \ldots) \tag{3.14}$$

where

$$c_n = \tfrac{1}{2}({}^nK_0 - {}^nI_0), \tag{3.15a}$$

$${}^n\mu_m = 2mi\{{}^nJ_{2m+1} - {}^nI_{2m+1} - (2m+1)({}^nK_{2m+1} - {}^nL_{2m+1})\}, \tag{3.15b}$$

$${}^n\nu_m = (2m+1)\{{}^nI_{2m+2} + {}^nJ_{2m+2} + 2m\,{}^nK_{2m+2} - (2m+2)\,{}^nL_{2m+2}\}. \tag{3.15c}$$

In general, when the curve of notch is unsymmetrical, these coefficients are complex. The quantities involved in the above expressions are integrals defined by

$${}^nI_m = \frac{1}{\beta-\alpha}\int_\alpha^\beta \frac{z_0'\bar{z}_0'}{z_0^m}\exp\left(-\frac{2n\pi i\phi}{\beta-\alpha}\right)\mathrm{d}\phi, \tag{3.16a}$$

$${}^nJ_m = \frac{1}{\beta-\alpha}\int_\alpha^\beta \frac{z_0'\bar{z}_0'}{\bar{z}_0^m}\exp\left(-\frac{2n\pi i\phi}{\beta-\alpha}\right)\mathrm{d}\phi, \tag{3.16b}$$

$${}^nK_m = \frac{1}{\beta-\alpha}\int_\alpha^\beta \frac{z_0'^2}{z_0^m}\exp\left(-\frac{2n\pi i\phi}{\beta-\alpha} - 2i\phi\right)\mathrm{d}\phi, \tag{3.16c}$$

$${}^nL_m = \frac{1}{\beta-\alpha}\int_\alpha^\beta \frac{z_0'^2\bar{z}_0}{z_0^{m+1}}\exp\left(-\frac{2n\pi i\phi}{\beta-\alpha} - 2i\phi\right)\mathrm{d}\phi, \tag{3.16d}$$

where

$$z_0 = \omega(i\,e^{-i\phi}), \qquad \bar{z}_0 = \bar{\omega}(-i\,e^{i\phi}),$$
$$z_0' = \omega'(i\,e^{-i\phi}), \qquad \bar{z}_0' = \bar{\omega}'(-i\,e^{i\phi}). \tag{3.17}$$

The first two integrals are connected by

$${}^{n}J_m = {}^{-n}\bar{I}_m. \tag{3.18}$$

Further relations may be found for the particular mapping function in question.

To solve the set of equations, it is necessary to split it first into two sets with real coefficients and then truncate them into finite sets corresponding to an index of truncation N. The sets can then be solved for A_m and B_m by matrix inversion or otherwise. Generally, there presents no particular difficulty in solving such sets. The subsequent calculation of the stresses is straightforward. However, it appears that the truncation always impairs accuracy of the results because the convergence of the series in the solution is in general exceedingly slow. A mere increase of the number of terms in the truncated sets does not appear to improve appreciably the accuracy of the results unless an unusually large number of terms are used, which are not practical. Furthermore, previous studies on the solutions of a semicircular notch, including Maunsell's, as well as those of other problems [3–6] reveal that there exists Mitchell–Ling effect in the values of stress. That is to say, the stress and particularly the maximum stress forms an oscillatory sequence with the index of truncation N. The elements of the sequence are alternately below and above, and as N increases converge very slowly to, the true stress, save in the neighborhood of concentrated load if any. As the present solution is formulated from Maunsell's solution, there is therefore ground to anticipate the same effect to occur in the values of stress even when the notch is now no longer semicircular in shape. By utilizing this effect, accurate values of stress can be obtained as the averages of two separate computations corresponding to two appropriate consecutive values of the index of truncation N. The method will be demonstrated by referring to a semicircular notch in Section 3.6.

It may be mentioned that several years ago, Bowie gave a different method of solution for an edge notch of general shape [7]. His solution was a complex variable solution derived by further utilizing the reflection argument and extension principle. He illustrated the solution by applying it to a semicircular notch and also to a semi-elliptic notch. However, since he used the image of the curve of notch to form the closed curve in the transformation, it seems that its applicability is more restrictive.

3.4 The symmetrical case

In the following, we shall consider in particular the special case in which the given notch is symmetrical with respect to the y-axis. It is seen that in this case, the portion of the circumference of the transformed unit circle in the ζ-plane is also symmetrical with respect to the imaginary axis. Hence, $\alpha = -\beta$ and the coefficient C_n becomes

$$C_n = \frac{1}{2\beta T}\int_{-\beta}^{\beta} [J^2(\sigma_{\rho\rho} + i\sigma_{\rho\phi})]_{\rho=1} \exp\left(-\frac{n\pi i\phi}{\beta}\right) \mathrm{d}\phi. \tag{3.19}$$

Again, by symmetry, this coefficient is necessarily real. Consequently, the coefficients involved in the set of equations in (3.14) are all real. The analysis is thus considerably simplified. Further, when $\alpha = -\beta$, the integrals in equations (3.16) besides ${}^nJ_m$ become

$${}^nI_m = \frac{1}{2\beta}\int_{-\beta}^{\beta} \frac{z'\bar{z}_0'}{z_0^m} \exp\left(-\frac{n\pi i\phi}{\beta}\right) \mathrm{d}\phi, \tag{3.20a}$$

$${}^nK_m = \frac{1}{2\beta}\int_{-\beta}^{\beta} \frac{z_0'^2}{z_0^m} \exp\left(-\frac{n\pi i\phi}{\beta} - 2i\phi\right) \mathrm{d}\phi, \tag{3.20b}$$

$${}^nL_m = \frac{1}{2\beta}\int_{-\beta}^{\beta} \frac{z_0'^2\bar{z}_0}{z_0^{m+1}} \exp\left(-\frac{n\pi i\phi}{\beta} - 2i\phi\right) \mathrm{d}\phi. \tag{3.20c}$$

By inferring from the relations in equations (3.15) the integrals are seen to be real when m is even or purely imaginary when m is odd. Corresponding to an index of truncation N, we suppose that the first N coefficients of both A_m and B_m are taken while the remaining coefficients are neglected. These $2N$ coefficients can be solved from an equal number of equations taken appropriately from the set.

When A_m and B_m are determined, the stresses in the plate can then be evaluated from equations (3.3) if referred to the rectangular coordinates (x, y) or from equations (3.8) if referred to the polar coordinates (ρ, ϕ). In the symmetrical case, the tangential stress σ_{xy} or $\sigma_{\rho\phi}$ vanishes identically along the line of symmetry $x = 0$. In particular, expressions for the stress along the curve of notch as well as along the line of symmetry are given below.

Since $\sigma_{\rho\rho}$ vanishes at $\rho = 1$, the stress along the curve of notch is therefore

$$[\sigma_{\phi\phi}]_{\rho=1} = [\sigma_{\rho\rho} + \sigma_{\phi\phi}]_{\rho=1}$$

$$= T - 4T \sum_{m=1}^{N} \mathrm{Re}\left[\frac{2miA_m}{z_0^{2m+1}} - \frac{(2m-1)B_{m-1}}{z_0^{2m}}\right], \tag{3.21}$$

where as before z_0 is the value of z at $\rho = 1$. The maximum stress occurs at the crown of notch where $z = iy_0$. The expression is

$$[\sigma_{\phi\phi}]_{\max} = T - 4T \sum_{m=1}^{N} (-1)^m \left\{\frac{2mA_m}{y_0^{2m+1}} - \frac{(2m-1)B_{m-1}}{y_0^{2m}}\right\}. \tag{3.22}$$

Besides, the stress σ_{xx} along the line of symmetry $x = 0$ for $y \geqq y_0$ is

$$[\sigma_{xx}]_{x=0} = \frac{1}{b^2}\frac{\partial^2}{\partial y^2}[\chi]_{x=0}$$

$$= T + 2T \sum_{m=1}^{N} (-1)^m \left\{\frac{4m^2A_m}{y^{2m+1}} - \frac{(2m-2)(2m-1)B_{m-1}}{y^{2m}}\right\}. \tag{3.23}$$

The other stress σ_{yy} is generally small. It is noted that the series involved in the preceding expressions of stress are shown corresponding to the index of truncation N.

In subsequence, we shall consider the following particular cases of the symmetrical notch:

(i) A semicircular notch. The curve of notch is a semicircle.
(ii) A circular notch. The curve of notch is a major or minor circular arc.
(iii) A semi-elliptic notch. The curve of notch is a semi-ellipse, deep or shallow.
(iv) A U-type notch. The curve of notch is defined by a similar transformation considered by Greenspan.
(v) A V-type notch. The curve of notch is a corner of a generalized hypotrochoid between the mid-points of two adjacent sides.

3.5 Mapping functions in symmetrical case

The geometries of the preceding five symmetrical notches in the z-plane are shown in Figures 3.2(i)–(v). In each case, AA' denotes the opening of the notch and C the crown of notch.

Before proceeding to treat each case separately, we first give here a summary of the mapping function $\omega(\zeta)$ to be used in each case. Here the mapping function transforms in all cases the exterior of a closed curve in the z-plane into the exterior of a unit circle in the ζ-plane. The portion of the closed curve in the upper half of the z-plane is the curve of notch under consideration.

(i) Semicircular notch, $z = \zeta$.
(ii) Circular notch, $z = \zeta + i\lambda$.
$\lambda > 0$, a major circular arc; $\lambda < 0$, a minor circular arc.

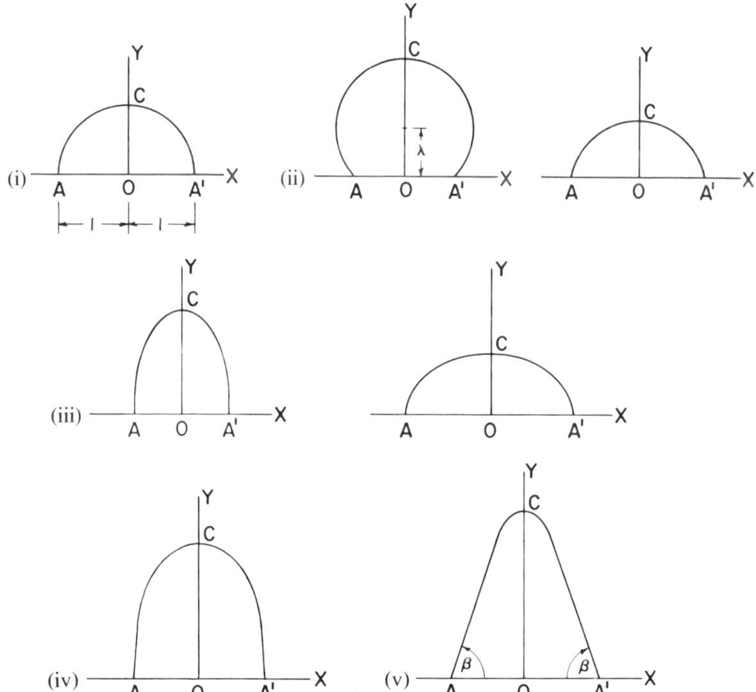

Figure 3.2. A symmetrical edge notch in a semi-infinite plate. (i) Semicircular notch; (ii) Circular notch; (iii) Semi-elliptic notch; (iv) U-type notch; (v) V-type notch.

(iii) Semi-elliptic notch,

$$z = \zeta - \frac{\lambda}{\zeta}.$$

$0 < \lambda < 1$, a deep notch; $-1 < \lambda < 0$, a shallow notch.

(iv) U-type notch,

$$z = \zeta + \frac{2pq}{\zeta} + \frac{q^2}{\zeta^3}, \quad (0 < |q^2| < 1,\ 0 < |pq| < 1).$$

(v) V-type notch,

$$z = \zeta + \frac{i^{s+1}}{s^2\zeta^2} - i\lambda \cot\beta. \qquad \lambda = \left(1 - \frac{1}{s^2}\right)\sin\beta, \qquad \beta = \frac{\pi}{s+1}, \quad (s > 1).$$

3.6 The semicircular notch

The mapping function for the semicircular notch takes the simplest form $z = \zeta$. Here, $\beta = \pi/2$, $z_0 = i\,e^{-i\phi}$ and $z_0' = 1$. The three integrals in equations (3.20) are therefore

$$^{n}I_m = \frac{1}{\pi i^m}\int_{-\pi/2}^{\pi/2} e^{(m-2n)i\phi}\,d\phi, \tag{3.24a}$$

$$^{n}K_m = \frac{1}{\pi i^m}\int_{-\pi/2}^{\pi/2} e^{(m-2n-2)i\phi}\,d\phi, \tag{3.24b}$$

$$^{n}L_m = -\frac{1}{\pi i^m}\int_{-\pi/2}^{\pi/2} e^{(m-2n)i\phi}\,d\phi. \tag{3.24c}$$

For positive or negative integral value of n including zero, we have

$$\frac{1}{\pi}\int_{-\pi/2}^{\pi/2} e^{-2ni\phi}\,d\phi = \begin{cases} 1, & (n = 0) \\ 0, & (n \neq 0) \end{cases} \tag{3.25a}$$

$$\frac{1}{\pi}\int_{-\pi/2}^{\pi/2} e^{-(2n+1)i\phi}\,d\phi = \frac{2(-1)^n}{(2n+1)\pi}. \tag{3.25b}$$

Consequently, we find the following values together with $^{n}J_m$, according to the lower subscript being even or odd:

$$
\begin{aligned}
{}^{n}I_{2m} &= (-1)^{m}\,\delta_{m,n}, & {}^{n}I_{2m+1} &= \frac{2(-1)^{n}i}{(2n-2m-1)\pi},\\
{}^{n}J_{2m} &= (-1)^{m}\,\delta_{m,-n}, & {}^{n}J_{2m+1} &= \frac{2(-1)^{n}i}{(2n+2m+1)\pi},\\
{}^{n}K_{2m} &= (-1)^{m}\,\delta_{m-1,n}, & {}^{n}K_{2m+1} &= -\frac{2(-1)^{n}i}{(2n-2m+1)\pi},\\
{}^{n}L_{2m} &= -(-1)^{m}\,\delta_{m,n}, & {}^{n}L_{2m+1} &= -\frac{2(-1)^{n}i}{(2n-2m-1)\pi},
\end{aligned}
\tag{3.26}
$$

where $\delta_{m,n}$ is the Kronecker delta. Hence, the set of equations is

$$\sum_{m=1}^{\infty} {}^{0}\mu_m A_m = -\tfrac{1}{2}, \tag{3.27}$$

and for $n \geqslant 1$,

$$(-1)^{n}(2n-1)(2n+1)B_{n-1} - (-1)^{n}2n(2n+1)B_n + \sum_{m=1}^{\infty} {}^{n}\mu_m A_m = 0, \tag{3.28}$$

$$-(-1)^{n}(2n+1)B_n + \sum_{m=1}^{\infty} {}^{-n}\mu_m A_m = \tfrac{1}{2}\delta_{1,n}, \tag{3.29}$$

where for positive or negative integral n including zero,

$${}^{n}\mu_m = (-1)^{n}\frac{16m(2m+1)(2n+1)}{(2n+2m+1)\{(2n-2m)^2-1\}}. \tag{3.30}$$

From the last two equations, we have by eliminating B_n, for $n \geqslant 1$,

$$B_{n-1} = \tfrac{1}{3}\delta_{1,n} + \frac{(-1)^{n}}{4n^2-1}\sum_{m=1}^{\infty}(2n\,{}^{-n}\mu_m - {}^{n}\mu_m)A_m. \tag{3.31}$$

Further, by eliminating B_n altogether from equations (3.29) and (3.31), we find for $n \geqslant 1$,

$$\sum_{m=1}^{\infty}\left({}^{-n}\mu_m + \frac{2n+2}{2n+3}\,{}^{-n-1}\mu_m - \frac{1}{2n+3}\,{}^{n+1}\mu_m\right)A_m = \tfrac{1}{2}\delta_{1,n}. \tag{3.32}$$

In the set of equations (3.27) and (3.32), we retain the first N

coefficients of A_m and solve for the N coefficients from the first N equations of the set. Subsequently, from equation (3.31) and the N coefficients of A_m, the first $N+1$ coefficients of B_n are obtained. The stress in the plate can then be calculated. In particular, the maximum stress at the crown of notch is given by equation (3.22).

To illustrate the Mitchell–Ling effect of truncation, the maximum stress at the crown of notch corresponding to several consecutive values of the index of truncation N are computed [4] and shown in Table 3.1.

TABLE 3.1

Maximum stress at crown of semicircular notch

N	$[\sigma_{\phi\phi}]_{\max}/T$	Average
19	3.0660	
		3.06535
20	3.0647	
		3.06530
21	3.0659	
		3.06535
22	3.0648	
		3.06530
23	3.0658	
		3.06535
24	3.0649	

Alternately, the maximum stress is given by the integral

$$\frac{16T}{\pi^3}\int_0^\infty \frac{\theta^2 \cosh\theta \, \mathrm{d}\theta}{\sinh^2\theta - (2\theta/\pi)^2}, \tag{3.33}$$

which is obtained from the solution in terms of bipolar coordinates to be described later in section 3.8. In evaluating this integral, it is found that the value accurate to $6D$ is 3.065336. Indeed, these computations carry more decimal places than generally necessary for practical purposes. It is seen that the values of stress shown above form an oscillatory sequence with N and each of the average values corresponding to two consecutive values of N is astonishingly close to the true value given by the integral. These results lead us to infer that accurate values of stress can be obtained by simply averaging the results corresponding to two properly chosen consecutive values of N.

3.7 The circular notch in general

The mapping function for a circular notch is $z = \zeta + i\lambda$, where

$$\lambda = -\cot\beta, \quad (0 < \beta < \pi). \tag{3.34}$$

The center of the circular notch is at $z = i\lambda$. The subtending angle of the circular notch at its center is 2β. When $\lambda > 0$, the notch is a major circular arc and when $\lambda < 0$, the notch is a minor circular arc. In the particular case $\lambda = 0$, the notch is a semicircle with $\beta = \pi/2$. Here, in general, $z_0 = i(\mathrm{e}^{-i\phi} + \lambda)$ and $z_0' = 1$. The three integrals in equations (3.20) are therefore

$$ {}^nI_m = \frac{1}{2\beta i^m}\int_{-\beta}^{\beta}\frac{1}{(1+\lambda\,\mathrm{e}^{i\phi})^m}\exp\left(-\frac{n\pi i\phi}{\beta}+mi\phi\right)\mathrm{d}\phi, \tag{3.35a} $$

$$ {}^nK_m = \frac{1}{2\beta i^m}\int_{-\beta}^{\beta}\frac{1}{(1+\lambda\,\mathrm{e}^{i\phi})^m}\exp\left(-\frac{n\pi i\phi}{\beta}+mi\phi-2i\phi\right)\mathrm{d}\phi, \tag{3.35b} $$

$$ {}^nL_m = -\frac{1}{2\beta i^m}\int_{-\beta}^{\beta}\frac{1+\lambda\,\mathrm{e}^{-i\phi}}{(1+\lambda\,\mathrm{e}^{i\phi})^{m+1}}\exp\left(-\frac{n\pi i\phi}{\beta}+mi\phi\right)\mathrm{d}\phi. \tag{3.35c} $$

It can be shown that these integrals are connected by

$$ {}^nL_m = -{}^nI_m + i\lambda({}^nI_{m+1} - {}^nK_{m+1}). \tag{3.36} $$

Hence, in this case, only two integrals need to be evaluated. To evaluate them, we expand $(1+\lambda\,\mathrm{e}^{i\phi})^{-m}$ into series by binomial theorem and make use of the relation,

$$ \frac{1}{2\beta}\int_{-\beta}^{\beta}\exp\left(-\frac{n\pi i\phi}{\beta}+mi\phi\right)\mathrm{d}\phi = \begin{cases} \dfrac{(-1)^n\sin m\beta}{m\beta - n\pi}, & (m\beta \neq n\pi) \\ 1. & (m\beta - n\pi) \end{cases} \tag{3.37} $$

The expansion is different according to λ being numerically less or greater than unity. We find when $|\lambda| < 1$,

$$ {}^nI_m = \frac{1}{2\beta i^m}\sum_{s=0}^{\infty}(-1)^{n+s}\binom{m+s-1}{s}\frac{\lambda^s\sin(m+s)\beta}{(m+s)\beta - n\pi}, \tag{3.38a} $$

$$ {}^nK_m = \frac{1}{2\beta i^m}\sum_{s=0}^{\infty}(-1)^{n+s}\binom{m+s-1}{s}\frac{\lambda^s\sin(m+s-2)\beta}{(m+s-2)\beta - n\pi}, \tag{3.38b} $$

and when $|\lambda| > 1$,

$$ {}^nI_m = \frac{1}{2\beta i^m}\sum_{s=0}^{\infty}\frac{(-1)^{n+s}}{\lambda^{m+s}}\binom{m+s-1}{s}\frac{\sin s\beta}{s\beta + n\pi}, \tag{3.39a} $$

$$ {}^nK_m = \frac{1}{2\beta i^m}\sum_{s=0}^{\infty}\frac{(-1)^{n+s}}{\lambda^{m+s}}\binom{m+s-1}{s}\frac{\sin(s+2)\beta}{(s+2)\beta + n\pi}. \tag{3.39b} $$

Note that in the preceding expressions, if the denominator of the term is zero, the sine function in the numerator is also zero. The limiting value of their ratio is $(-1)^n$. The convergence of the series becomes slower as λ is nearer to unity numerically. Further, both expansions fail when $|\lambda| = 1$. In such instances, the integrals can be evaluated directly by using Simpson rule or otherwise. If Simpson rule is used, accuracy of the results can be ascertained by repeating the process with a smaller increment until the values are stable to the desired degree of accuracy.

When the coefficients are evaluated, the rest of the solution is merely a routine. It is mentioned that in the case of circular notch, an alternate solution is available in Airy's stress function in terms of bipolar coordinates. Such a solution together with the numerical results will be described in subsequence.

3.8 Solution of circular notch in bipolar coordinates

An alternate method of solution of the circular notch is to find the Airy's stress function in terms of bipolar coordinates. Such a solution was given independently by Isibasi [8], Weinel [9] and Ling [10]. A general solution in bipolar coordinates admitting of wider application was given earlier by Jeffery [11]. The relevant solution for the circular notch is outlined here according to the formulation by Ling. It is mentioned that there are minor variations in the definition of bipolar coordinates. In Ling's solution, the pair of bipolar coordinates (ξ, η) is defined by

$$x + iy = -\coth \tfrac{1}{2} i(\xi + i\eta), \tag{3.40}$$

so that the two poles of the coordinates are situated on the x-axis at the points $z = \pm 1$ or the opening AA' of the circular notch as shown in

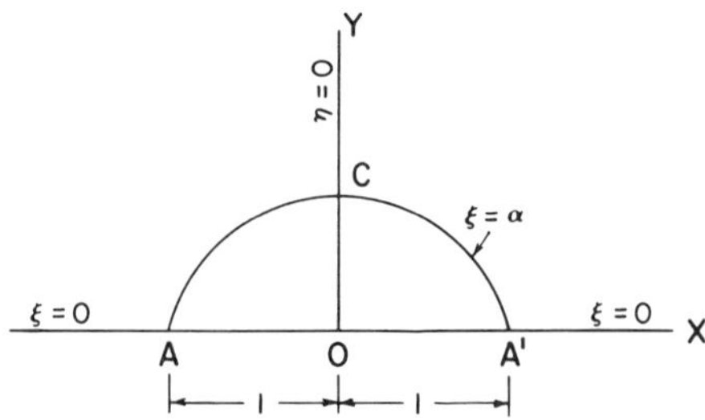

Figure 3.3. A circular notch in bipolar coordinates.

Figure 3.3. By separating the real and imaginary parts, we have

$$x = J \sinh \eta, \quad y = J \sin \xi, \tag{3.41}$$

where

$$J = 1/(\cosh \eta - \cos \xi). \tag{3.42}$$

The biharmonic equation $\nabla^4 \chi = 0$ to be satisfied by the Airy's stress function is then transformed to

$$\left(\frac{\partial^4}{\partial \xi^4} + 2\frac{\partial^4}{\partial \xi^2\, \partial \eta^2} + \frac{\partial^4}{\partial \eta^4} + 2\frac{\partial^2}{\partial \xi^2} - 2\frac{\partial^2}{\partial \eta^2} + 1\right)\frac{\chi}{J} = 0. \tag{3.43}$$

Now suppose that the solution is given as the sum of two biharmonic functions by

$$\chi = \chi_0 + \chi_1, \tag{3.44}$$

where χ_0 provides a uniform unit longitudinal tension T in an otherwise unnotched plate and χ_1 is added to allow for the effect of the notch. χ_0 is

$$\chi_0 = \tfrac{1}{2} T b^2 y^2, \tag{3.45}$$

or, in bipolar coordinates,

$$\frac{\chi_0}{J} = \tfrac{1}{2} T b^2 \frac{\sin^2 \xi}{\cosh \eta - \cos \xi}. \tag{3.46}$$

An integral solution even in η satisfying the transformed differential equation is found in the form:

$$\frac{\chi_1}{J} = T b^2 \int_0^\infty F_n(\xi) \cos n\eta \, dn, \tag{3.47}$$

where

$$F_n(\xi) = (f_n \sin \xi + h_n \cos \xi) \sinh h\xi + (g_n \sin \xi + k_n \cos \xi) \cosh n\xi, \tag{3.48}$$

in which f_n, h_n, g_n and k_n are parametric functions of n to be determined.

The stresses referring to the bipolar coordinates can then be derived from χ/J [12] as follows:

$$\sigma_{\xi\xi} = \left\{(\cosh\eta - \cos\xi)\frac{\partial^2}{\partial\eta^2} - \sinh\eta\frac{\partial}{\partial\eta} - \sin\xi\frac{\partial}{\partial\xi} + \cos\xi\right\}\left(\frac{\chi}{J}\right), \tag{3.49a}$$

$$\sigma_{\eta\eta} = \left\{(\cosh\eta - \cos\xi)\frac{\partial^2}{\partial\xi^2} - \sinh\eta\frac{\partial}{\partial\eta} - \sin\xi\frac{\partial}{\partial\xi} + \cosh\eta\right\}\left(\frac{\chi}{J}\right), \tag{3.49b}$$

$$\sigma_{\xi\eta} = -(\cosh\eta - \cos\xi)\frac{\partial^2}{\partial\xi\,\partial\eta}\left(\frac{\chi}{J}\right). \tag{3.49c}$$

The boundary conditions at infinity require that when $(\xi, \eta) = 0$, the stresses are not disturbed by the presence of χ_1. This requirement leads to

$$\int_0^\infty F_n(0)\,\mathrm{d}n = 0. \tag{3.50}$$

The boundary conditions of no traction along the straight portions of the edge require that when $\xi = 0$, $\sigma_{\xi\xi} = 0$ and $\sigma_{\xi\eta} = 0$. We have, respectively,

$$-(\cosh\eta - 1)\int_0^\infty n^2F_n(0)\cos n\eta\,\mathrm{d}n + \sinh\eta\int_0^\infty nF_n(0)\sin n\eta\,\mathrm{d}n + \int_0^\infty F_n(0)\cos n\eta\,\mathrm{d}n = 0, \tag{3.51a}$$

$$(\cosh\eta - 1)\int_0^\infty nF_n'(0)\sin n\eta\,\mathrm{d}n = 0, \tag{3.51b}$$

where the prime denotes derivative. All the three equations are satisfied if

$$F_n(0) = 0, \qquad F_n'(0) = 0. \tag{3.52}$$

We thus find

$$k_n = 0, \qquad g_n = -nh_n. \tag{3.53}$$

The curve of notch is represented by $\xi = \alpha$, where $\alpha = \pi - \beta$; β denoting the half measure of the circular arc as before. The boundary conditions of no traction along the curve of notch require that $\sigma_{\xi\xi} = 0$

and $\sigma_{\xi\eta} = 0$ again when $\xi = \alpha$. Similarly, we have

$$\frac{\sin^2\alpha \sinh^2\eta}{(\cosh\eta - \cos\alpha)^2} - (\cosh\eta - \cos\alpha)\int_0^\infty n^2 F_n(\alpha)\cos n\eta \, dn + \sinh\eta \int_0^\infty nF_n(\alpha)\sin n\eta \, dn + \cos\alpha \int_0^\infty F_n(\alpha)\cos n\eta \, dn - \sin\alpha \int_0^\infty F_n'(\alpha)\cos n\eta \, dn = 0, \tag{3.54a}$$

$$-\frac{\sin\alpha \sinh\eta(1 - \cos\alpha\cosh\eta)}{(\cosh\eta - \cos\alpha)^2} + (\cosh\eta - \cos\alpha)\int_0^\infty nF_n'(\alpha)\sin n\eta \, dn = 0. \tag{3.54b}$$

Here, the second equation is satisfied if

$$\int_0^\infty nF_n'(\alpha)\sin n\eta \, dn = \frac{\sin\alpha \sinh\eta(1 - \cos\alpha\cosh\eta)}{(\cosh\eta - \cos\alpha)^3}. \tag{3.55}$$

Fourier sine transform then gives

$$nF_n'(\alpha) = \frac{2\sin\alpha}{\pi}\int_0^\infty \frac{(1 - \cos\alpha\cosh\eta)\sinh\eta \sin n\eta \, d\eta}{(\cosh\eta - \cos\alpha)^3}. \tag{3.56}$$

The integral can be evaluated by Cauchy theorem of residues. The result is

$$F_n'(\alpha)\sinh n\pi = n\cosh n(\pi - \alpha)\sin\alpha - \sinh n(\pi - \alpha)\cos\alpha. \tag{3.57}$$

To find $F_n(\alpha)$, we differentiate the first equation with respect to η and obtain with the aid of the second equation,

$$\int_0^\infty n(n^2 + 1)F_n(\alpha)\sin n\eta \, dn = -\frac{3\sin^2\alpha \sinh\eta(1 - \cos\alpha\cosh\eta)}{(\cosh\eta - \cos\alpha)^4}. \tag{3.58}$$

Integration with respect to η yields

$$\int_0^\infty (n^2 + 1)F_n(\alpha)\cos n\eta \, dn = \frac{3\sin^2\alpha\cos\alpha}{2(\cosh\eta - \cos\alpha)^2} - \frac{\sin^4\alpha}{(\cosh\eta - \cos\alpha)^3}. \tag{3.59}$$

Here the constant of integration is zero since the integral vanishes when η tends to ∞. Fourier cosine transform then gives

$$F_n(\alpha) = \frac{2}{(n^2+1)\pi}\int_0^\infty \left\{\frac{3\sin^2\alpha\cos\alpha}{2(\cosh\eta-\cos\alpha)^2} - \frac{\sin^4\alpha}{(\cosh\eta-\cos\alpha)^3}\right\}\cos n\eta \, d\eta$$

$$= -\frac{\sinh n(\pi-\alpha)\sin\alpha}{\sinh n\pi}. \tag{3.60}$$

Here, the integral is evaluated by Cauchy theorem of residues. We thus have

$$f_n = \coth n\pi - \frac{\sinh 2n\alpha - n\sin 2\alpha}{2(\sinh^2 n\alpha - n^2\sin^2\alpha)}, \tag{3.61a}$$

$$h_n = -\frac{n\sin^2\alpha}{\sinh^2 n\alpha - n^2\sin^2\alpha}. \tag{3.61b}$$

Hence, with these values of f_n and h_n, we have

$$F_n(\xi) = f_n \sinh n\xi \sin\xi - h_n(n\cosh n\xi\sin\xi - \sinh n\xi\cos\xi), \tag{3.62}$$

so that the stress function is fully determined. In particular, the stress along the curve of notch $\xi = \alpha$ is given by

$$[\sigma_{\eta\eta}]_{\xi=\alpha} = 2T(\cosh\eta - \cos\alpha) \times \int_0^\infty \frac{n(n\cosh n\alpha\sin\alpha - \sinh n\alpha\cos\alpha)}{\sinh^2 n\alpha - n^2\sin^2\alpha}\cos n\eta \, dn. \tag{3.63}$$

The maximum stress occurs at the crown of notch where $\eta = 0$. By further putting $\theta = n\alpha$, the maximum stress is

$$[\sigma_{\eta\eta}]_{\max} = \frac{2T(1-\cos\alpha)\sin\alpha}{\alpha^3}\int_0^\infty \frac{\theta(\theta\cosh\theta - \alpha\cot\alpha\sinh\theta)\,d\theta}{\sinh^2\theta - (\sin\alpha/\alpha)^2\theta^2}. \tag{3.64}$$

In particular, when $\alpha = \pi/2$, the notch is semicircular and the integral becomes

$$[\sigma_{\eta\eta}]_{\max} = \frac{16T}{\pi^3}\int_0^\infty \frac{\theta^2\cosh\theta\,d\theta}{\sinh^2\theta - (2\theta/\pi)^2}. \tag{3.65}$$

Again, when $\alpha = 0$, the notch becomes an indefinitely large circle tangential to the edge and the integral takes the limiting value

$$[\sigma_{\eta\eta}]_{\max} = T\int_0^\infty \frac{\theta(\theta \cosh \theta - \sinh \theta)\, d\theta}{\sinh^2 \theta - \theta^2}. \tag{3.66}$$

Besides, the stress $\sigma_{\eta\eta}$ along the line of symmetry $\eta = 0$ is

$$[\sigma_{\eta\eta}]_{\eta=0} = T + T\int_0^\infty \left\{(1 - \cos \xi)\frac{d^2}{d\xi^2} - \sin \xi \frac{d}{d\xi} + 1\right\} F_n(\xi)\, dn. \tag{3.67}$$

The preceding integrals may be evaluated numerically by Simpson rule or otherwise. A method of evaluating the integrals of maximum stress with the aid of tabulated values is given later. The resulting values of the maximum stress at the crown of notch is shown versus β in Table 3.2 and also graphically in Figure 3.4. The stresses along the curve of

TABLE 3.2

Maximum stress at the crown of circular notch

β	$[\sigma_{\eta\eta}]_{\max}/T$	Remark
0°	1.000	Unnotched plate
30	1.707	
60	2.424	
90	3.065	Semicircular notch
120	3.568	
150	3.882	
180	3.999	Tangential hole

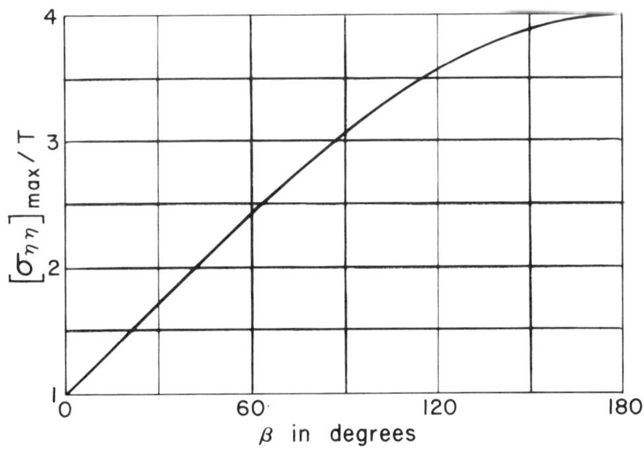

Figure 3.4. Maximum stress at crown of circular notch.

notch $\xi = \alpha$ and the line of symmetry $\eta = 0$ for several typical values of β are shown graphically in Figures 3.5 and 3.6. Note that in the former, the value of η is given by

$$\eta = \cosh^{-1}\left(\frac{1 - \cos\beta\cos\phi}{\cos\phi - \cos\beta}\right), \tag{3.68}$$

and in the latter, the value of ξ is given by

$$\xi = 2\cot^{-1} y. \tag{3.69}$$

To evaluate the integral in equation (3.64), we separate the integrand into two parts, expand each part into series and integrate the first part by parts. We find

$$[\sigma_{\eta\eta}]_{\max} = 2T(1 - \cos\alpha)\sum_{s=0}^{\infty}\frac{\sin^{2s+1}\alpha}{\alpha^{2s+3}}\left(\frac{2s+2}{2s+1} - \alpha\cot\alpha\right)\int_0^{\infty}\frac{\theta^{2s+1}\,d\theta}{\sinh^{2s+1}\theta}. \tag{3.70}$$

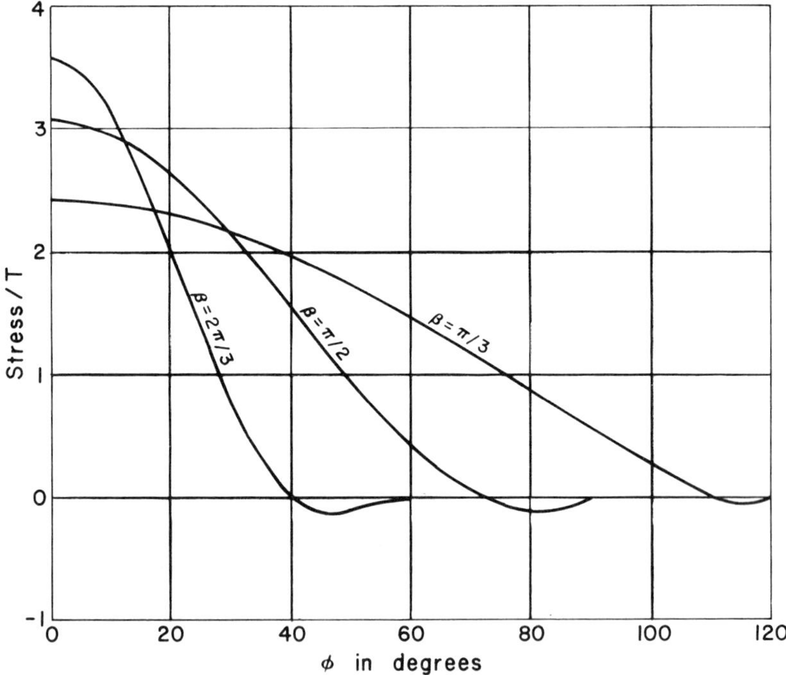

Figure 3.5. Stress along circular notch.

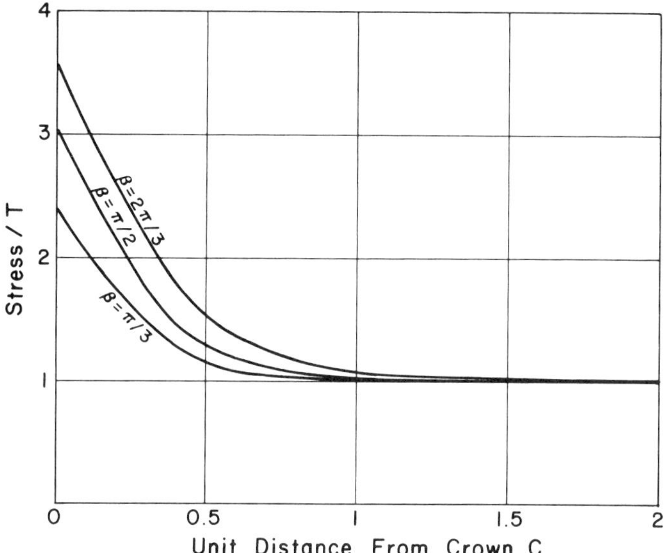

Figure 3.6. Stress along line of symmetry.

The integral involved in the series was tabulated together with a factor $\frac{1}{2}$ to $7D$ by Ling and Nelson [13] and also tabulated to $8D$ as a special case of a more general integral by Ling and Wu [14]. With these values, the series can readily be computed. The integral in equation (3.65) is a particular case in which $\alpha = \pi/2$. Similarly, the integral in equation (3.66) can be developed into the series:

$$\int_0^{\infty} \frac{\theta(\theta \cosh \theta - \sinh \theta)\, \mathrm{d}\theta}{\sinh^2 \theta - \theta^2} = \sum_{s=1}^{\infty} 2s(I_{2s+1}^* - I_{2s+1}), \tag{3.71}$$

where

$$I_s = \frac{1}{2(s!)} \int_0^{\infty} \frac{\theta^s\, \mathrm{d}\theta}{\sinh \theta + \theta}, \qquad I_s^* = \frac{1}{2(s!)} \int_0^{\infty} \frac{\theta^s\, \mathrm{d}\theta}{\sinh \theta - \theta}, \tag{3.72}$$

which are two of the four Howland integrals. They were tabulated with high precision to $18D$ when s is an even integer by Nelson [15] and to $25D$ when s is an odd integer by Ling and Lin [16]. Thus, the value of the series is readily obtained.

3.9 The semi-elliptic notch

The mapping function

$$z = \zeta - \frac{\lambda}{\zeta} \tag{3.73}$$

transforms the exterior region of an ellipse in the z-plane into the exterior of a unit circle in the ζ-plane with its center at the origin. The curve of notch in the upper half of the z-plane is a semi-ellipse with $\beta = \pi/2$. When $0 < \lambda < 1$, it is a deep semi-elliptic notch and when $-1 < \lambda < 0$, it is a shallow semi-elliptic notch. The depth ratio γ is

$$\gamma = \frac{1+\lambda}{1-\lambda}, \tag{3.74}$$

which is the ratio of the depth of the crown to one-half of the opening of the notch. For a prescribed value of γ, the value of λ is therefore

$$\lambda = \frac{\gamma - 1}{\gamma + 1}. \tag{3.75}$$

Here, with the polar coordinates (ρ, ϕ) defined in the ζ-plane by (3.7),

$$z_0 = i(\mathrm{e}^{-i\phi} + \lambda\,\mathrm{e}^{i\phi}), \qquad z_0' = 1 - \lambda\,\mathrm{e}^{2i\phi}. \tag{3.76}$$

The two integrals ${}^nI_0$ and ${}^nK_0$ are readily integrable. Their values are

$$\begin{aligned} {}^nI_0 &= (1+\lambda^2)\delta_{n,0} - \lambda\delta_{n,1} - \lambda\delta_{n,-1}, \\ {}^nK_0 &= -2\lambda\delta_{n,0} + \lambda^2\delta_{n,1} + \delta_{n,-1}, \end{aligned} \tag{3.77}$$

where $\delta_{n,m}$ is Kronecker delta. Hence, by equation (3.15a),

$$c_n = -\frac{(1+\lambda)^2}{2}\delta_{n,0} + \frac{\lambda(1+\lambda)}{2}\delta_{n,1} + \frac{1+\lambda}{2}\delta_{n,-1} \tag{3.78}$$

With the preceding values of z_0 and z_0', it is further found that the last two integrals in equations (3.20) are connected by

$$^{n}L_m = \lambda\ ^{n-1}K_{m+2} + (1+\lambda^2)\ ^{n}K_{m+2} + \lambda\ ^{n+1}K_{m+2}. \tag{3.79}$$

The remaining two integrals can be expressed in terms of an auxiliary integral defined by

$$^{n}H_m = \frac{1}{\pi}\int_{-\pi/2}^{\pi/2} \frac{1}{z_0^m}\, e^{-2ni\phi}\, d\phi. \tag{3.80}$$

The relations are

$$\begin{aligned} ^{n}I_m &= -\lambda\ ^{n-1}H_m + (1+\lambda^2)\ ^{n}H_m - \lambda\ ^{n+1}H_m, \\ ^{n}K_m &= \lambda^2\ ^{n-1}H_m - 2\lambda\ ^{n}H_m + {}^{n+1}H_m. \end{aligned} \tag{3.81}$$

The auxiliary integral can be evaluated by using the binomial expansion

$$\frac{1}{z_0^m} = \frac{1}{i^m}\, e^{mi\phi} \sum_{p=0}^{\infty} (-1)^p \binom{m+p-1}{p} \lambda^p\, e^{2pi\phi}, \quad (m \geqslant 1) \tag{3.82}$$

and then applying equations (3.25). The results are

$$^{n}H_{2m} = \begin{cases} (-1)^n \dbinom{n+m-1}{n-m} \lambda^{n-m}, & (n \geqslant m) \\ 0, & (n \leqslant m-1) \end{cases} \tag{3.83a}$$

$$^{n}H_{2m+1} = -\frac{2i}{\pi}(-1)^n \sum_{p=0}^{\infty} \binom{2m+p}{p} \frac{\lambda^p}{2m+2p-2n+1}. \tag{3.83b}$$

The following recurrence relation may be mentioned:

$$^{n+1}H_m + 2\lambda\ ^{n}H_m + \lambda^2\ ^{n-1}H_m + {}^{n}H_{m-2} = 0, \tag{3.84}$$

which may be used for checking purpose or be used to evaluate the auxiliary integral from the values of $^{0}H_m$, $^{1}H_m$ and $^{n}H_1$ when m is an odd integer. When the integrals are evaluated, the coefficients $^{n}\mu_m$ and $^{n}\nu_m$ in equation (3.14) can be calculated. They are all real. In particular, $^{0}\nu_m$ vanishes for all m and both $^{n}\nu_m$ and $^{-n}\nu_m$ vanish for $m \geqslant n+1$ when $n \geqslant 1$. The following truncated set of equations is obtained:

$$\sum_{m=1}^{N} {}^{0}\mu_m A_m = c_0, \tag{3.85}$$

$$\sum_{m=1}^{N} {}^{n}\mu_m A_m + \sum_{m=0}^{n} {}^{n}\nu_m B_m = c_n, \quad (n = 1, 2, 3, \ldots, N) \tag{3.86}$$

$$\sum_{m=1}^{N} {}^{-n}\mu_m A_m + \sum_{m=0}^{n} {}^{-n}\nu_m B_m = c_{-n}, \quad (n = 1, 2, 3, \ldots, N) \tag{3.87}$$

Here, the integer N is the index of truncation.

The preceding set of equations may be solved as follows: (i) Eliminate B_n from the n-th equations in (3.86) and (3.87) so that the resulting set of N equations can be solved for B_m from B_0 up to B_{N-1} by forward substitution in terms of A_m. (ii) Substitute B_m into the first $(N-1)$ equations in (3.86) or (3.87), or a combination of them, so as to eliminate B_m altogether. (iii) Solve for A_m by matrix inversion or otherwise from the set of N equations formed by equation (3.85) and the $(N-1)$ equations thus obtained. (iv) Compute B_m up to B_{N-1} from the values of A_m. In this manner, the first N coefficients of both A_m and B_m are solved.

Subsequently, the values of maximum stress at the crown of notch are computed as the averages of two separate computations corresponding to $N = 15$ and $N = 16$. The results are shown in Table 3.3 and also graphically in Figure 3.7. The case $\gamma = 1$ is for a semicircular notch.

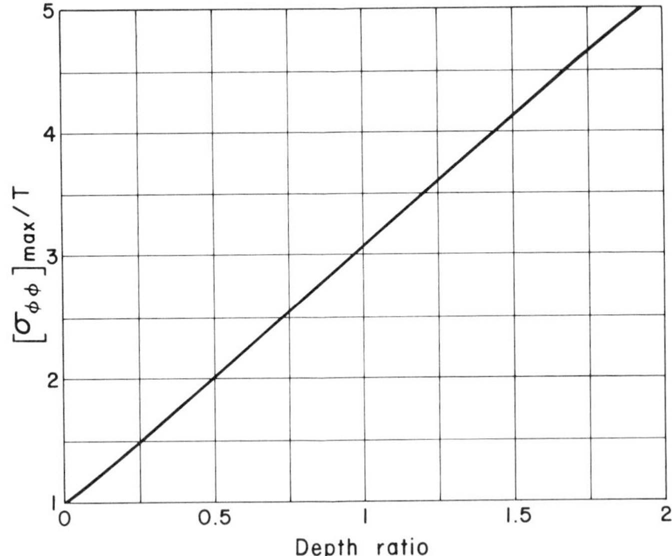

Figure 3.7. Maximum stress at crown of semi-elliptic notch.

TABLE 3.3

Maximum stress at crown of semi-elliptic notch

γ	$[\sigma_{\phi\phi}]_{max}/T$
0.9	2.861
1.0	3.065
1.1	3.278
1.2	3.492
1.3	3.706
1.4	3.921
1.5	4.136
1.6	4.341

3.10 The U-type notch

Consider a mapping function in the form:

$$z = \zeta + \frac{2pq}{\zeta} + \frac{q^2}{\zeta^3}, \quad (0 < |q^2| < 1, \ 0 < |pq| < 1) \tag{3.88}$$

where p and q are two parameters, either both real or both purely imaginary. The function transforms conformally the exterior region of a closed curve in the z-plane into the exterior of a unit circle in the ζ-plane with its center at the origin. The closed curve in the z-plane is symmetrical with respect to both the x-axis and y-axis. The U-type notch under consideration is formed by the upper half of the closed curve with $\beta = \pi/2$. Such a transformation with a particular pair of values for the parameters was used by Greenspan [17] to approximate an ovaloid hole in a plate, which is in the form of a square with a semicircle erected on each of the two opposite sides.

Likewise, a pair of polar coordinates (ρ, ϕ) is defined in the ζ-plane by equation (3.7) so that on the unit circle $\rho = 1$ the values of z and its derivative with respect to ζ are

$$z_0 = i\,\mathrm{e}^{-i\phi}(1 - 2pq\,\mathrm{e}^{2i\phi} + q^2\,\mathrm{e}^{4i\phi}), \tag{3.89a}$$

$$z_0' = 1 + 2pq\,\mathrm{e}^{2i\phi} - 3q^2\,\mathrm{e}^{4i\phi}. \tag{3.89b}$$

As shown in Figure 3.2(iv), the opening of the notch AA' corresponds to the points on the unit circle $\rho = 1$ at $\phi = \mp\pi/2$. The crown of notch C is at $z = i(1 - 2pq + q^2)$. The depth ratio is given by

$$\gamma = \frac{1 - 2pq + q^2}{1 + 2pq + q^2}. \tag{3.90}$$

Consequently, when $\gamma > 1$, it is necessary that $pq < 0$ and $|pq| < (1 + q^2)/2$. The radius of curvature of the closed curve is found equal to

$$R = \frac{\{1 + 4p^2q^2 + 9q^4 + 4pq(1 - 3q^2)\cos 2\phi - 6q^2 \cos 4\phi\}^{\frac{3}{2}}}{1 - 4p^2q^2 - 27q^4 + 24pq^3 \cos 2\phi + 6q^2 \cos 4\phi}. \tag{3.91}$$

Suppose that the curve of notch is straight or of zero curvature at the opening AA'. By equating the denominator of the preceding fraction to zero at $\phi = \mp\pi/2$, we have

$$(1 - 2pq - 3q^2)(1 + 2pq + 9q^2) = 0. \tag{3.92}$$

The quadratic equation suitable for our purpose is

$$1 + 2pq + 9q^2 = 0. \tag{3.93}$$

With this relation of p and q, the parameters can be expressed in terms of the depth ratio γ as follows:

$$q^2 = -\frac{1}{4\gamma + 5}, \qquad pq = -\frac{2(\gamma - 1)}{4\gamma + 5}. \tag{3.94}$$

Also, the radius of curvature at the crown of notch C where $\phi = 0$ becomes

$$R_0 = -\frac{9}{4(\gamma - 1)}. \tag{3.95}$$

The negative sign indicates that the curve is concave downward.

With the values of z_0 and z_0' in equations (3.89), it is found that the last two integrals in equations (3.20) are connected by

$$\begin{aligned} {}^nL_m = (1 + 4p^2q^2 + q^4)\,{}^nK_{m+2} + q^2({}^{n-2}K_{m+2} + {}^{n+2}K_{m+2}) \\ - 2pq(1 + q^2)({}^{n-1}K_{m+2} + {}^{n+1}K_{m+2}). \end{aligned} \tag{3.96}$$

The first two integrals can be expressed in terms of an auxiliary integral defined similarly by equation (3.80). The relations are

$$
\begin{aligned}
{}^{n}I_m &= (1+4p^2q^2+9q^4)\,{}^{n}H_m + 2pq(1-3q^2)({}^{n-1}H_m + {}^{n+1}H_m) \\
&\quad - 3q^2({}^{n-2}H_m + {}^{n+2}H_m),
\end{aligned} \tag{3.97a}
$$

$$
\begin{aligned}
{}^{n}K_m &= {}^{n}H_m + 4pq\,{}^{n-1}H_m - 2q^2(3-2p^2)\,{}^{n-1}H_m \\
&\quad - 12pq^3\,{}^{n-3}H_m + 9q^4\,{}^{n-1}H_m.
\end{aligned} \tag{3.97b}
$$

To evaluate the auxiliary integral, we expand

$$(1-2pq\,\mathrm{e}^{2i\phi} + q^2\,\mathrm{e}^{4i\phi})^{-m} = \sum_{s=0}^{\infty} C_s^m(p)q^s\,\mathrm{e}^{2si\phi}, \tag{3.98}$$

where C_s^m is a Gegenbauer polynomial of order m and degree s [18]. By further applying equations (3.25), the following results are found according to m being even or odd:

$$
{}^{n}H_{2m} = \begin{cases} (-1)^m\lambda^{-2m}C_{n-m}^{2m}(p)q^{n-m}, & (n \geqslant m) \\ 0, & (n \leqslant m-1) \end{cases} \tag{3.99a}
$$

$$
{}^{n}H_{2m+1} = \frac{2i}{\pi\lambda^{2m+1}} \sum_{s=0}^{\infty} \frac{(-1)^{n+s}q^s}{2m-2n+2s+1} C_s^{2m+1}(p). \tag{3.99b}
$$

The following recurrence relation may be mentioned:

$$
\begin{aligned}
{}^{n+1}H_m - 4pq\,{}^{n}H_m + 2q^2(1+2p^2)\,{}^{n-1}H_m \\
- 4pq^3\,{}^{n-2}H_m + q^4\,{}^{n-3}H_m + {}^{n}H_{m-2} = 0.
\end{aligned} \tag{3.100}
$$

Note that the Gegenbauer polynomial is given by

$$C_n^m(p) = \frac{1}{(m-1)!} \sum_{s=0}^{[n/2]} \frac{(-1)^s(m+n-s-1)!}{s!(n-2s)!}(2p)^{n-2s}, \tag{3.101}$$

where $[n/2]$ is the greatest integer not exceeding $n/2$. Two recurrence relations of the polynomial may be mentioned:

$$C_n^m(p) = (n+2m-2+p)C_{n-1}^m(p), \tag{3.102a}$$

$$(n+m)C_{n+1}^{m-1}(p) = (m-1)\{C_{n+1}^m(p) - C_{n-1}^m(p)\}. \tag{3.102b}$$

It is mentioned that the problem of a U-type notch was also considered by Seika [19]. His solution was also formulated by using complex variable method but based on a different process. However, in his mapping function, the values of p and q were not connected so that the curve of notch was not straight at the opening. He solved the resulting set of equations for the parametric coefficients by using a method of perturbation. This method is favorable only when the parameters are numerically small.

3.11 The V-type notch

Consider a mapping function in the form

$$z = \zeta + \frac{p}{a^{s+1}\zeta^{s}} + q, \tag{3.103}$$

where a, p and q are parameters. Likewise, define a pair of polar coordinates (ρ, ϕ) in the ζ-plane by equation (3.7). When the parameters are properly chosen, the mapping function transforms conformally the exterior region of a hypotrochoid in the z-plane with its center at $z = q$ into the exterior of a unit circle $\rho = 1$ in the ζ-plane. The hypotrochoid is a regular curvilinear polygon of $(s+1)$ sides if s is an integer not less than 2. If $s = 1$, it becomes an ellipse. When ϕ varies from $-\pi$ to π on the unit circle, the corresponding curve in the z-plane is a complete hypotrochoid.

In particular, when $|p| = 1/s$ and $a = 1$, the hypotrochoid becomes a hypocycloid with $(s+1)$ cusps. We shall henceforth let $|p| = 1/s$ and $a > 1$ so that the curve is a hypotrochoid with rounded vertices. The sides of the hypotrochoid are slightly curved. The curvature at the midpoint of each side changes gradually from concavity to convexity as a increases from unity up to a certain value. The value of a will be adjusted so that each side is straight or flattened at the midpoint with zero curvature.

Further, consider a corner of the hypotrochoid consisting of a rounded vertex and two adjacent sides up to the midpoint of each side. Let 2β be the subtending angle of the corner at the center of the hypotrochoid. Then

$$\beta = \frac{\pi}{s+1}. \tag{3.104}$$

Let the curve of the V-type notch be formed by such a corner. By referring to Figure 3.2(v), let the opening of the notch AA' be at $z = \mp\lambda$. The curve of notch is symmetrical with respect to the y-axis and each side intersects the x-axis symmetrically at an angle β. The center O' of the hypotrochoid is on the y-axis at $z = -i\lambda \cot \beta$. Consequently, $q = -i\lambda \cot \beta$. Let the three corresponding points A, A' and C on the unit circle $\rho = 1$ be at $\phi = -\beta, \beta$ and 0, respectively. Note that C is the crown of notch. Then the mapping function becomes

$$z = \zeta + \frac{i^{s+1}}{sa^{s+1}\zeta^s} - i\lambda \cot \beta, \tag{3.105}$$

where

$$\lambda = \left(1 - \frac{1}{sa^{s+1}}\right) \sin \beta. \tag{3.106}$$

The radius of curvature of the curve of notch is found equal to

$$R = -\frac{\{a^{2s+2} + 1 - 2a^{s+1} \cos (s+1)\phi\}^{\frac{3}{2}}}{a^{s+1}\{a^{2s+2} - s + (s-1)a^{s+1} \cos(s+1)\phi\}}. \tag{3.107}$$

Hence, the requirement of zero curvature at A and A' is satisfied if the denominator of the preceding fraction vanishes at $\phi = \mp\beta$. This leads to

$$a^{s+1} = s. \tag{3.108}$$

Consequently, the mapping function becomes

$$z = \zeta + \frac{i^{s+1}}{s^2\zeta^s} - i\lambda \cot \beta, \tag{3.109}$$

where λ now is

$$\lambda = \left(1 - \frac{1}{s^2}\right) \sin \beta. \tag{3.110}$$

Also, the radius of curvature becomes

$$R = -\frac{\{s^2 + 1 - 2s \cos(s+1)\phi\}^{\frac{3}{2}}}{s^2(s-1)\{1 + \cos(s+1)\phi\}}. \tag{3.111}$$

At the crown of notch C, where $\phi = 0$, it is

$$R_0 = -\frac{(s-1)^2}{2s^2}. \tag{3.112}$$

Besides, at C the value of z is

$$z = i\left\{1 + \frac{1}{s^2} - \left(1 - \frac{1}{s^2}\right)\cos\beta\right\}. \tag{3.113}$$

Hence, the depth ratio of the notch is

$$\gamma = \frac{s^2+1}{s^2-1}\csc\beta - \cot\beta. \tag{3.114}$$

It is noted that the foregoing results are obtained under the supposition that s is an integer. We may now relax this restriction and consider s to be a real number greater than unity so as to provide a continuous variation for the curve of notch. It is seen that under such an extension all the preceding relations remain valid. However, the hypotrochoid is no longer a complete hypotrochoid if s is not an integer. The curve intersects itself on the y-axis to form a closed curve as ϕ varies correspondingly from $-\pi$ to π on the unit circle $\rho = 1$. Such an incomplete hypotrochoid may be called a generalized hypotrochoid. It still has rounded corners and flattened sides. With one corner of the generalized hypotrochoid as the V-type notch, the curve of notch in the z-plane is transformed by the mapping function into a portion of the circumference of the unit circle $\rho = 1$ in the ζ-plane from $\phi = -\beta$ to $\phi = \beta$.

When $\rho = 1$, we have from the mapping function

$$z_0 = i\,\mathrm{e}^{-i\phi}\left(1 + \frac{1}{s^2}\,\mathrm{e}^{\pi i\phi/\beta}\right) - i\lambda\cot\beta, \tag{3.115a}$$

$$z_0' = 1 - \frac{1}{s}\,\mathrm{e}^{\pi i\phi/\beta}. \tag{3.115b}$$

As here s in general is not an integer, it appears more convenient to evaluate the integrals in equations (3.20) directly by using Simpson rule or other numerical method.

3.12 A single notch in a strip

In the foregoing analysis, we deal with the fundamental problem of a single edge notch in a semi-infinite plate under longitudinal tension. In what follows in this Chapter, we shall extend the analysis to deal with further problems of the notched plate in which neighboring boundaries are present such that their interaction on the stresses has to be considered. The neighboring boundaries may be in the form of an additional straight edge or one or more edge notches. However, for shortness, the analysis will be carried just to the step at which the two analytic functions $\Phi(z)$ and $\Psi(z)$ are found. The remaining steps of the analysis consist of adjusting the boundary conditions along the notches and then leading to the determination of the parametric coefficients involved in the analytic functions. There seems no particular difficulty in carrying out these steps in the manner described previously.

Suppose first that there exists in the notched plate a second edge parallel to the first. We thus face a problem of analyzing the stresses in an infinite strip containing a single edge notch. Such a problem was considered by Tamate [20] and Ling [21] independently for both longitudinal tension and transverse bending when the notch is semicircular. Both solutions were expressed in Airy's stress function. To extend the tension problem to an edge notch of general shape, the relevant Airy's stress function is first converted into the two analytic functions by the method described in section 3.2.

Let the two edges of the infinite strip be represented in the z-plane by the lines $y=0$ and $y=2$, respectively. In Ling's solution, the Airy's stress function for longitudinal tension is expressed in the form:

$$\chi = Tb^2\left\{\chi_{00} + \sum_{k=0}^{\infty} (A_{2k}\chi_{2k} + B_{2k}\chi'_{2k}\right\}, \tag{3.116}$$

where T is the uniform unit tension in the longitudinal direction and b is one half of the width of the strip. The factor b^2T is introduced in order to render the real parametric coefficients A_{2k} and B_{2k} dimensionless. The functions involved are

$$\chi_{00} = \tfrac{1}{2}y^2, \tag{3.117a}$$

$$\begin{aligned}\chi_{2k} = \frac{1}{(2k+1)!}\Bigg[2\int_0^\infty m^{2k}(1+my)\,\mathrm{e}^{-my}\cos mx\,\mathrm{d}m \\ + \int_0^\infty m^{2k}\{f_1(m)(\sinh my - my\cosh my) \\ + f_2(m)m^2y\sinh my\}\cos mx\,\mathrm{d}m\Bigg],\end{aligned} \tag{3.117b}$$

$$\chi'_{2k} = \frac{1}{(2k)!}\Bigg[2\int_0^\infty m^{2k} y\, \mathrm{e}^{-my} \cos mx \,\mathrm{d}m$$

$$+ \int_0^\infty m^{2k}\{f_2(m)(\sinh my - my \cosh my)$$

$$- f_3(m) y \sinh my\} \cos mx \,\mathrm{d}m\Bigg], \tag{3.117c}$$

where

$$f_1(m) = \frac{1+2m-\mathrm{e}^{-2m}}{\sinh 2m + 2m} - \frac{1+2m+\mathrm{e}^{-2m}}{\sinh 2m - 2m}, \tag{3.118a}$$

$$f_2(m) = \frac{2}{\sinh 2m + 2m} - \frac{2}{\sinh 2m - 2m}, \tag{3.118b}$$

$$f_3(m) = \frac{1-2m+\mathrm{e}^{-2m}}{\sinh 2m + 2m} - \frac{1-2m-\mathrm{e}^{-2m}}{\sinh 2m - 2m}. \tag{3.118c}$$

This stress function is biharmonic as required and possesses a singularity at the origin. It satisfies all the given boundary conditions along the edges as well as at the ends of the strip, save at the origin which is a singularity. Moreover, it is even in x so that it is suitable for solving a symmetrical notch in a strip under longitudinal tension. The first integral in χ_{2k} and χ'_{2k} is integrable. The two corresponding analytic functions are found as follows:

$$\Phi(z) = \frac{1}{4} + \sum_{k=0}^{\infty} (-1)^k \left\{\frac{(2k+2)iA_{2k}}{z^{2k+3}} + \frac{(2k+1)B_{2k}}{z^{2k+2}}\right\}$$

$$+ \sum_{k=0}^{\infty} \frac{A_{2k}}{2(2k+1)!} \int_0^\infty m^{2k+2}\{mf_2(m) \cos mz + if_1(m) \sin mz\} \,\mathrm{d}m$$

$$- \sum_{k=0}^{\infty} \frac{B_{2k}}{2(2k)!} \int_0^\infty m^{2k+1}\{f_3(m) \cos mz - imf_2(m) \sin mz\} \,\mathrm{d}m, \tag{3.119a}$$

$$\Psi(z) = -\frac{1}{2} + \sum_{k=0}^{\infty} (-1)^k \left\{\frac{(2k+2)(2k+3)iA_{2k}}{z^{2k+3}} + \frac{2k(2k+1)B_{2k}}{z^{2k+2}}\right\}$$

$$+ \sum_{k=0}^{\infty} \frac{A_{2k}}{2(2k+1)!} \int_0^\infty m^{2k+3}\{f_2(m)(mz \sin mz - 2 \cos mz)$$

$$- if_1(m) z \cos mz\} \,\mathrm{d}m$$

$$- \sum_{k=0}^{\infty} \frac{B_{2k}}{2(2k)!} \int_0^\infty m^{2k+1}\{f_3(m)(mz \sin mz - 2 \cos mz)$$

$$- im^2 f_2(m) z \cos mz\} \,\mathrm{d}m. \tag{3.119b}$$

It is seen that the functions apart from the integrals give precisely the Maunsell's solution in equations (3.5) though the notation is slightly different. Therefore, the integrals in the functions may be regarded as additional terms introduced to account for the interaction due to the presence of the opposite edge. The adjustment of the boundary conditions on the notch leads to a similar set of equations as in (3.14).

3.13 A pair of symmetrical notches in a strip

Suppose that the infinite strip contains a second notch located symmetrically on the opposite edge. The two notches are of equal size and symmetrical shape. In the particular case when the notches are semicircular, the problem was solved earlier by Ling for both longitudinal tension and transverse bending [22], [23] and [5]. Likewise, let the two edges of the strip be represented in the z-plane by the lines $y = 0$ and $y = 2$. Furthermore, let the Airy's stress function be given by equation (3.116). Whereas the function χ_{00} is identical, the other two functions in the series that follow are now given by

$$\chi_{2k} = \frac{4}{(2k+1)!}\int_0^\infty \frac{m^{2k}\sinh m}{\sinh 2m + 2m}\{(1 + m\coth m)\cosh m(y-1) - m(y-1)\sinh m(y-1)\}\cos mx \,\mathrm{d}m, \tag{3.120a}$$

$$\chi'_{2k} = \frac{4}{(2k)!}\int_0^\infty \frac{m^{2k}\cosh m}{\sinh 2m + 2m}\{\tanh m\cosh m(y-1) - (y-1)\sinh m(y-1)\}\cos mx \,\mathrm{d}m. \tag{3.120b}$$

Both of them are even in x and also even with respect to the line $y = 1$. Each such integral possesses two singularities on the y-axis, one on the lower edge at the origin and the other on the upper edge at the point $(x, y) = (0, 2)$. The terms singular at the origin can be isolated from each integral and be integrated. The two corresponding analytic functions are found as follows:

$$\begin{aligned}\Phi(z) = {} & \frac{1}{4} + \sum_{k=0}^{\infty}(-1)^k\left\{\frac{(2k+2)iA_{2k}}{z^{2k+3}} + \frac{(2k+1)B_{2k}}{z^{2k+2}}\right\} \\ & + \sum_{k=0}^{\infty}\frac{A_{2k}}{(2k+1)!}\int_0^\infty \frac{m^{2k+2}}{\sinh 2m + 2m}\{2m\,\mathrm{e}^{miz} + i(1 - \mathrm{e}^{-2m})\sin mz\}\,\mathrm{d}m \\ & - \sum_{k=0}^{\infty}\frac{B_{2k}}{(2k)!}\int_0^\infty \frac{m^{2k+1}}{\sinh 2m + 2m}\{2m\,\mathrm{e}^{miz} - (1 + \mathrm{e}^{-2m})\cos mz\}\,\mathrm{d}m,\end{aligned} \tag{3.121a}$$

$$\begin{aligned}\Psi(z) = &-\frac{1}{2} + \sum_{k=0}^{\infty} (-1)^k \left\{\frac{(2k+2)(2k+3)iA_{2k}}{z^{2k+3}} + \frac{2k(2k+1)B_{2k}}{z^{2k+2}}\right\} \\ &- \sum_{k=0}^{\infty} \frac{A_{2k}}{(2k+1)!} \int_0^{\infty} \frac{m^{2k+3}}{\sinh 2m + 2m} \{4 \cos mz + 2imz\, \mathrm{e}^{miz} \\ &+ i(1 - \mathrm{e}^{-2m}) z \cos mz\}\, \mathrm{d}m \\ &- \sum_{k=0}^{\infty} \frac{B_{2k}}{(2k)!} \int_0^{\infty} \frac{m^{2k+1}}{\sinh 2m + 2m} \{4m \cos mz + 2im^2 z\, \mathrm{e}^{miz} \\ &+ (1 + \mathrm{e}^{-2m})(mz \sin mz - 2 \cos mz)\}\, \mathrm{d}m. \end{aligned} \tag{3.121b}$$

Again, the functions apart from the integrals give the Maunsell's solution in equations (3.5). Consequently, the integrals in the functions may be regarded as the additional terms introduced to account for the interaction due to the presence of the opposite edge and the opposite notch. The adjustment of the boundary conditions of the notch on the lower edge leads to a similar set of equations as in equation (3.14). By symmetry, if the boundary conditions of the notch on the lower edge are satisfied, those of the notch on the upper edge are automatically satisfied.

3.14 A pair of staggered notches in a strip

Suppose that the pair of edge notches is located in staggered positions on the opposite edges of the strip as shown in Figure 3.8. The notches are of equal size and symmetrical shape. Let the edges be likewise represented by the lines $y = 0$ and $y = 2$ in the z-plane. Consider two points of the strip, one on the lower edge at the origin and the other on the upper edge at $z = c$, where $c = a + 2i$. It appears that the solution of the present problem can be constructed by superposing the solution in section 3.12 for a single notch in a strip. However, additional functions

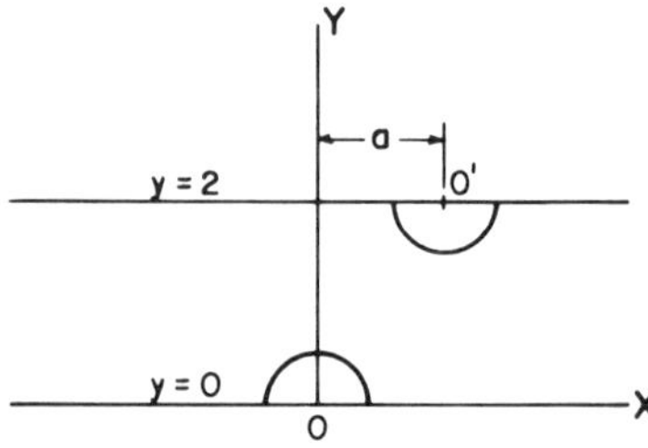

Figure 3.8. A pair of staggered notches in an infinite strip.

odd in x are needed since the solution in section 3.12 is even in x and suitable only when the state of stress in the strip is symmetrical with respect to the y-axis. The state of stress in the present problem does not possess such a symmetry but possesses a center of symmetry at the point $z = c/2$. The functions odd in x can be obtained as follows:

$$\chi_{2k+1} = -\frac{1}{2k+2}\frac{\partial \chi_{2k}}{\partial x}, \qquad \chi'_{2K+1} = -\frac{1}{2k+1}\frac{\partial \chi'_{2k}}{\partial x}, \tag{3.122}$$

where χ_{2k} and χ'_{2k} are given in equations (3.117). Further, denote

$$\chi^* = Tb^2 \sum_{k=0}^{\infty} (A_{2k+1}\chi_{2k+1} + B_{2k+1}\chi'_{2k+1}), \tag{3.123}$$

where A_{2k+1} and B_{2k+1} are real parametric coefficients. The two analytic functions corresponding to χ^* are found as

$$\begin{aligned}\Phi_1(z) = &\sum_{k=0}^{\infty} (-1)^k \left\{\frac{(2k+3)iA_{2k+1}}{z^{2k+4}} + \frac{(2k+2)B_{2k+1}}{z^{2k+3}}\right\} \\ &+ \sum_{k=0}^{\infty} \frac{A_{2k+1}}{2(2k+2)!}\int_0^{\infty} m^{2k+3}\{mf_2(m)\sin mz - if_1(m)\cos mz\}\,dm \\ &- \sum_{k=0}^{\infty} \frac{B_{2k+1}}{2(2k+1)!}\int_0^{\infty} m^{2k+2}\{f_3(m)\sin mz + imf_2(m)\cos mz\}\,dm,\end{aligned} \tag{3.124a}$$

$$\begin{aligned}\Psi_1(z) = &\sum_{k=0}^{\infty} (-1)^k \left\{\frac{(2k+3)(2k+4)iA_{2k+1}}{z^{2k+4}} + \frac{(2k+2)(2k+3)B_{2k+1}}{z^{2k+3}}\right\} \\ &- \sum_{k=0}^{\infty} \frac{A_{2k+1}}{2(2k+2)!}\int_0^{\infty} m^{2k+4}\{f_2(m)(mz\cos mz + 2\sin mz) \\ &+ if_1(m)z\sin mz\}\,dm \\ &+ \sum_{k=0}^{\infty} \frac{B_{2k+1}}{2(2k+1)!}\int_0^{\infty} m^{2k+3}\{f_3(m)(mz\cos mz + 2\sin mz) \\ &- imf_2(m)z\sin mz\}\,dm.\end{aligned} \tag{3.124b}$$

where f_1, f_2 and f_3 are given in equations (3.118).

Write each analytic function in equations (3.119) as the sum of two parts in the form:

$$\Phi(z) = \tfrac{1}{4} + \Phi_2(z), \tag{3.125a}$$

$$\Psi(z) = \tfrac{1}{2} + \Psi_2(z). \tag{3.125b}$$

By superposition, the two analytic functions of the present problem are constructed by

$$\Phi(z) = \tfrac{1}{4} + \Phi_1(z) + \Phi_2(z) + \Phi_1(c - z) + \Phi_2(c - z), \tag{3.126a}$$

$$\Psi(z) = -\tfrac{1}{2} + \Psi_1(z) + \Psi_2(z) + \Psi_1(c - z) + \Psi_2(c - z). \tag{3.126b}$$

These two functions possess two singularities on the edges, one at $z = 0$ and the other at $z = c$. Besides, they possess a center of symmetry at $z = c/2$. In adjusting the boundary conditions of the notches, the functions are expanded near the origin such that the singular terms are isolated. Again, by symmetry, if the boundary conditions of the notch on the lower edge are satisfied, those of the notch on the upper edge are automatically satisfied.

3.15 Multiple edge notches

When there are more than one notch on the edge of a semi-infinite plate or on either or both edges of an infinite strip, the solution can be constructed similarly by superposition from the fundamental solution of a single notch in the respective plate or strip. When the notches are not of the same size, the boundary conditions of each notch are to be adjusted individually. The resulting solution is thus exceedingly complicated. However, if the notches are all of the same shape and size and in addition they are distributed along the edge or edges in such a manner that the state of stress toward each notch is identical or invariant through appropriate rectilinear transformations, the solution then becomes greatly simplified. Such a distribution of notches may be referred to as invariant distribution. When the notches are in invariant distribution, it is seen that if the boundary conditions on one of the notches are satisfied, those on all the other notches are automatically satisfied.

Two of the cases of invariant distribution of notches were solved by Atsumi [24] and [25], one concerning two pairs of symmetrically-located semicircular notches and the other concerning infinite pairs of such notches on the edges of a strip. It may be mentioned that the entire class of invariant distribution of notches consists of altogether 16 cases, in which four cases are for a semi-infinite plate. The construction of the Airy's stress function in each case together with the evaluation of the coefficients involved was discussed systematically by Ling in a previous paper [26]. The two corresponding analytic functions can then be found accordingly.

References

[1] F. G. Maunsell, Stresses in a notched plate under tension, *Philosophical Magazine*, 21, pp. 765–773 (1936).

[2] N. I. Muskhelishvili, *Some Basic Problems of Mathematical Theory of Elasticity*, 4th edition, English translation, Noordhoff, Groningen, 1963.

[3] L. H. Mitchell, Stress concentration at semicircular notch, *Journal of Applied Mechanics*, 32, *Trans. ASME*, 87, series E, pp. 938–939 (1965).

[4] C. B. Ling, On stress concentration at semicircular notch, *Journal of Applied Mechanics*, 34, *Trans. ASME*, 89, series E, p. 522 (1967).

[5] C. B. Ling, On stress concentration factor in a notched strip, *Journal of Applied Mechanics*, 35, *Trans. ASME*, 90, series E, pp. 833–835 (1968).

[6] C. B. Ling and F. H. Cheng, Stress systems in a rectangular plate, *Applied Scientific Research, series A*, 25, pp. 97–112 (1971).

[7] O. L. Bowie, Analysis of edge notch in a semi-infinite region, *Journal of Mathematics and Physics*, 45, pp. 356–366 (1966).

[8] T. Isibasi, Stresses in a semi-infinite plate with a circular notch under uniform tension, *Memoirs of Faculty of Engineering, Kyushu Imperial University, Fukuoka, Japan*, IX, no. 2, pp. 131–143 (1940).

[9] E. Weinel, Die Spannungserhoehung durch Kreisbogenkerben, *Zeitschrift für Angewandte Mathematik und Mechanik*, 21, pp. 228–230 (1941).

[10] C. B. Ling, On the stresses in a notched plate under tension, *Journal of Mathematics and Physics*, 26, pp. 184–198 (1948).

[11] G. B. Jeffery, Plane stress and plane strain in bipolar coordinates, *Philosophical Transactions of Royal Society of London, series A*, 221, pp. 265–293 (1921).

[12] E. G. Coker and L. N. G. Filon, *A Treatise on Photo-Elasticity*, Cambridge University Press, 2nd edition, 1957, p. 163.

[13] C. B. Ling and C. W. Nelson, On evaluation of Howland's integrals, *Annals of Academia Sinica, no. 2*, part 2, Taiwan, Republic of China, pp. 45–50 (1955).

[14] C. B. Ling and H. C. Wu, Tables of values of three infinite integrals, *Mathematics of Computation*, 19, pp. 360–365 (1965).

[15] C. W. Nelson, New tables of Howland's and related integrals, *Mathematics of Computation*, 15, pp. 12–18 (1961).

[16] C. B. Ling and J. Lin, A new method of evaluation of Howland integrals, *Mathematics of Computation*, 25, pp. 331–337 (1971).

[17] M. Greenspan, Effect of a small hole on the stresses in a uniformly loaded plate, *Quarterly of Applied Mathematics*, 2, pp. 60–71 (1944).

[18] W. Magnus, F. Oberhettinger and R. P. Soni, *Formulas and Theorems for Special Functions of Mathematical Physics*, 3rd edition, Springer-Verlag, 1966.

[19] M. Seika, Stresses in a semi-infinite plate containing a U-type notch under uniform tension, *Ingenieur-Archiv*, 27, pp. 285–294 (1960).

[20] O. Tamate, Stresses in an infinite strip with a semicircular notch under uniform tension and pure bending, *Technology Reports of Tohoku University, Japan*, 16, pp. 34–53 (1952).

[21] C. B. Ling, On the stresses in a strip having a single semicircular notch, *Proceedings of Ninth Japan National Congress for Applied Mechanics*, pp. 171–181 (1959).

[22] C. B. Ling, Stresses in a notched strip under tension, *Journal of Applied Mechanics*, 14, *Trans. ASME*, 69, pp. A275–280 (1947).

[23] C. B. Ling, On the stresses in a notched strip, *Journal of Applied Mechanics*, 19, *Trans. ASME*, 74, pp. 141–146 (1952).

[24] A. Atsumi, Stress concentrations in a strip under tension and containing two pairs of semicircular notches placed on the edges symmetrically, *Journal of Applied Mechanics*, 24, *Trans. ASME*, 79, pp. 565–573 (1957).

[25] A. Atsumi, Stress concentrations in a strip under tension and containing an infinite row of semicircular notches, *Quarterly Journal of Mechanics and Applied Mathematics*, 11, pp. 478–490 (1958).

[26] C. B. Ling, On invariant distribution of notches in an infinite strip, *Collected Papers of Chih-Bing Ling, Institute of Mathematics, Academia Sinica, Taiwan, Republic of China*, pp. 246–270 (1963).

M. K. Kassir

4 Three dimensional notch problems

4.1 Introduction

This chapter is concerned with the nature of the stress field in the neighborhood of a three dimensional notch embedded in an infinite, homogeneous and isotropically elastic medium. The analysis is based upon Eshelby's work [1], which contains formulas for the cartesian components of displacement and stress induced in an ellipsoidal inclusion and in its surrounding solid when the inclusion undergoes arbitrary transformation in strain. These formulas involve various derivatives of a harmonic function, often used in many branches of mathematical physics and, in particular, in potential theory to determine the potential of a solid homogeneous ellipsoid [2]. In the region outside the inclusion, the elastic field can be expressed in terms of the cartesian coordinates (x, y, z) and the ellipsoidal coordinate ξ [3]. In order to determine explicit expressions for the elastic field at a given point, a cubic equation (see equation 4.5) has to be solved for ξ and a sizeable amount of algebraic work is involved.

In this study, the asymptotic form of ξ near the boundary of the inclusion is determined. With the aid of this result, the local displacement and stress fields are derived in terms of a convenient set of spherical coordinates r, θ and ϕ measured from the edge of a focal plane of the ellipsoid. Such information provide the necessary tools to extend the energy-based theories of fracture mechanics [4, 5] to three dimensional blunted notches and cavities. As the ellipsoid shrinks to an elliptical lamina with sharp edges, the singular stress field derived by Kassir and Sih [6] for the case of a flat elliptical crack is recovered.

Section 2 contains a brief outline of Eshelby's solution to the ellipsoidal inclusion and inhomogeneity problems. The functional relations between the variable ξ and the coordinates (r, θ, ϕ) are derived in section 3. Two sets of relations valid for small values of r are presented.

The first form describes the position of an arbitrary external point with reference to the boundary of a 'limiting ellipse' obtained as a result of shrinking the triaxial inclusion into a biaxial one. These relations and their associated elastic field are utilized to recover the singular stress field associated with sharp notches and cavities [6, 7]. The second set of relations determines the position of the external point relative to the edge of the horizontal plane of the ellipsoid and are used to determine the elastic fields around notches with smooth boundaries (i.e. non-singular stress fields). The limiting forms of ξ are then employed in section 4 to determine the local displacement field. As compared to the simple expressions obtained in case of the elliptical flat crack [6], the resulting displacement field is quite involved in its angular and radial distributions. It is also a function of the location around the periphery of a focal plane of the inclusion. Finally, section 5 contains asymptotic expansions of the components of stress near the notch surface.

4.2 Ellipsoidal inclusions and inhomogeneities

In this section, Eshelby's solution of the ellipsoidal inclusion and inhomogeneity problems is briefly summarized [1]. The relevant portions needed for determining the local stress distribution are only given. For additional information the reader is referred to Eshelby's original work.

Eshelby considers two closely related problems in the infinitesimal theory of three-dimensional elasticity. The first problem deals with the elastic field in an ellipsoidal region (the inclusion) and in its surrounding medium (the matrix) when the inclusion undergoes a transformation that, in the absence of the constraint imposed by the matrix, would be equivalent to a prescribed uniform strain (e_{ij}^T). Both the inclusion and the matrix are assumed to be made of the same homogeneous and isotropic material and to be perfectly bonded to each other. It turns out that the resulting stress is constant throughout the inclusion. In the second problem, an ellipsoidal elastic body (the inhomogeneity) in an otherwise infinite solid with different elastic properties disturbs a uniform stress applied at remote distances from the inhomogeneity. It is required to find the induced stress field everywhere in the medium. The important case of an ellipsoidal cavity [8] is obtained by setting the elastic constants of the inhomogeneity to be zero. For this problem, Eshelby's formulation provides a set of linear algebraic equations which relate the components of stress within the body to the loads applied at infinity (see also ref. [9]).

Since the state of stress in an ellipsoidal inclusion which has undergone uniform transformation is constant, one can easily relate the inclusion and inhomogeneity problems. Thus, in order to obtain the elastic field around an ellipsoidal cavity which perturb uniform traction at far away distances, a constant stress equal and opposite to the one induced in the inclusion is superimposed on the elastic field of the inclusion. This leaves the region of the inclusion free of stress and gives rise to uniform stress at infinity as required. Because of this analogy, attention is focused on Eshelby's formulation of the ellipsoidal inclusion problem for purposes of determining the local stress field around three-dimensional notches in solids.

Consider an ellipsoidal inclusion having three unequal principal semi-axes (a, b, c) and surrounded by an infinite, homogeneous and isotropic elastic medium made of an identical material. Referring to a system of cartesian coordinates with the origin located at the center, the surface of the inclusion is described by the ellipsoid

$$x^2/a^2 + y^2/b^2 + z^2/c^2 = 1, \tag{4.1}$$

Suppose that the inclusion is subjected to an arbitrary homogeneous transformation which, in the absence of the matrix, can be characterized by the constant strains $(e_{xx}^T, e_{yy}^T, e_{zz}^T, e_{xy}^T, e_{yz}^T, e_{zx}^T)$. Eshelby [1] has determined the induced displacements in the medium

$$\begin{aligned}
8\pi(1-\nu)u_x^c = {} & \frac{e_{yy}^T - e_{xx}^T}{a^2 - b^2}\frac{\partial}{\partial y}\left(a^2 y\frac{\partial f}{\partial x} - b^2 x\frac{\partial f}{\partial y}\right) \\
& + \frac{e_{zz}^T - e_{xx}^T}{c^2 - a^2}\frac{\partial}{\partial z}\left(c^2 x\frac{\partial f}{\partial z} - a^2 z\frac{\partial f}{\partial x}\right) \\
& - 2[(1-\nu)e_{xx}^T + \nu(e_{yy}^T + e_{zz}^T)]\frac{\partial f}{\partial x} \\
& - 4(1-\nu)\left(e_{xy}^T\frac{\partial f}{\partial y} + e_{xz}^T\frac{\partial f}{\partial z}\right) \\
& + \frac{\partial}{\partial x}\left[\frac{2e_{xy}^T}{a^2 - b^2}\left(a^2 y\frac{\partial f}{\partial x} - b^2 x\frac{\partial f}{\partial y}\right)\right. \\
& + \frac{2e_{yz}^T}{b^2 - c^2}\left(b^2 z\frac{\partial f}{\partial y} - c^2 y\frac{\partial f}{\partial z}\right) \\
& \left. + \frac{2e_{zx}^T}{c^2 - a^2}\left(c^2 x\frac{\partial f}{\partial z} - a^2 z\frac{\partial f}{\partial x}\right)\right],
\end{aligned} \tag{4.2}$$

where (u_x^c, u_y^c, u_z^c) designate the displacement components and ν is Poisson's ratio of the material. u_y^c and u_z^c are found from equation (4.2) by cyclic permutation of the letters (x, y, z) and (a, b, c). In equation (4.2), $f(x, y, z)$ stands for the potential [2]

$$f = \pi abc \int_\xi^\infty [1 - x^2/(a^2+s) - y^2/(b^2+s) - z^2/(c^2+s)] \frac{\mathrm{d}s}{\sqrt{Q(s)}}, \tag{4.3}$$

in which

$$Q(s) = (a^2+s)(b^2+s)(c^2+s), \tag{4.4}$$

and ξ is the variable in the system of ellipsoidal coordinates (ξ, η, ζ) which describes a family of confocal ellipsoids in space. In fact, ξ is the greatest (and only positive) root of the cubic equation in s [3]

$$x^2/(a^2+s) + y^2/(b^2+s) + z^2/(c^2+s) = 1, \tag{4.5}$$

and can assume any value in the interval $(-c^2, \infty)$. The exterior- and surface-points of the inclusion are reached by setting $\xi = 0$. Upon employing the substitutions

$$c^2 + s = (a^2 - c^2)(cn^2t/sn^2t), \tag{4.6a}$$

$$c^2 + \xi = (a^2 - c^2)(cn^2u/sn^2u), \tag{4.6b}$$

the integration in equation (4.3) can be performed and f takes the form

$$\begin{aligned} f = \frac{2\pi abc}{k^2 l^3} \Big[& (k^2 l^2 - x^2 + y^2)u \\ & + k'^{-2}(k'^2 x^2 - y^2 + k^2 z^2)E(u) \\ & + \frac{lk^2}{k'^2 A}\left(\frac{C}{B}y^2 - \frac{B}{C}z^2\right)\Big]. \end{aligned} \tag{4.7}$$

In equations (4.6) and (4.7), u and $E(u)$ are, respectively, the normal elliptic integrals of the first and second kinds of modulus k and argument am. u, where

$$\left.\begin{aligned} & E(u) = \int_0^u \mathrm{d}n^2 t \, \mathrm{d}t, \qquad k^2 = 1 - k'^2 = \frac{a^2 - b^2}{l^2} \\ & \text{am. } u = \sin^{-1}\left(\frac{l}{A}\right), \quad 0 < k^2 < 1, \quad 0 < \text{am. } u < \frac{\pi}{2} \end{aligned}\right\}, \tag{4.8}$$

while $\mathrm{sn}\, u$, $\mathrm{cn}\, u$ and $\mathrm{dn}\, u$ are the Jacobian elliptic functions, and the following abbreviations have been adopted

$$\left.\begin{aligned} A &= (a^2+\xi)^{\frac{1}{2}} \\ B &= (b^2+\xi)^{\frac{1}{2}} \\ C &= (c^2+\xi)^{\frac{1}{2}} \\ l &= (a^2-c^2)^{\frac{1}{2}} \end{aligned}\right\} . \tag{4.9}$$

It is also assumed that $c^2 < b^2 < a^2$.

The corresponding state of stress in the matrix (σ_{ij}^c) can be computed by applying the usual Hooke's law to equations (4.2), i.e., in indicial notations

$$\left.\begin{aligned} &\sigma_{ij}^c = \lambda e_{mm}^c \delta_{ij} + 2\mu e_{ij}^c \\ &(i \text{ and } j) = (x, y \text{ and } z) \end{aligned}\right\} \tag{4.10}$$

Here, λ and μ denote Lame's constants. These stresses vanish at infinity. The various derivatives of f which are needed for determining the elastic field are listed in Appendix A.

In the interior region of the inclusion ($\xi = 0$), the potential in equation (4.7) reduces to

$$\begin{aligned} f = \frac{2\pi abc}{k^2 l^3} \Big[&(k^2 l^2 - x^2 + y^2)u + k'^{-2}(k'^2 x^2 - y^2 + k^2 z^2)E(u) \\ &+ \frac{lk^2}{abck'^2}(c^2 y^2 - b^2 z^2) \Big], \end{aligned} \tag{4.11}$$

and the stresses are found from the relations

$$\sigma_{ij} = \sigma_{ij}^c - \sigma_{ij}^T, \quad \xi = 0, \tag{4.12}$$

where σ_{ij}^T are the stresses corresponding to the strains e_{ij}^T. It is clear from equations (4.2), (4.10), (4.11) and (4.12) that the stresses in the inclusion are constants [1].

From the viewpoint of fracture mechanics, a great deal of information may be gained from a study of the stress distribution around a three dimensional flaw or cavity in an elastic solid. The local field can be relied upon to predict the failure loads and fracture trajectories in solids. In order to determine the stresses around an ellipsoidal cavity perturbing uniform tractions (σ_{ij}^∞) at distant boundaries, the negative of the stresses in equation (4.12) are superimposed on the stresses computed by means of equations (4.10) and (4.12). This leaves the inclusion region free of stress (i.e. a cavity) and the stresses at a general point in the matrix become

$$\sigma_{ij} = \sigma_{ij}^{\infty} + \sigma_{ij}^{c}. \tag{4.13}$$

Since σ_{ij}^{∞} are constants, it suffices to examine σ_{ij}^{c} for purposes of investigating the local distribution of stress around the notch. This is done in the sequel.

4.3 Relations between ξ and local coordinates

The derivation of the stress state near the notch hinges upon a knowledge of the asymptotic expansion of ξ in terms of some local coordinates (r, θ, ϕ) measured from the boundary of the notch. Consider the surfaces described by equation (4.5) with $s = \xi$, where $-c^2 \leqslant \xi < \infty$. As long as $\xi > -c^2$, these surfaces form a family of confocal ellipsoids. The ellipsoid tends to become large and nearly spherical in shape for large and positive values of ξ and, when $\xi = 0$, it reduces to the one in equation (4.1). A thin ellipsoid is obtained by allowing ξ to assume negative values and, in the limiting case of $\xi \to -c^2$, the ellipsoid shrinks to a flat surface given by the relations

$$x^2/l^2 + y^2/(k'l)^2 \leqslant 1, \qquad z = 0. \tag{4.14}$$

Also, upon setting $c = 0$ in equation (4.14), the following confocal ellipse is obtained

$$x^2/a^2 + y^2/b^2 \leqslant 1, \qquad z = 0. \tag{4.15}$$

With a view toward determining a local stress field which would be valid for sharp ellipsoidal notches (i.e. for $c \to 0$), it is expedient at first to expand ξ near the periphery of the ellipse (4.14). A point (Q_1) on this periphery can be represented by the parametric relations

$$x = l \cos \phi_1, \qquad y = k'l \sin \phi_1. \tag{4.16}$$

The location of a point (P_1) in the layer surrounding the notch can be described by the coordinates (r_1, θ_1) located in a plane normal to the

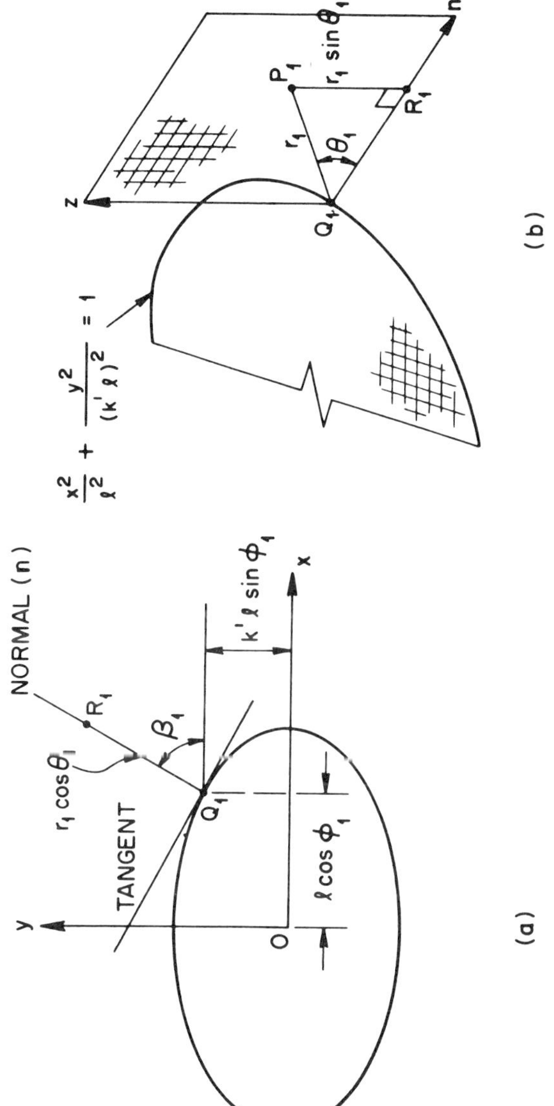

Figure 4.1. Local coordinates near periphery of ellipse in equation (4.14).

ellipse (4.14) as shown in Figure (4.1). In this figure, the segment $Q_1R_1 = r_1 \cos\theta_1$ is the projection of the radial distance r_1 along the direction of the outward normal to the ellipse. This outward normal makes an angle β_1 with the positive x-axis which is related to ϕ_1 by the transformations

$$\sin\phi_1 = \Phi_1^{\frac{1}{2}} \sin\beta_1, \tag{4.17a}$$

$$k' \cos\phi_1 = \Phi_1^{\frac{1}{2}} \cos\beta_1, \tag{4.17b}$$

where the following contraction has been employed

$$\Phi_1 = 1 - k^2 \cos^2\phi_1. \tag{4.18}$$

In equations (4.17) and (4.18) k and k' are defined in equations (4.8) and (4.9). It follows that the coordinates of point (P_1) can be expressed by

$$x = l\cos\phi_1 + r_1\cos\theta_1\cos\beta_1, \tag{4.19a}$$

$$y = k'l\sin\phi_1 + r_1\cos\theta_1\sin\beta_1, \tag{4.19b}$$

$$z = r_1\sin\theta_1, \tag{4.19c}$$

For the parameter ξ, let

$$\xi = a_0 + a_1 r_1 + a_2 r_1^2 + O(r_1^3) \tag{4.20}$$

where a_n $(n = 0, 1, 2)$ are, in general, functions of θ_1 and ϕ_1, and higher order terms in r_1 are neglected. Since $r_1 = 0$ corresponds to $\xi = -c^2$, it follows that

$$a_0 = -c^2. \tag{4.21}$$

Inserting equations (4.19)–(4.21) into expression (4.5) with $s = \xi$, expanding for small values of r_1 and making use of equations (4.17) and (4.18), the following result is obtained

$$\left[a_1\Phi_1 - 2k'l\cos\theta_1\Phi_1^{\frac{1}{2}} - \frac{k'^2l^2}{a_1}\sin^2\theta_1\right] r_1$$

$$+\left[a_2\left(\Phi_1 + \frac{k'^2l^2}{a_1^2}\sin^2\theta_1\right) - \frac{k'^4\cos^2\phi_1 + \sin^2\phi_1}{(k'l)^2}\right.$$

$$\left.\times\left(a_1^2 - 2k'la_1\Phi_1^{-\frac{1}{2}}\cos\theta_1 + k'^2l^2\frac{\cos^2\theta_1}{\Phi_1}\right)\right] r_1^2 + O(r_1^3), \tag{4.22}$$

On account of equation (4.22), it is readily confirmed that the values of a_1 and a_2 are

$$a_1 = 2k'l\Phi_1^{-\frac{1}{2}}\cos^2(\theta_1/2), \tag{4.23a}$$

$$a_2 = \Phi_1^{-2}(\Phi_1 - k^2k'^2\cos^2\phi_1)\cos^2(\theta_1/2). \tag{4.23b}$$

Knowing the asymptotic expansion of ξ, the expression for the displacement and stress fields in terms of the coordinates (r_1, θ_1, ϕ_1) may now be obtained. Appendix B contains the necessary expansions of various terms appearing in Eshelby's formulas for the displacements and stresses in terms of the coordinates (r_1, θ_1, ϕ_1).

On the other hand, for triaxial notches (i.e. where $c \neq 0$), the position of point (P) may be described by a system of spherical coordinates (r, θ, ϕ) measured from the boundary of the ellipse (4.15) as illustrated in Figure (4.2). Referring to this figure, the coordinates of point Q are

$$x = a\cos\phi, \qquad y = b\sin\phi, \tag{4.24}$$

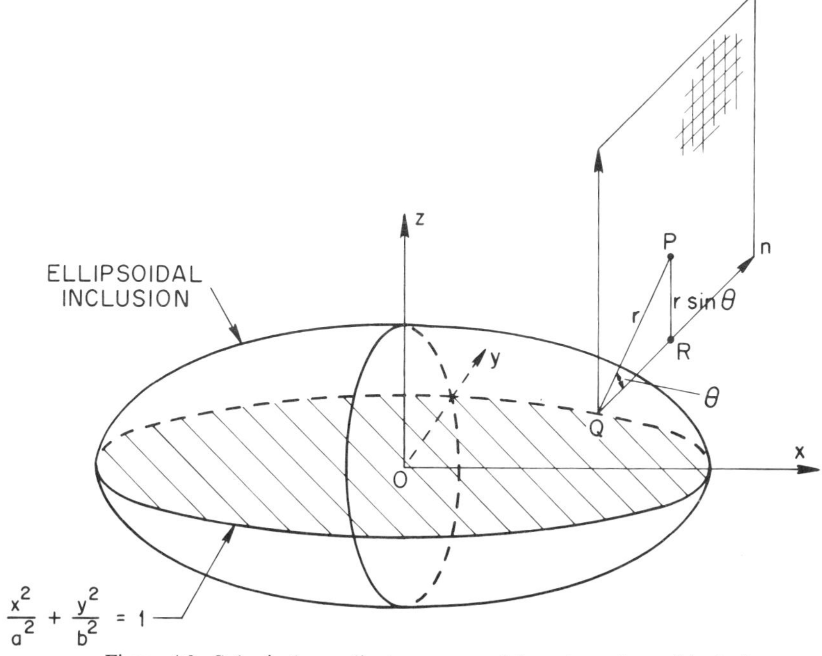

Figure 4.2. Spherical coordinates measured from boundary of inclusion.

while the distance QR (i.e. $r\cos\theta$) is the projection of the radial vector (r) in the direction of the outward normal to the ellipse (4.15) at Q. In this case, it can be confirmed that the position of point P is given by

$$x=\cos\phi\left(a+\frac{b}{a}r\cos\theta\,\Phi^{-\frac{1}{2}}\right), \tag{4.25a}$$

$$y=\sin\phi(b+r\cos\theta\,\Phi^{-\frac{1}{2}}), \tag{4.25b}$$

$$z=r\sin\theta, \tag{4.25c}$$

in which

$$\Phi=1-(1-b^2/a^2)\cos^2\phi. \tag{4.26}$$

In addition, the local expansion of ξ can be derived in a manner similar to the one used previously. It turns out to be

$$\xi=b_1r+b_2r^2+O(r^3), \tag{4.27}$$

where higher order terms in r have been neglected and the functions b_1 and b_2 designate

$$b_1=2b\Phi^{-\frac{1}{2}}\cos\theta, \tag{4.28a}$$

$$b_2=\left(\frac{b}{c}\right)^2\frac{\sin^2\theta}{\Phi}+\frac{\cos^2\theta}{\Phi^2}\left(\sin^2\phi+\frac{b^4}{a^4}\cos^2\phi\right). \tag{4.28b}$$

Note that when $r\to 0$, $\xi=0$ as mentioned previously. The relations (4.25)–(4.28) are used in the sequel to obtain the displacement and stress fields in a volume element of the matrix situated at the point P. Some asymptotic expressions required for the latter purpose are given in appendix B.

4.4 Expansion of the displacement field

In the region outside the notch, the expressions for the displacements are obtained by inserting the appropriate derivatives of f from appendix A into equations (4.2). This results into

$$
\begin{aligned}
u_x^c = {} & (\varepsilon_{yy} - \varepsilon_{xx})x\left[\left(\frac{k'^2a^2 + b^2}{k^2l^2}\right)E(u) - \left(\frac{a^2 + b^2}{k^2l^2}\right)k'^2u - \frac{b^2C}{lAB} + \frac{Hy^2}{A^2B^4}\right] \\
& + (\varepsilon_{zz} - \varepsilon_{xx})x\left[\left(\frac{k'^2a^2 - k^2c^2}{l^2}\right)E(u) - \frac{k'^2a^2}{l^2}u + \frac{k^2c^2B}{lAC} + \frac{Hz^2}{A^2C^4}\right] \\
& - 2k'^2x[(1-\nu)\varepsilon_{xx} + \nu(\varepsilon_{yy} + \varepsilon_{zz})][E(u) - u] \\
& + 2y\varepsilon_{xy}\left\{-\left[2(1-\nu) + \frac{a^2 + b^2}{k^2l^2}\right]k'^2u + \left[2(1-\nu) + \frac{a^2k'^2 + b^2}{k^2l^2}\right]E(u)\right. \\
& \left. - \left[2(1-\nu) + \frac{b^2}{k^2l^2}\right]\frac{k^2lC}{AB} + \frac{Hx^2}{A^4B^2}\right\} \\
& + 2z\varepsilon_{xz}\left\{-\frac{k'^2a^2}{l^2}u - \left[2(1-\nu) + \frac{k^2c^2 - k'^2a^2}{k^2l^2}\right]k^2E(u)\right. \\
& \left. + \left[2(1-\nu) + \frac{c^2}{l^2}\right]\frac{k^2lB}{AC} + \frac{Hx^2}{A^4C^2}\right\} + 2\varepsilon_{yz}\frac{xyzH}{A^2B^2C^2},
\end{aligned}
\tag{4.29a}
$$

$$
\begin{aligned}
u_y^c = {} & (\varepsilon_{zz} - \varepsilon_{yy})y\left[\frac{b^2}{l^2}u - \left(\frac{b^2 + c^2k^2}{k'^2l^2}\right)E(u) + \frac{k^2}{k'^2l}\left(\frac{b^2C}{AB} + \frac{c^2B}{AC}\right) + \frac{Hz^2}{B^2C^4}\right] \\
& + (\varepsilon_{xx} - \varepsilon_{yy})y\left[-\frac{k'^2(a^2 + b^2)}{k^2l^2}u + \left(\frac{k'^2a^2 + b^2}{k^2l^2}\right)E(u) - \frac{b^2C}{lAB} + \frac{Hx^2}{A^4B^2}\right] \\
& - 2y[(1-\nu)\varepsilon_{yy} + \nu(\varepsilon_{zz} + \varepsilon_{xx})]\left[k'^2u - E(u) + \frac{k^2lC}{AB}\right] \\
& + 2z\varepsilon_{yz}\left\{\frac{b^2}{l^2}u - \left[2(1-\nu)k^2 + \frac{b^2 + k^2c^2}{k'^2l^2}\right]E(u) + \frac{b^2k^2C}{k'^2lAB}\right. \\
& \left. + \left[2(1-\nu) + \frac{c^2}{k'^2l^2}\right]\frac{k^2lB}{AC} + \frac{Hy^2}{B^4C^2}\right\} \\
& + 2x\varepsilon_{xy}\left\{\left[2(1-\nu) - \frac{a^2 + b^2}{k^2l^2}\right]k'^2u - \left[2(1-\nu)k'^2 - \frac{b^2 + a^2k'^2}{k^2l^2}\right]E(u)\right. \\
& \left. - \frac{b^2C}{lAB} + \frac{Hy^2}{A^2B^4}\right\} + 2\varepsilon_{xz}\frac{xyzH}{A^2B^2C^2},
\end{aligned}
\tag{4.29b}
$$

$$
\begin{aligned}
u_z^c = {} & (\varepsilon_{xx} - \varepsilon_{zz})z\left[-\frac{k'^2a^2}{l^2}u + \left(\frac{k'^2a^2 - k^2c^2}{l^2}\right)E(u) + \frac{k^2c^2B}{lAC} + \frac{Hx^2}{A^4C^2}\right] \\
& + (\varepsilon_{yy} - \varepsilon_{zz})z\left[\frac{b^2}{l^2}u - \left(\frac{b^2 + k^2c^2}{k'^2l^2}\right)E(u) + \frac{k^2}{k'^2l}\left(\frac{b^2C}{AB} + \frac{c^2B}{AC}\right) + \frac{Hy^2}{B^4C^2}\right] \\
& - 2k^2z[(1-\nu)\varepsilon_{zz} + \nu(\varepsilon_{xx} + \varepsilon_{yy})]\left[E(u) - \frac{lB}{AC}\right]
\end{aligned}
$$

$$+2x\varepsilon_{xz}\left\{\left[2(1-\nu)-\frac{a^2}{l^2}\right]k'^2u-\left[2(1-\nu)k'^2+\frac{c^2k^2-a^2k'^2}{l^2}\right]E(u)\right.$$
$$\left.+\frac{k^2c^2B}{lAC}+\frac{Hz^2}{A^2C^4}\right\}$$
$$+2y\varepsilon_{yz}\left\{-\left[2(1-\nu)-\frac{b^2}{k'^2l^2}\right]k'^2u+\left[2(1-\nu)-\frac{b^2+c^2k^2}{k'^2l^2}\right]E(u)\right.$$
$$\left.-2(1-\nu)\frac{k^2lC}{AB}+\frac{k^2}{k'^2l}\left(\frac{b^2C}{AB}+\frac{c^2B}{AC}\right)+\frac{Hz^2}{B^2C^4}\right\}+2\varepsilon_{xy}\frac{xyzH}{A^2B^2C^2}, \quad (4.29c)$$

where the following abbreviations have been introduced

$$\epsilon_{ij}=\frac{abce_{ij}^T}{2(1-\nu)k^2k'^2l^3}, \quad (i \text{ and } j=x,\ y \text{ and } z), \quad (4.30a)$$

$$H=\frac{l^3k^2k'^2\xi}{ABC\left[\dfrac{x^2}{A^4}+\dfrac{y^2}{B^4}+\dfrac{z^2}{C^4}\right]}. \quad (4.30b)$$

Note that for the problem of the ellipsoidal inhomogeneity, the specified strains e_{ij}^T, i and $j=x$, y and z, in equations (4.29) and (4.30) can be related to the applied loading at infinity by means of certain sets of algebraic equations [1].

First, let us obtain the local form of the displacement field with reference to the boundary of the ellipse (4.14). Without loss in generality and for the sake of brevity, let us assume further that $e_{ij}^T=0$, for $i\neq j$. Inserting the results given in equations (4.18)–(4.23) into equations (4.29), expanding for small values of r_1, utilizing the expressions listed in appendix B (equations B4.2 through B4.9) and simplifying the algebra, it is found that

$$u_x^c=\frac{k^2}{l}\left(\frac{2r_1}{k'l}\right)^{\frac{1}{2}}\cos\phi_1\cos\frac{\theta_1}{2}\,\Phi_1^{-\frac{1}{4}}\left\{(\varepsilon_{xx}-\varepsilon_{yy})\left[c^2+\frac{\sin^2\phi_1}{8\Phi_1}\left(4k'l\Phi_1^{\frac{1}{2}}\frac{c^2}{r_1}\right.\right.\right.$$
$$\left.\left.-c^2(5+k'^2)-8(k'l)^2\cos^2\frac{\theta_1}{2}-\left(3+4\cos^2\frac{\theta_1}{2}\right)\frac{(k'c)^2}{\Phi_1}\right)\right]$$
$$+(\varepsilon_{zz}-\varepsilon_{xx})\left[(k'a)^2\left(1+\sin^2\frac{\theta_1}{2}\right)+\frac{c^2}{8}\left(4k'l\frac{\Phi_1^{\frac{1}{2}}}{r_1}+3-k'^2\right.\right.$$
$$\left.\left.+8k'^2\cos^2\frac{\theta_1}{2}+\left(1+4\sin^2\frac{\theta_1}{2}\right)\frac{k'^2}{\Phi_1}\right)\right]$$
$$\left.-2(k'l)^2[(1-\nu)\varepsilon_{xx}+\nu(\varepsilon_{yy}+\varepsilon_{zz})]\right\}+O(r_1^{\frac{3}{2}}), \quad (4.31a)$$

$$u_y^c = \frac{k^2 k'}{l}\left(\frac{2r_1}{k'l}\right)^{\frac{1}{2}} \sin \phi_1 \cos \frac{\theta_1}{2} \Phi_1^{-\frac{1}{4}} \Big\{ (\varepsilon_{yy} - \varepsilon_{xx}) \Big[c^2 + \frac{\cos^2 \phi_1}{8\Phi_1} \Big(4k'l\Phi_1^{\frac{1}{2}} \frac{c^2}{r_1}$$

$$- c^2(1 + 5k'^2) - 8(k'l)^2 \cos^2 \frac{\theta_1}{2} - \Big(3 + 4\cos^2 \frac{\theta_1}{2}\Big) \frac{(k'c)^2}{\Phi_1} \Big) \Big]$$

$$+ k'^{-2}(\varepsilon_{zz} - \varepsilon_{yy}) \Big[b^2\Big(1 + \sin^2 \frac{\theta_1}{2}\Big) + \frac{c^2}{8} \Big(4k'l \frac{\Phi_1^{\frac{1}{2}}}{r_1} + 3k'^2 - 1$$

$$+ 8\cos^2 \frac{\theta_1}{2} + \Big(1 + 4\sin^2 \frac{\theta_1}{2}\Big) \frac{k'^2}{\Phi_1} \Big) \Big]$$

$$- 2l^2[(1 - \nu)\varepsilon_{yy} + \nu(\varepsilon_{zz} + \varepsilon_{xx})] \Big\} + O(r_1^{\frac{3}{2}}), \qquad (4.31b)$$

$$u_z^c = \left(\frac{k}{l}\right)^2 (2k'lr_1)^{\frac{1}{2}} \sin \frac{\theta_1}{2} \Phi_1^{\frac{1}{4}} \Big\{ (\varepsilon_{xx} - \varepsilon_{zz}) \Big[c^2 + \frac{\cos^2 \phi_1}{8\Phi_1} \Big(-4k'l\Phi_1^{\frac{1}{2}} \frac{c^2}{r_1}$$

$$+ 8(k'a)^2 \cos^2 \frac{\theta_1}{2} + \frac{(k'c)^2}{\Phi_1} \Big(1 + 4\cos^2 \frac{\theta_1}{2}\Big) - c^2 \Big(1 + k'^2 - 8k'^2 \sin^2 \frac{\theta_1}{2}\Big) \Big) \Big]$$

$$+ k'^{-2}(\varepsilon_{yy} - \varepsilon_{zz}) \Big[c^2 + \frac{\sin^2 \phi_1}{8\Phi_1} \Big(-4k'l\Phi_1^{\frac{1}{2}} \frac{c^2}{r_1} + 8b^2 \cos^2 \frac{\theta_1}{2}$$

$$+ \frac{(k'c)^2}{\Phi_1} \Big(1 + 4\cos^2 \frac{\theta_1}{2}\Big) - c^2 \Big(1 + k'^2 - 8\sin^2 \frac{\theta_1}{2}\Big) \Big) \Big]$$

$$+ 2l^2[(1 - \nu)\varepsilon_{zz} + \nu(\varepsilon_{xx} + \varepsilon_{yy})] \Big\} + O(r_1^{\frac{3}{2}}). \qquad (4.31c)$$

In equations (4.31), the higher order terms in r_1 have been neglected. As mentioned previously, the displacement field in equations (4.31) is measured with reference to the 'limiting' ellipse (4.14). In order for the triaxial inclusion to assume the region occupied by this ellipse, it must sweep-up all points in space originally occupied by it as it shrinks to a flat lamina ($c = 0$). It is evident from equations (4.8), (4.9) and (4.31) that as $c \to 0$; $l = a$, $k' = b/a$, $k = (1 - b^2/a^2)^{\frac{1}{2}}$ and the displacement field in eq. (4.31) assumes the well-known form [6]

$$u_x^c = \frac{b^{\frac{3}{2}} k^2}{a} \sqrt{2r_1} \cos \varphi_1 \cos \frac{\theta_1}{2} \Phi_1^{-\frac{1}{4}} \Big\{ (\varepsilon_{yy} - \varepsilon_{xx}) \cos^2 \frac{\theta_1}{2} \frac{\sin^2 \phi_1}{\Phi_1}$$

$$+ (\varepsilon_{zz} - \varepsilon_{xx}) \Big(1 + \sin^2 \frac{\theta}{2}\Big)$$

$$- 2[(1 - \nu)\varepsilon_{xx} + \nu(\varepsilon_{yy} + \varepsilon_{zz})] \Big\} + O(r_1^{\frac{3}{2}}), \qquad (4.32a)$$

$$u_y^c = b^{\frac{1}{2}} k^2 \sqrt{2r_1} \sin \phi_1 \cos \frac{\theta_1}{2} \Phi_1^{-\frac{1}{4}} \left\{ (\varepsilon_{xx} - \varepsilon_{yy}) k'^2 \cos^2 \frac{\theta_1}{2} \frac{\cos^2 \phi_1}{\Phi_1} \right.$$

$$+ (\varepsilon_{zz} - \varepsilon_{yy}) \left(1 + \sin^2 \frac{\theta_1}{2}\right)$$

$$\left. - 2[(1-\nu)\varepsilon_{yy} + \nu(\varepsilon_{zz} + \varepsilon_{xx})] \right\} + O(r_1^{\frac{3}{2}}), \tag{4.32b}$$

$$u_z^c = b^{\frac{1}{2}} k^2 \sqrt{2r_1} \sin \frac{\theta_1}{2} \Phi_1^{\frac{1}{4}} \left\{ \varepsilon_{xx} \left[2\nu + \cos^2 \frac{\theta_1}{2} \frac{k'^2 \cos^2 \phi_1}{\Phi_1} \right] \right.$$

$$+ \varepsilon_{yy} \left[2\nu + \cos^2 \frac{\theta_1}{2} \frac{\sin^2 \phi_1}{\Phi_1} \right]$$

$$\left. + \varepsilon_{zz} \left[2(1-\nu) - \cos^2 \frac{\theta_1}{2} \right] \right\} + O(r_1^{\frac{3}{2}}). \tag{4.32c}$$

By way of further comparison, referring to a triply orthogonal system of coordinates (n, t, z) such that the axes n-, t- and z- are always described along the binormal, tangent and principal normal to the boundary of the ellipse at Q_1, u_z^c in equations (4.32) remains invariant and u_x^c and u_y^c combine to yield

$$u_n^c = u_x^c \cos \beta_1 + u_y^c \sin \beta_1$$

$$= k^2 (2br_1)\, \Phi_1^{-\frac{3}{4}} \cos \frac{\theta_1}{2} \left[\left(1 - 2\nu + \sin^2 \frac{\theta_1}{2}\right) \Phi_1 \varepsilon_{zz} \right.$$

$$- \left(2\nu k'^2 \cos^2 \phi_1 + \left(3 - 2\nu + \sin^2 \frac{\theta_1}{2}\right) \sin^2 \phi_1 \right) \varepsilon_{yy}$$

$$\left. - \left(2\nu \sin^2 \phi_1 + \left(3 - 2\nu + \sin^2 \frac{\theta_1}{2}\right) k'^2 \cos^2 \phi_1 \right) \varepsilon_{xx} \right] + O(r_1^{\frac{3}{2}}), \quad c = 0, \tag{4.33a}$$

$$u_t^c = -u_x^c \sin \beta_1 + u_y^c \cos \beta_1$$

$$= 4(1-\nu) k^2 k' (2br_1)^{\frac{1}{2}} \Phi_1^{-\frac{3}{4}} \cos \frac{\theta_1}{2} \sin \phi_1 \cos \phi_1$$

$$\times (\varepsilon_{xx} - \varepsilon_{yy}) + O(r_1^{\frac{3}{2}}), \quad c = 0. \tag{4.33b}$$

It is also of interest to examine the components of the displacement near the border of non-sharp notches ($c \neq 0$). For this purpose, the coordinates r, θ and ϕ are used. Inserting equations (4.25)–(4.28) into the basic formulas in equations (4.29), making use of the asymptotic results in appendix B (equations (B4.11) through (B4.20)) and expanding for small values of r leads to the relations

$$u_x^c = \frac{bc\cos\phi}{al}\Big\{a[\varepsilon_{xx} - \varepsilon_{yy} + k^2(\varepsilon_{zz} - \varepsilon_{xx})]$$

$$+ \frac{b\cos\theta}{a\Phi^{\frac{1}{2}}} r\Big[k^2(\varepsilon_{zz} - \varepsilon_{xx}) + (\varepsilon_{xx} - \varepsilon_{yy})\Big(1 - \frac{2k^2k'^2l^4}{b^2c^2\Phi}\sin^2\phi\Big)$$

$$- \frac{2k^2k'^2l^4}{b^2c^2}(\varepsilon_{xx} - \nu\varepsilon_{xx} + \nu\varepsilon_{yy} + \nu\varepsilon_{zz})\Big]\Big\} - O(r^2), \tag{4.34a}$$

$$u_y^c = \frac{bc\sin\phi}{al}\Big\{b\Big[\varepsilon_{yy} - \varepsilon_{xx} + 2\Big(\frac{k}{k'}\Big)^2(\varepsilon_{zz} - \varepsilon_{yy})$$

$$-2\Big(\frac{kl}{b}\Big)^2(\varepsilon_{yy} - \nu\varepsilon_{yy} + 2\nu\varepsilon_{xx} + 2\nu\varepsilon_{zz})\Big]$$

$$+ \frac{k^2\cos\theta}{\Phi^{\frac{1}{2}}} r\Big[2k'^{-2}(\varepsilon_{zz} - \varepsilon_{yy}) + k^{-4}(\varepsilon_{xx} - \varepsilon_{yy})\Big(\frac{2k'^2k^4l^4}{a^2c^2\Phi}\cos^2\phi - 1\Big)$$

$$-2\Big(\frac{l}{c}\Big)^2(\varepsilon_{yy} - \nu\varepsilon_{yy} + \nu\varepsilon_{zz} + \nu\varepsilon_{xx})\Big]\Big\} + O(r^2), \tag{4.34b}$$

$$u_z^c = \frac{bck^2}{al} r\sin\theta\Big\{\varepsilon_{xx} - \varepsilon_{zz} + 2k'^{-2}(\varepsilon_{yy} - \varepsilon_{zz})$$

$$+2\Big(\frac{l}{c}\Big)^2[(1-\nu)\varepsilon_{zz} + \nu(\varepsilon_{xx} + \varepsilon_{yy})]\Big\} + O(r^2). \tag{4.34c}$$

In equations (4.34), ε_{ij}, i and $j = x$, y and z, stand for the quantities in equation (4.30a) and terms containing r^2 and higher order values in r have been neglected.

4.5 Local stress distribution

The general character of the three-dimensional stresses near the border of the inclusion is examined in this section. Inasmuch as the details have been presented in previous sections, the emphasis here is confined to the nature of the results. The displacement expressions in equations (4.31) can be utilized to derive the pertinent stress field when measured with reference to the border of ellipse (4.14). Denoting the shear modulus of the solid by μ and carrying out the necessary computations, it is found that near the apex of the major axis of the latter ellipse

$$\frac{\sigma_{xx}}{2\mu}=\frac{k^2\cos\frac{\theta_1}{2}}{k'l^{\frac{3}{2}}\sqrt{2r_1}}\Big\{(\varepsilon_{xx}-\varepsilon_{yy})c^2-2(\varepsilon_{xx}+\nu\varepsilon_{yy})k'^2l^2$$
$$+(\varepsilon_{zz}-\varepsilon_{xx})\Big[\frac{k'^2lc^2}{2r_1}(1-2\cos\theta_1)+\frac{15}{8}k'^2c^2$$
$$+\left(k'^2l^2+\frac{c^2}{2}\right)\left(1-\sin\frac{\theta_1}{2}\sin\frac{3\theta_1}{2}\right)\Big]\Big\}+O(r_1^0),\quad \phi_1=0, \tag{4.35a}$$

$$\frac{\sigma_{yy}}{2\mu}=\frac{k^2\cos\frac{\theta_1}{2}}{k'l^{\frac{3}{2}}\sqrt{2r_1}}(\varepsilon_{zz}-\varepsilon_{xx})(c^2+2\nu k'^2l^2)+O(r_1^0),\quad \phi=0, \tag{4.35b}$$

$$\frac{\sigma_{zz}}{2\mu}=\frac{k^2\cos\frac{\theta_1}{2}}{k'l^{\frac{3}{2}}\sqrt{2r_1}}\Big\{(\varepsilon_{yy}-\varepsilon_{zz})c^2+2(\varepsilon_{zz}+\nu\varepsilon_{yy})k'^2l^2$$
$$+(\varepsilon_{xx}-\varepsilon_{zz})\Big[\frac{k'^2lc^2}{2r_1}(1-2\cos\theta_1)+\frac{15}{8}k'^2c^2$$
$$+\left(k'^2l^2+\frac{c^2}{2}\right)\left(1-\sin\frac{\theta_1}{2}\sin\frac{3\theta_1}{2}\right)\Big]\Big\}+O(r_1^0),\quad \phi_1=0, \tag{4.35c}$$

$$\frac{\sigma_{xz}}{2\mu}=\frac{k^2\sin\frac{\theta_1}{2}}{k'l^{\frac{3}{2}}\sqrt{2r_1}}\Big\{\frac{c^2}{2}(\varepsilon_{xx}-2\varepsilon_{yy}+\varepsilon_{zz})-(\varepsilon_{xx}+2\nu\varepsilon_{yy}+\varepsilon_{zz})k'^2l^2$$
$$+(\varepsilon_{zz}-\varepsilon_{xx})\Big[-\frac{k'^2lc^2}{2r_1}(1+2\cos\theta_1)+k'^2c^2\left(\frac{7}{8}-\cos\frac{\theta_1}{2}\cos\frac{3\theta_1}{2}\right)$$
$$+\left(a^2k'^2+\frac{c^2}{2}\right)\left(1+\cos\frac{\theta_1}{2}\cos\frac{3\theta_1}{2}\right)\Big]\Big\}+O(r_1^0),\quad \phi_1=0, \tag{4.35d}$$

$$\frac{\sigma_{xy}}{2\mu}=O(r_1^0),\quad \phi_1=0, \tag{4.35e}$$

$$\frac{\sigma_{yz}}{2\mu}=O(r_1^0),\quad \phi=0. \tag{4.35f}$$

Upon setting $c\to 0$ in equations (4.35) the following relations are reached

$$\frac{\sigma_{xx}}{2\mu}=\frac{k^2b\cos\frac{\theta_1}{2}}{a^{\frac{1}{2}}\sqrt{2r_1}}\left[-2(\varepsilon_{xx}+\nu\varepsilon_{yy})+(\varepsilon_{zz}-\varepsilon_{xx})\left(1-\sin\frac{\theta_1}{2}\sin\frac{3\theta_1}{2}\right)\right]+O(r_1^0), \tag{4.36a}$$

$$\frac{\sigma_{yy}}{2\mu}=\frac{k^2b\cos\frac{\theta_1}{2}}{a^{\frac{1}{2}}\sqrt{2r_1}}2\nu(\varepsilon_{zz}-\varepsilon_{xx})+O(r_1^0), \tag{4.36b}$$

$$\frac{\sigma_{zz}}{2\mu}=\frac{k^2b\cos\frac{\theta_1}{2}}{a^{\frac{1}{2}}\sqrt{2r_1}}\left[2(\varepsilon_{zz}+\nu\varepsilon_{yy})+(\varepsilon_{xx}-\varepsilon_{zz})\left(1-\sin\frac{\theta_1}{2}\sin\frac{3\theta_1}{2}\right)\right]+O(r_1^0), \tag{4.36c}$$

$$\frac{\sigma_{xz}}{2\mu}=\frac{k^2b\sin\frac{\theta_1}{2}}{a^{\frac{1}{2}}\sqrt{2r_1}}\left[-(\varepsilon_{xx}+2\nu\varepsilon_{yy}+\varepsilon_{zz})+(\varepsilon_{zz}-\varepsilon_{xx})\right.$$

$$\left.\times\left(1+\cos\frac{\theta_1}{2}\cos\frac{3\theta_1}{2}\right)\right]+O(r_1^0), \tag{4.36d}$$

$$\frac{\sigma_{xy}}{2\mu}=O(r_1^0), \tag{4.36e}$$

$$\frac{\sigma_{y3}}{2\mu}=O(r_1^0). \tag{4.36f}$$

The expressions in equations (4.36) can be identified with the stress field around an elliptical crack under arbitrary loading [6]. Moreover, the stress σ_{yy} ($\phi_1=0$) is observed to be equal to Poissin's ratio ν times the sum of σ_{xx} and σ_{zz}, a well-known relationship in two-dimensional problems of plane strain.

In the vicinity of the horizontal section of a triaxial notch ($c\neq 0$), the stress state can be derived from the displacement components in equations (4.34). Making use of these expressions and the identities

$$\frac{\partial r}{\partial x}=\frac{b_1\cos\phi}{2a},\qquad \frac{\partial r}{\partial y}=\frac{b_1\sin\varphi}{2b},\qquad \frac{\partial r}{\partial z}=\sin\theta, \tag{4.37a}$$

$$\left.\begin{aligned}&\frac{\partial\theta}{\partial x}=-\frac{b\cos\phi}{a\Phi^{\frac{1}{2}}}\frac{\sin\theta}{r},\\ &\frac{\partial\theta}{\partial y}=-\frac{\sin\phi}{\Phi^{\frac{1}{2}}}\frac{\sin\theta}{r},\qquad \frac{\partial\theta}{\partial z}=\frac{\cos\theta}{r}\end{aligned}\right\} \tag{4.37b}$$

and

$$\left.\begin{aligned}&\frac{\partial\phi}{\partial x}=-\frac{\sin\phi}{a\Phi}+\cdots\\ &\frac{\partial\phi}{\partial y}=\frac{b\cos\phi}{a^2\Phi}+\cdots\\ &\frac{\partial\phi}{\partial z}=0\end{aligned}\right\}, \tag{4.37c}$$

the stress field at an arbitrary point near the border of the notch assumes the form

$$\frac{\sigma_{xx}}{2\mu} = \frac{bc}{al}\left[\varepsilon_{xx} - \varepsilon_{yy} + k^2(\varepsilon_{zz} - \varepsilon_{xx}) + \frac{2k^2k'^2l^4}{a^2c^2}(\varepsilon_{yy} - \varepsilon_{xx})\frac{\sin^2\phi\cos^2\phi}{\Phi^2} - \frac{2k^2k'^2l^4}{b^2c^2\Phi}(\varepsilon_{xx} - \nu\varepsilon_{xx} + \nu\varepsilon_{yy} + \nu\varepsilon_{zz})\right] + O(r) \tag{4.38a}$$

$$\frac{\sigma_{yy}}{2\mu} = \frac{bc}{al}\left[\varepsilon_{yy} - \varepsilon_{xx} + 2\left(\frac{k}{k'}\right)^2(\varepsilon_{zz} - \varepsilon_{yy}) - 2\left(\frac{kk'l^2\sin\phi\cos\phi}{ac\Phi}\right)^2(\varepsilon_{yy} - \varepsilon_{xx}) - 2\left(\frac{kl}{ac}\right)^2\frac{a^2\sin^2\phi + c^2\cos^2\phi}{\Phi}(\varepsilon_{yy} - \nu\varepsilon_{yy} + \nu\varepsilon_{zz} + \nu\varepsilon_{xx})\right] + O(r) \tag{4.38b}$$

$$\frac{\sigma_{zz}}{2\mu} = \frac{bc}{al}\left[k^2(\varepsilon_{xx} - \varepsilon_{zz}) + 2\left(\frac{k}{k'}\right)^2(\varepsilon_{yy} - \varepsilon_{zz}) + 2\left(\frac{kl}{c}\right)^2(\varepsilon_{zz} - \nu\varepsilon_{zz} + \nu\varepsilon_{xx} + \nu\varepsilon_{yy})\right] + O(r) \tag{4.38c}$$

$$\frac{\sigma_{xy}}{2\mu} = \frac{2k^2k'^2l^3\sin\phi\cos\phi}{a^2c\Phi}\left[-\varepsilon_{xx} - \nu\varepsilon_{zz} + (\varepsilon_{xx} - \varepsilon_{yy})\frac{b^2\cos^2\phi}{a^2\Phi}\right] + O(r) \tag{4.38d}$$

$$\frac{\sigma_{xz}}{2\mu} = O(r), \tag{4.38e}$$

$$\frac{\sigma_{yz}}{2\mu} = O(r). \tag{4.38f}$$

The expressions in equations (4.38) are functions of the locations of point Q as it traces the periphery of ellipse (4.15). Also, it appears that there are no counterparts to these expressions in the two-dimensional theory of elasticity even when they are transformed to a new system of coordinates (n, t, z) chosen such that the axes n, t, z are always directed along the binormal, tangent, and principal normal of the curve (4.15), respectively.

4.6 Appendix A: Derivatives of $f(x, y, z; \xi)$

Referring to equation (4.7), the first partial derivatives of f with respect to x, y and z are computed by regarding ξ as a constant. This follows immediately from equations (4.3) and (4.5). In forming the higher derivatives, the following results are needed

$$\frac{\partial \xi}{\partial x} = \frac{2x}{A^2h^2}, \qquad \frac{\partial \xi}{\partial y} = \frac{2y}{B^2h^2} \qquad \frac{\partial \xi}{\partial z} = \frac{2z}{C^2h^2} \tag{A.41}$$

and

$$\frac{\partial u}{\partial \xi} = -\frac{l}{2ABC}, \qquad \frac{\partial E(u)}{\partial \xi} = -\frac{lB}{2A^3C}, \tag{A4.2}$$

where A, B, C and l are given in equation (4.9), and

$$h^2 = x^2/A^4 + y^2/B^4 + z^2/C^4. \tag{A4.3}$$

Making use of the results in equations (A4.1)–(A4.3), it is found that

$$\left.\begin{aligned} \frac{\partial u}{\partial x} &= -\frac{lx}{A^3BCh^2}, \quad \frac{\partial u}{\partial y} = -\frac{ly}{AB^3Ch^2}, \quad \frac{\partial u}{\partial z} = -\frac{lz}{ABC^3h^2} \\ \frac{\partial E}{\partial x} &= -\frac{lBx}{A^5Ch^2}, \quad \frac{\partial E}{\partial y} = -\frac{ly}{A^3BCh^2}, \quad \frac{\partial E}{\partial z} = -\frac{lBz}{A^3C^3h^2} \end{aligned}\right\}, \tag{A4.4}$$

and the partial derivatives of f are:

$$\frac{\partial f}{\partial x} = \frac{4\pi abc}{l^3k^2} x[E(u) - u], \tag{A4.5}$$

$$\frac{\partial f}{\partial y} = \frac{4\pi abc}{l^3k^2k'^2} y\left[k'^2u - E(u) + k^2\frac{lC}{BA}\right], \tag{A4.6}$$

$$\frac{\partial f}{\partial z} = \frac{4\pi abc}{l^3k'^2} z\left[E(u) - \frac{lB}{AC}\right], \tag{A4.7}$$

$$\frac{\partial^2 f}{\partial x\, \partial y} = 4\pi abc\frac{xy}{A^3B^3Ch^2}, \tag{A4.8}$$

$$\frac{\partial^2 f}{\partial y\, \partial z} = 4\pi abc\frac{yz}{AB^3C^3h^2}, \tag{A4.9}$$

$$\frac{\partial^2 f}{\partial z\,\partial x} = 4\pi abc\frac{zx}{A^3BC^3h^2}, \tag{A4.10}$$

$$\frac{\partial^2 f}{\partial x^2} = \frac{4\pi abc}{l^3k^2}\left[E(u) - u + \frac{l^3k^2x^2}{A^5BCh^2}\right], \tag{A4.11}$$

$$\frac{\partial^2 f}{\partial y^2} = \frac{4\pi abc}{l^3k^2k'^2}\left[k'^2u - E(u) + \frac{k^2lC}{BA} + \frac{l^3k^2k'^2y^2}{AB^5Ch^2}\right], \tag{A4.12}$$

$$\frac{\partial^2 f}{\partial z^2} = \frac{4\pi abc}{l^3k'^2}\left[E(u) - \frac{lB}{AC} + \frac{l^3k'^2z^2}{ABC^5h^2}\right]. \tag{A4.13}$$

4.7 Appendix B: Limiting forms of some expressions in terms of local coordinates

Near the vicinity of the edge of ellipse (4.14), the asymptotic expansion of ξ is

$$\xi = -c^2 + a_1r_1 + a_2r_1^2 + \cdots, \tag{B4.1a}$$

where

$$a_1 = 2k'l\Phi_1^{-\frac{1}{2}}\cos^2\frac{\theta_1}{2}, \tag{B4.1b}$$

$$a_2 = \Phi_1^{-2}(\Phi_1 - k^2k'^2\cos^2\phi_1)\cos^2\frac{\theta_1}{2}, \tag{B4.1c}$$

$$\Phi_1 = 1 - k^2\cos^2\phi_1. \tag{B4.1d}$$

Utilizing these relations in conjunction with equations (4.19), the expressions listed below are generated for small values of r_1

$$u = -(a_1r_1)^{\frac{1}{2}}/(k'l) + O(r_1^{\frac{3}{2}}), \tag{B4.2}$$

$$E(u) = -\frac{k'}{l}(a_1r_1)^{\frac{1}{2}} + O(r_1^{\frac{3}{2}}) \tag{B4.3}$$

$$\frac{B}{AC} = \frac{k'}{(a_1r_1)^{\frac{1}{2}}}\left[1 + \left(\frac{k^2a_1}{k'^2l^2} - \frac{a_2}{a_1}\right)\frac{r_1}{2}\right] + O(r_1^{\frac{3}{2}}), \tag{B4.4}$$

$$\frac{C}{AB} = \frac{(a_1r_1)^{\frac{1}{2}}}{k'l^2}\left[1 + \left(\frac{a_2}{a_1} - \frac{1+k'^2}{k'^2l^2}a_1\right)\frac{r_1}{2}\right] + O(r_1^{\frac{5}{2}}), \tag{B4.5}$$

$$(x/A)^2 = \cos^2\phi_1\left[1 - 2\frac{k'}{l}\Phi_1^{-\frac{1}{2}}\left(\sin^2\frac{\theta_1}{2}\right)r_1\right] + O(r_1^2), \tag{B4.6}$$

$$(y/B)^2 = \sin^2\phi_1\left[1 - \frac{2\sin^2\left(\frac{\theta_1}{2}\right)}{k'l\Phi_1^{\frac{1}{2}}}r_1\right] + O(r_1^2), \tag{B4.7}$$

$$(z/C)^2 = \frac{\sin^2\theta_1}{a_1}\left[r_1 - \frac{1}{2k'l\Phi_1^{\frac{1}{2}}}\left(1 - \frac{k^2k'^2}{\Phi_1}\cos^2\phi_1\right)r_1^2 + O(r_1^3)\right] \tag{B4.8}$$

$$H = \frac{l}{2}(k^2k'^2)\left(\frac{a_1}{\Phi_1}\right)^{\frac{1}{2}}\left\{-\frac{c^2}{r_1^{\frac{1}{2}}} + r_1^{\frac{1}{2}}\left[a_1 + \frac{c^2}{4k'l\Phi_1^{\frac{3}{2}}}\right.\right.$$
$$\left.\left.\times\left(3k'^2 + 4k'^2\cos^2\frac{\theta_1}{2} - 3(1+k'^2)\Phi_1\right)\right]\right\} + O(r_1^{\frac{3}{2}}). \tag{B4.9}$$

Similarly, in terms of the coordinates (r, θ, ϕ) displayed in Figure (4.2), ξ takes the form

$$\xi = b_1 r + b_2 r^2 + O(r^3), \tag{B4.10a}$$

with

$$b_1 = 2b\Phi^{-\frac{1}{2}}\cos\theta, \tag{B4.10b}$$

$$b_2 = \left(\frac{b}{c}\right)^2\frac{\sin^2\theta}{\Phi} + \frac{\cos^2\theta}{\Phi^2}\left(\sin^2\phi + \frac{b^4}{a^4}\cos^2\phi\right) \tag{B4.10c}$$

$$\Phi = 1 - \left(1 - \frac{b^2}{a^2}\right)\cos^2\phi \tag{B4.10d}$$

and the following relations are obtained in the neighborhood of the notch boundary $(c \neq 0)$

$$u = -\frac{lb_1}{2abc}r + O(r^2), \tag{B4.11}$$

$$E(u) = -\frac{lbb_1}{2a^3c}r + O(r^2), \tag{B4.12}$$

$$\frac{C}{AB} = \frac{c}{ab}\left[1 + \left(\frac{1}{c^2} - \frac{1}{a^2} - \frac{1}{b^2}\right)\frac{b_1 r}{2}\right] + O(r^2), \tag{B4.13}$$

$$\frac{B}{AC} = \frac{b}{ac}\left[1 + \left(\frac{1}{b^2} - \frac{1}{a^2} - \frac{1}{c^2}\right)\frac{b_1 r}{2}\right] + O(r^2), \tag{B4.14}$$

$$\frac{x^2}{A^4B^2} = \frac{\cos^2\phi}{(ab)^2}\left[1 - \frac{a^2+b^2}{a^2b^2}b_1r\right] + O(r^2), \tag{B4.15}$$

$$\frac{x^2}{A^4C^2} = \frac{\cos^2\phi}{(ac)^2}\left[1 - \frac{a^2+c^2}{a^2c^2}b_1r\right] + O(r^2), \tag{B4.16}$$

$$\frac{y^2}{A^2B^4} = \frac{\sin^2\phi}{(ab)^2}\left[1 - \frac{a^2+b^2}{a^2b^2}b_1r\right] + O(r^2), \tag{B4.17}$$

$$\frac{y^2}{B^4C^2} = \frac{\sin^2\phi}{(bc)^2}\left[1 - \frac{b^2+c^2}{b^2c^2}b_1r\right] + O(r^2), \tag{B4.18}$$

$$\frac{xyz}{(ABC)^2} = \frac{\sin\theta\sin\phi\cos\phi}{abc^2}r + O(r^2), \tag{B4.19}$$

$$H = \frac{bl^3k^2k'^2}{ac\Phi}b_1r + O(r^2). \tag{B4.20}$$

References

[1] Eshelby, J.D., *Elastic Inclusions and Inhomogeneities, progress in solid mechanics*, 2, Ed. by I. N. Sneddon and R. Hill, Interscience, pp. 89–140 (1961).

[2] Kellogg, O. D., *Potential Theory*, Dover Publication, New York, p. 196 (1953).

[3] Whittaker, E. T. and Watson, G. N., *Modern Analysis*, Cambridge University Press, Cambridge, England, p. 548 (1962).

[4] Griffith, A. A., *Theory of Rupture, Proc. 1st. International Congress of Applied Mechanics, Delft*, pp. 55–63 (1924).

[5] Sih, G. C., *A Special Theory of Crack Propagation, Methods of Analysis and Solutions to Crack Problems*, Ed. by G. C. Sih, Wolters-Noordhoff, pp. xxi-xLv (1973).

[6] Kassir, M. K. and Sih, G. C., Three Dimensional Stress Distribution Around an Elliptical Crack Under Arbitrary Loadings, *Journal of Applied Mechanics*, 33, pp. 601–611 (1966).

[7] Kassir, M. K. and Sih, G. C., Some Three-Dimensional Inclusion Problems in Elasticity, *Int. J. Solids and Structures*, 4, pp. 225–241 (1968).

[8] Sadowsky, M. A. and Sternberg, E., Stress Concentration Around a Triaxial Cavity, *Journal of Applied Mechanics*, 16, pp. 149–157 (1949).

[9] Mirandy, L. and Paul, B., Stresses at the Surface of a Flat Three-Dimensional Ellipsoidal Cavity, *Journal of Engineering Materials and Technology, Trans. ASME*, 98, Series H, 1976, pp. 164–172 (1976).

[10] Kassir, M. K. and Sih, G. C., *Three Dimensional Crack Problems*, Ed. by G. C. Sih, Vol. 2 of a series on mechanics of fracture, Noordhoff International Publishing, Leyden, Netherlands (1975).

Author's index

Subject index